Indigenous Knowledge

compiled by

Sarah Johnson

Themes in Environmental History, 3

'Themes in Environmental History' is a series of readers for students and researchers. Each volume aims to cover a prominent subject in the discipline, combining theoretical chapters and case studies. All chapters have been previously published in the White Horse Press journals *Environment and History* and *Environmental Values*.

1. *Bioinvaders* (2010) ISBN 978-1-874267-55-3
2. *Landscapes* (2010) ISBN 978-1-874267-60-7
3. *Indigenous Knowledge* (2012) ISBN 978-1-874267-68-3

British Library Cataloguing in Publication Data
A catalogue record for this book is available from the British Library

ISBN   978-1-874267-68-3 (PB)

Contents

Publisher's Introduction

Sarah Johnson

WHOSE INDIGENOUS KNOWLEDGE?

The title of this volume immediately raises problems of definition. The exact meaning of 'indigenous' is contested – Nygren, for example, draws attention to the status of well-established but non-native settlers. Goodall locates the concept in the 'interaction between the self-representation of the individuals and groups asserting their indigeneity on the one hand and, on the other, the pressures and goals of allies and enemies, whether within the nation state or internationally'. Even having established who is indigenous, it is far from straightforward to decide what constitutes indigenous knowledge and a wary eye must be kept on the differences between outsiders' and insiders' understandings of that knowledge or those knowledges. A seminal example of the reflexive nature of such understandings is provided by the war of words between anthropologists Sahlins and Obseyekere[1] about whether, in fact, the Hawaians really perceived Captain Cook as the incarnation of their god Lono. The questions of how far outsiders can truly claim to understand indigenous knowledge, and how that knowledge is inevitably altered by contact with the outside, remain topics of vigorous contemporary debate.

In the environmental domain, with global claims on resources and their conservation, the interaction of indigenous and outside knowledges and outlooks is both complex and consequential. Should the knowledge be taken to relate only to and derive only from the culture under study, or to be more global in its application and its sources? How might one indigenous person's knowledge differ from another's and to what extent are proponents – within and outside a culture – of indigenous views influenced by other interests, if indeed such interests are extricable from 'knowledge'? Such questions are not idle since, as Goodall points out,

> Indigenous people's knowledge of their environments ... is widely invoked today in many arenas of environmental analysis and natural resource management as a potential source of beneficial approaches to sustainability.

---

1. See Marshall Sahlins, *Islands of History* (University of Chicago Press, 1985); Gananath Obeyesekere, *The Apotheosis of Captain Cook: European Mythmaking in the Pacific* (Princeton University Press, 1992).

Introduction

However, such invocations can often be fuzzy or unthinking and in conservation politics it often arises that

> indigenous knowledge is celebrated but at the same time called on to carry the burden of finding solutions to major environmental crises, without allowing such reflection on how such knowledge might be constructed and transmitted. (Goodall)

Outsiders' perceptions of indigenous people and their knowledge inevitably mediate the application of such knowledge to issues constructed in or understood through global knowledge systems. As Nygren observes, post-imperialist Western observers are no longer generally guilty of defining indigenous peoples as savages 'eking out a miserable hand-to-mouth existence through hunting and gathering and ... doomed to give way to more progressive ways of exploiting the tropical resources', but unnuanced essentialisms remain rife, often tending these days (as in the age of Rousseau) to idealism: indigenous peoples may be 'depicted as guardians of the forest, blessed with inherent environmental wisdom', as part of threatened Nature but also part of the solution, if only we would stop and listen to their simple, holistic wisdom. Nygren's paper challenges at the outset of this volume the tendency to see through such a lens of idealism, to confuse indigenous peoples and their knowledges with 'the imaginative spaces that the tropical peoples occupy in the travel writers' [or any observers'] minds' and to fit the observed cultures into narratives arising from observers' acquired ideas of moral hierarchies, environmental concern or yearning for fragile paradises. Later in the volume, Kull and Henley both challenge the tendency to allow myths – whether of indigenous people's 'virtue' or 'vice' – to disproportionately inform forest conservation policy, in Madagascar and Indonesia respectively.

Nygren, analysing the depiction of South American rainforest dwellers in travel accounts published in the *National Geographic*, argues that writers tend to represent such people in accordance with the dominant ideas of their time and milieu – the march of progress and Christian missionising in the Victorian era, environmental awareness and the 'mission' to save threatened ecosystems in the twentieth century – and, over and above this cultural currency, to inscribe 'their own feelings and their own reactions' onto what is encountered, eliding the feelings and motivations of the encountered. Nygren points out that, in the texts she has studied, native conceptions of nature rarely become an object of study and that definitions of indigenous people tend to be monolithic, to take little account of diversity between members of a culture. It is as spurious to say that 'Amazonian Indians think' as to say 'Europeans think': while true on many partial levels, generalisations about indigenous environmental knowledge are unsubtle at best, ignoring, for example, the dynamic nature of cultural systems in contact with history and the ability of every individual to interpret, besides which, as Mukamuri points out, 'Though individuals share an environment, they compete for natural resources: they have well defined personal interests as well as group concerns'. Later essays demonstrate that crude generalisations are rife in the portrayal of indigenous peoples' relationships with their environment: Kull draws attention, for example, to the 'schizophrenic' portrayal of Malagasy farmers 'as both ignorant, backward farmers without a care for biodiversity and as potentially wise indigenous resource managers' who must be worked with to safeguard nature through traditional means. The reality, as so often in these cases, is more complex:

> The *tantsaha* are experienced resource managers, constrained by socio- political struggles over resource control, by market demands, and by government policies, not ignorant destroyers pushed by hunger and poverty.

Nygren begins her essay by probing such dichotomies as 'The West' and 'the rest', the 'othering' tendency of travel-writing since exotic travel began, and ends by critiquing

> the artificiality of making categorical distinctions between what is 'authentic' and what is 'spurious'. In today's world of globalisation and hybridisation, one can no longer predict who will put on a loincloth and lift up a blowgun and who will slip into jeans and pick up a mobile telephone

This observation highlights the difficulty of pinning down any absolute 'indigenous knowledge', or indeed indigenous persona, in isolation from other factors.

Karjalainen and Habeck in their study of Northern Russia push the idea of hybridisation further, insisting that exogenous factors are actually necessary to creating 'indigenous knowledge' as opposed to 'perception':

> an individual's perception of the environment is embedded in his/her everyday engagement with the surroundings ('the environment' as seen from within). Environmental knowledge is of more cognitive character: it originates mainly from outside the context of everyday life and is imparted via various forms of communication ('The Environment' as seen from the outside). From the interplay of these two levels arises what we call local environmental knowledge, a kind of knowledge which has its own moral and symbolic dimension within the social, cultural and political setting.

They link environmental knowledge to concern about the environment, something they argue comes from communication with other cultures and systems, in practice often media, science and education; environmental perception, by contrast, refers to experience of surroundings, to 'milieu-specific life-worlds' (which may be urban as well as rural); knowledge may grow out of and be enriched by perception of the home environment but the authors argue that 'knowledge is often at odds with "perception"', in that people's life-world might predispose them to be unable or unwilling to 'reconcile their personal experiences with scientifically established knowledge'. For example, when widely publicised environmental disasters bring environmental affairs to prominence in communities where they were not previously discussed, local people may be convinced, against scientific evidence, that these have had detrimental effects on health or on livestock or fish. Thus the process of a group of people developing environmental knowledge becomes embroiled with that of criticising the validity of knowledge disseminated by other groups.

TRANSACTIVE AND INTERACTIVE: THE FLUIDITY OF KNOWLEDGE

Karjalainen and Habeck conclude that, since 'local environmental knowledge is the result of the transactions and interactions, within a local context, between environmental perception and environmental knowledge', it is impossible simply to 'harvest' facts from such knowledge without careful consideration of the contexts and practices that have formed it and the 'symbolic dimension' these may have imparted. Local knowledge may include 'statements that are not simply descriptive, but moral and performative'. This issue is taken up by Goodall in her study of Aboriginal knowledge about water. She concedes that some fundamental knowledge may be 'transmitted unchanged over many generations' but claims that,

> Most oral knowledge is passed on in the far more flexible conditions of performance, often … in participatory and interactive settings. Here there are opportunities to engage apparently unalterable narratives with the historical changes in both environment and social life.

In other words, indigenous environmental knowledge is subject to change over time, and to shaping by the subjectivities of those who perform it, an issue defines Middleton's paper about evolving memories of Malagasy cactus extermination in Madagascar. Goodall argues that '"indigenous knowledge" can be more effectively understood as a process rather than as an archive, both before and after colonisation', not least because of the orality that is an important feature of most understandings of indigenous environmental knowledge. Oral traditions may assert their unchanging truthfulness but in reality they 'have always been a dynamic form, which engaged with and reflected changing social and environmental circumstances'. Indeed, as Middleton notes of rural Madagascar, 'in a non-literate society … where transmitted memory is oral rather than inscribed, important narrational shifts can occur without the narrators themselves being necessarily conscious of the fact'. Thus, Goodall argues, Aboriginal environmental knowledge was not some precious artefact rudely shattered by colonisation; instead, it has evolved in tandem with colonisation's unfolding. In a sense, Goodall is being controversial here, since she reports not only the non-indigenous tendency to view indigenous knowledge as a sacred fossil but its depiction by Aboriginal writer Tex Skuthorpe 'as a timeless, static and "intact" pre-invasion knowledge system which can be viewed whole and in opposition to western land management'. She describes, as an example of a living tradition's greedy absorption of the new, a ceremonial performance based on 'the rescue of a World War 2 American bomber pilot whose plane had crashed nearby in 1942' – in form the dance and song were wholly traditional, but the subject matter reveals a close integration of tradition and the modern. Mukamuri notes similarly of southern Zimbabwe that 'most of the *mitoro* [rainmaking ceremonies] being held in this region were actually established during the colonial period, as created rather than maintained traditions'.

Of course such created traditions spice acquired experience with indigenous sensibility. There is also the question of whether the indigenous knowledge that 'survives' contact with other cultures was inherently the most important or whether the very corpus of indigenous knowledge, even if not betraying outside influence, is dependent on relationships with the outside. Goodall notes that Aboriginal conceptions of environment can offer an 'alternative geography', in which water becomes highly significant. She describes the case of Boobera lagoon, disastrously silted by various settler activities that transgressed the prohibitions laid down in local myth. 'Aboriginal people are arguing that their knowledge contained an approach which would have protected an important resource'. Here, perhaps, is a prototypical case of indigenous knowledge encouraging sustainable behaviour (or prohibiting unsustainable). However, was water historically at the heart of Aboriginal conceptions of the environment? Since Aboriginal people found themselves restricted by settlers from access to many places, did watercourses take on a special significance in their traditions; is it possible that 'other special places, about which environmental knowledge might have been retold and learnt have not been visited so often or recently because the access to them has been closed off' and thus the knowledge system appears disproportionately to focus on water?

As the corpus of indigenous environmental knowledge evolves, its shape may alter considerably and so too the way people see the environment. The stories that survive serve to construct the next generation's sense of place and the sense of place in turn constructs the local subjects. But which stories survive is by no means random, being related to continuing relevance, social, political and cultural:

> Indigenous knowledge has been sustained since the invasion although in substantially altered forms, at some times reflecting pre-invasion conditions and at others reflecting newly emerging content arising from traditional bases but in engagement with very changed conditions. The capacity of indigenous people in conditions of historical change to identify and reflect on environmental changes is an important dimension of the broader value of indigenous knowledge.

Introduction

PLACE-MAKING, PLACE-MANAGING

Goodall ends her essay with an account of an Aborginal man's quest to make sense of a vastly changed landscape by recording and amplifying narratives, 'drawing together his memories of traditional stories and performance, his historical knowledge and awareness of change' and, more importantly teaching these stories and thereby 'producing locality, making indigenous knowledge live on'. This elderly individual was 'place-making' in the symbolic sense, and the other papers in the first section also probe the reflexive constitution of people, knowledge and environment, subjects that will resurface in later case studies. The following section grounds these relationships in how indigenous practice physically affects the local environment.

Chandran and Hughes' study of sacred groves in India and the Ancient Mediterranean explores the 'roles of these refugia in maintaining a balance between human groups and the ecosystems of which they are part'. The groves were dedicated to tree-dwelling deities and both cultures deemed that the gods would be honoured by 'keeping the original forest as undisturbed as possible', which had the desirable side-effect of conservation of the ecosystem. It seems highly likely that the habitations of the Gods were not randomly ascribed: as with the Aborigines exercising prohibitions over a lagoon that housed a vital freshwater spring, Mediterranean and Indian groves often protected watersheds, preventing rain runoff and thereby limiting both flood and drought. The authors report that, in India, changes in religion and the move to a less localised economy were 'deleterious to preservation of the groves' and that 'the traditional form of forest preservation' was replaced by heavy-handed colonial conservation (and resource exploitation) practices that 'were seen as attacks on the community-based system of use'. Later in this volume Damodaran investigates 'the complexities of the relationship between the spiritual memory of remembered landscapes and modern politics', noting that the sacred groves have been taken up again in an altered context, the resurgence of festivals surrounding them, despite a despoiled landscape, coming to stand for memory of 'an idyllic environmental past' and protest about 'contemporary political and economic problems'.

Chandran and Hughes argue that community-based conservation practices rooted in community need to husband resources might still be the key to sustainable co-habitation of humans and forests, an ideal that the sacred groves embodied. Other authors are considerably less hopeful about traditional practices holding the key to conservation, with Henley going so far as to state baldly that, 'The conservation of tropical rainforests is incompatible with agriculture – traditional or otherwise, and notwithstanding insistent claims to the contrary'. Virtanen emphasises far more than Chandran and Hughes the erasure of traditional conservation sites with the weakening of the religious beliefs that demanded their protection and disputes, linked to political changes, over who has the right to impose behavioural prohibitions. She reports 'signs that respect for customary norms is weakening with regard to sacred forests, which are threatened by uncontrolled bush-fires and illegal cutting of trees'. Indigenous beliefs that sufficed in the past to protect the forest are now incapable of doing so. She does note that the relative inaccessibility and low population density of Chôa were factors in its forest preservation, but argues that,

> it is unlikely that geographical factors alone would have been sufficient to preserve the sacred forests of Chinda and Mungwa in their present form. The sacredness of the forests has contributed substantially to the preservation of their relatively high biological diversity.

It is one thing, though, to observe that this is so – as Chandran and Hughes do with Indian sacred groves – and another to machinate in favour of such 'management' strategies. Virtanen argues that

> if local institutions are promoted from above merely as a tool to preserve bio-diversity for its own sake, separating the institutions from their socio-cultural bases, they will soon lose their legitimacy. Sacredness is a powerful means of conservation only when it is linked to a broadly respected belief system, with adequate normative controls and means for their enforcement.

When norms break down, eroded by competing knowledge and value systems, they are no longer adequate alone to control environmental practices. Meddling with indigenous knowledge can be counter-productive if the local legitimacy of this knowledge is cast into doubt by cultural and social change (as in the power changes of Madagascar's Civil War); Virtanen thus offers a corrective to the view that local management practices (especially if conservation is not the primary stated aim) are necessarily adequate to cope with times of upheaval during which the environment may be most at risk.

Mukamuri goes further, arguing that 'Researchers should be more sceptical when they talk about a "ritually controlled ecosystem"' and open their eyes to the fact that

> local religious institutions are used by ruling lineages for political control, to grant preferential access to particular resources and to enhance political hegemony. The symbolism expressed in the rituals and environmental taboos is more powerful than the idiom of 'conservation'.

Furthermore, he suggests that Karanga rituals do not primarily emphasise the environment, though they may have 'major ecological effects'. Both he and Virtanen warn that the ecological effects of indigenous practices are probably not their primary intentions. Furthermore, indigenous environmental thinking is far from homogeneous, as 'Chiefs, lineage elders, the young and old, rich and poor, men and women all try to manipulate the supposedly shared belief system to their own advantage', for example in chiefs' securing for themselves 'preferential access to rural resources'. He observes, for example, that some sacred pools acquire 'sacredness' in distinctly dubious ways and may even be 'made by people and belong to them for which they get economic benefits'; sacred places thus 'act as supporting pillars to the political hegemony of ruling lineages'. Likewise, it being the case that 'the mountain should burn at the start of a good rainy season', supposedly set alight by the spirit Zame, it seems that,

> some of the chiefs secretly start fires so that they can legitimise their power and hence maintain recognition. If the mountain does not burn, the chief's rule is questioned: Zame is saying that he should not be chief and therefore the rains will be poor. Of late the mountain has burnt but the rains have been poor, and this reinforces the questioning of local cynics as to whether it is being illicitly burnt.

As Virtanen argues, once the cultural traditions that demand certain environmental practices are called into question, the ecological regimes they promote are vulnerable. However, the process works two ways: if local people have other reasons to continue these practices, religion may be called into service to legitimise them. Mukamuri reports local people refusing to fell fruit trees, despite government policy, because they wished (quite reasonably) to continue to enjoy them: here 'people were able to mobilise environmental religion to better husband their natural resources'. Thus the degree to which indigenous ecological knowledge is promoted must be viewed in the context of other factors tending to its survival or occlusion. Mukamuri is not the first to warn that those seeking to intervene in such communities must be aware that 'groups and individuals have divergent interests', some of which may be served by husbanding traditional ecological practices and others of which may not. It is not enough to regard traditional ecological knowledge as intrinsically 'virtuous'.

Castro explores the complex motivations behind indigenous management strategies in the Amazonian floodplain, tracing the 'mosaics of prescriptions created throughout the history of the users according to a range of incentives and goals'. He divides such prescriptions into 'ecological', based on direct needs and resource characteristics; 'cultural', based on 'social interactions that may indirectly enhance or maintain resource productivity'; and 'political', 'developed from a "conscious" problem-solving process carried out by a user group'. He regards resource-use prescriptions as gradually evolving from the simplest – ecological – to the most complex – political. He notes that fear of mythical creatures seems in the past to have preserved the productivity of floodplain lakes in the Amazon, though he cautions against assuming 'conservation functionality of the ecological and cultural prescriptions'; as with Virtanen's observations about geographical factors limiting forest exploitation in Chinda and Mungwa, it is difficult to gauge exactly in which proportions factors influence a conservation outcome.

Gradually evolving from the subsistence-based Amerindian period, 'the maintenance of ecological and cultural prescriptions was subordinated to a higher social structure', but Castro recognises that this structure has itself altered from a top-down 'patronage' system to more recent emphasis on local stakeholder participation: the wording of prescriptions regarding local resources has changed but, by negotiating with the wider society, local actors have created fishing accords, for example, that 'are closely related to former LMS [local management systems], as the prescriptions capture ecological and social influences from the past. After all, fishing accords represent a social innovation to overcome an old problem – ensuring resource access'. Balali similarly describes how social conditions have altered to such an extent that traditional paradigms are insufficient alone to control resource management – in this case of water in Iran – but have been integrated with the 'paradigm of industrial modernity' to create solutions that partake of the best features of both, a combination he describes as 'reflexive modernity'. He traces the change from co-operatively managed *Qanat* irrigation systems to the imposition of modern wells and dams, to a system

> characterised (1) by the integration of traditional (indigenous, small scale) and modern (scientific, large scale) technological infrastructure, (2) by participatory water resources management in the form of multi-stakeholder platforms or water user associations, and (3) by its recognition of the diversity of cultural values of water

As with the forest-use prescriptions identified by Virtanen or the fishing restrictions of the Amerindian period described by Castro, the *Qanat* system is a social and ethical phenomenon, tied to culture and religion; it differs from the former practices in also representing a technological system that was specifically undermined by the promulgators of modernity:

> To pave the way for industrial modernity, the Iranian authorities tried to belittle all traditional irrigation and production systems. Most Iranian scholars and politicians exaggerated the technical deficiencies of the *Qanats* to justify their own programmes and to convince farmers to use pump extraction instead of *Qanat*

More recently, however, and in the face of problems thrown up by unbridled use of industrial technology, the *Qanat* system been to an extent rehabilitated, its inherent sustainability in tap the groundwater potential only up to, and never beyond the limits of natural replenishment tapping 'the groundwater potential only up to, and never beyond the limits of natural replenishment' being recognised. However, the *Qanat* system as it previously functioned cannot truly be restored in the absence of the traditional *Buneh* social organisation that provided its original context; this context having disappeared in a public revulsion against feudalism, an ersatz version of the previous local management system must be developed, more in keeping with the political mores of modernity, 'with a participatory rather than a hierarchical character'; one factor Balali cites as working against this is the perception, in an Iran embracing modernity, that *Qanats* are old fashioned and aligned with a system that the country wishes now to turn its back on. Chandran and Hughes, Virtanen, Mukamuri, Castro and Balali all trace in varying contexts and with varying emphases the dynamism of local resource management strategies, their ability to merge with strategies born of different motivations, the risks to conservation practices if local actors lose faith in the original motivations and no convincing new ones are developed and their survival if old ways are invested with new meaning.

MYTH, MEMORY AND RHETORIC

'It is a matter of perspective'. So Kull articulates this section's central concern, in his discussion of how the rural Malagasy do not perceive their tavy agriculture as 'sacrificing nature for short-term needs' but 'instead transforming nature to be of more use to them'. Logically, as farmers, land use strategies such as tavy, characterised by slash-and-burn, make sense, if there is an apparently inexhaustible supply of land and scant population. Kull argues that conservation strategies in Madagascar – developed by outside agencies, even if they attempt to work with local actors – are frequently based on environmental myths, of the Malagasy both as wanton destroyers and as potential allies in managing reserves, that have unintended and pernicious consequences in terms 'of decoupling local people from their livelihoods and cultural resource base instead of reinforcing their relations with their environment'. Middleton also deals with Madagascar but her myths are primarily those generated by the local people, not by colonial botanists and global environmentalists. She explores memories of Malagasy Cactus (eradicated by the French in the 1920s) and how these have changed over two decades in the light of changing local circumstances, most interestingly the controversy over whether to eradicate another troublesome prickly pear, rakatemena. Middleton found that, over time, the widespread local story of a great famine caused by the eradication of Malagasy cactus 'no longer monopolised the imagination' and that 'In less than a generation, widely shared memory and interpretation had given way to surprisingly diverse narrative about the past.' In her more recent fieldwork, many interviewees claimed that only vagabonds on the margins of society had depended on the cactus for subsistence and she links this to debate over whether now to eradicate rakatemena. Perhaps seeing the effects of infestation at first hand had prompted her respondents to remember how things really were before Malagasy cactus was eradicated, or perhaps those with a vested interest in rakatemena eradication were rewriting history to justify their wish? Poorer respondents seemed to cling to the famine narrative, as if 'by emphasising how intensely the ancestors had suffered when their staple was taken from them, they hoped to underscore the food security risks that raketamena eradication would pose'.

Middleton is interested in the circulation and modification of rhetoric about Malagasy cactus in local policy debates and the extent to which this has caught bodies from outside (such as WWF) unawares; and also in an intriguing hybridisation of local and outside knowledge: the Karembola are appropriating colonial tropes and metaphors and increasingly repeat an 'evolutionary narrative that represented the eradication of Malagasy Cactus as a civilising event that transformed primitive, anarchic forest-dwellers, ignorant of money and agriculture, into clean, industrious farmers settled in an open, domesticated landscape'. All accounts of perception and memory are at risk of being affected by the lens through which the researcher (as well as her subjects) looks and Middleton is aware that her questioning may have altered over the decades, her following up of cues and her changing research interests affecting the responses elicited. This is as valid a caveat in research regarding indigenous knowledge as in any sociological or anthropological study and returns our attention again to the difficulty of identifying any 'pure' indigenous knowledge', not hybridised with or modified by Western forms and assumptions. She does insist, though, that the assumption that folk memory gradually becomes more 'stylised, homogenised, dare I say fossilised', losing 'the immediacy of original experience' is here problematised by its being reinjected with vigour by the new context, so that her subjects 'zigzagged back from communal narrative to recover original experience, finding all kinds of fresh connections between the present and the past'.

During the colonial period in Chotanagpur, administrators argued 'that traditional forest use seriously exacerbated the destruction of the forest' (though in fact the effects of colonial forest exploitation would be infinitely greater) and must be curtailed; the relationship between local people and their environment – their dependence on the forest for food and in terms of religion – was altered by new proscriptions. Where previously the people had a 'lived relationship' with Nature, dancing, singing and sharing cigarettes with the forest spirits, the forest now came under the jurisdiction of 'colonial scientists and policy makers and later of a modernising nationalist elite'. But, as Damodaran explores, the memory of that past relationship survived and came to stand nostalgically for the lost past of Chotanagpur, 'to be revived in complex oppositional contexts'; even if the original functions have been largely lost, indigenous tradition is invoked to protest against change or to assert rights over ancestral lands. For example, protests against dam building at Koel Karo or teak planting in ancient sal forest have invoked the sacredness of groves, habitations of Gods, even though this may no longer literally be believed in; and these claims have been hybridised with 'reference to a global environmentalism (or environmental religion). This involves arguing that the local people are the best stewards of the landscape and have the best claims to control it'. Thus indigenous knowledge looks to memories of the past but also ensures its own future by assimilating contemporary Western claims about its superiority.

Introduction

COLLISIONS AND COLLABORATIONS

The final section deals explicitly with an issue that has been lurking throughout – the collisions and collaborations of different knowledge systems, with effects that are often surprising and contrary to assumptions about the stifling of indigenous meaning by global hegemonies. In Low's account of buchu, a herb traditionally used by the Khoisan, and its adoption by Western alternative medicine enthusiasts who have 'dumbed-down' its original complex associations, the Western users make the fallacious assumption that they are continuing an indigenous traditional use – which indeed is the selling point of the plant – whereas their terms of reference are totally different. They believe they have appropriated Khoisan knowledge but 'If the understanding of illness, how it is caused, spread and healed is different there is historical slippage in justifying parity of current European medical use with past Khoisan use'. Buchu has a long history of Western use, but as its efficacy has become doubted in orthodox medicine, the 'indigenous historical usage of buchu has become all the more emphasised'; it is sold as 'an implicitly healthy, indigenous alternative medicine', its use and marketing a combination of idealism and gullibility, a misunderstanding rather than an abuse of indigenous knowledge.

Dahlberg and Blaikie offer a more straightforward example of divergent understandings in their exploration of Botswanan villagers' perceptions of rainfall and drought as viewed beside scientific data. Scientists find no long-term change in rainfall while almost all villagers report a decline; scientists perceive overstocking with livestock as an environmental problem, locals do not. Memory comes under scrutiny again here, in that 'Initially many stated there were no droughts in 'the old days'', but would later recount stories told by their parents and grandparents, of droughts so bad that 'people had to eat the skins of animals', and 'there was one drought that was extremely bad and it was called "don't ask me"'. Changes in local activities and needs – in terms of roads, settlement, employment and 'how people move through the landscape' – may partly explain the discrepancy, in that 'rainfall may have declined when seen in relation to its relative sufficiency for an altered set of requirements'. The authors shrewdly observe that 'rainfall is often used as a metaphor for a general change in livelihood circumstances'; in the same way that Damodaran describes sacred groves becoming a symbol of political dissatisfaction and a locus of nostalgia for a past pristine environment, villagers here reporting declining rainfall 'may not primarily reflect a particular environmental change, but instead a feeling of regret for "the good old days"'. We all, including scientists, invent narratives that serve to make sense of a complex and often ambivalent reality and necessarily foreground the factors that seem of most relevance to particular circumstances and perceptions – in so doing, 'reality is simplified'. Since 'Local narratives are not told for the same reasons as those told by scientists' – they are experiential rather than analytical – it is not surprising that they are different. One of the many thorny issues in dealing with indigenous knowledge(s) is how to manage with mutual respect when it becomes necessary to privilege one narrative over another, in policymaking, for example.

Nag's short chapter turns the commonly assumed power-structure of us and other on its head by reporting how colonial administrators used predictions of famine made by tribal people in the Indo-Burmese Mizo Hills, based on signs observed in bamboo and rats. Instead of a tale of local knowledge being crushed or disregarded under colonialism, this short essay reports how the people 'were not used to ... being assisted in times of crisis' and their delight when the colonial government gave them aid, 'effecting a metamorphosis of the image of the Raj in the minds of the tribals' who had previously loathed the British and resisted their incursions into the Mizo Hills. A small dose of respect for and response to local environmental knowledge dramatically altered local perceptions of the colonists whose 'benevolence and kindness' they lauded and would come to look back on nostalgically; and of course the colonists benefited from the locals' ability to predict famine, which allowed time for preparation before things got too bad.

Henley ends the volume on a cautionary note against idealising indigenous knowledge systems and imputing to them an environmental benefit that is not verifiable in the field. He states:

> The view of traditional swidden farming as an environmentally benign practice is an idealised one, and should not be allowed to obscure the fundamental incompatibility of agriculture with nature conservation.

In analysing historical data, he finds that those who claim that Indonesian swidden farming is environmentally sound because its long fallow periods allow forest to regrow are ignoring or unaware of considerable historical evidence of unsustainable short fallow periods. That conservationists have encouraged the development of 'swidden-derived commercial arboriculture', perceiving it as 'ethnoconservation' and its practitioners as 'managers of the forest', based on an idealised notion of indigenous harmony with the environment, has, says Henley, potentially devastating environmental consequences. He states:

> traditional swidden farming practices in northern Sulawesi clearly did not involve any kind of symbiosis with the natural forest. On the contrary, they involved the destruction of that forest and its replacement by completely different, man-made ecosystems that were much less rich and diverse.

Potentially misguided conservation strategy has arisen from a romantic 'insistence on the ecological virtue' of tribal people and blindness to the destructive consequences of traditional resource use. As Henley argues, the fallow periods in even the longest swidden cycles are degrees of magnitude too short to allow the reestablishment of centuries-old climax rainforest vegetation. It would be rose-tinted in the extreme to hope for more than that

The support or consent of nearby swidden farming populations, if it can be obtained, may facilitate the protection of nature reserves. But what is certain is that in the area they actually farm, there will be no more rainforest.

As Kull's essay probes, the question of who has the right to assert the primacy of their goals is a delicate one: in the local context, unsustainable practices may seem desirable and their policing by the state or global agencies may lead to human hardship at worst, a feeling of disenfranchisement at best. The Madagascar case is just one among thousands where the balance between indigenous and global perspectives must be carefully managed; and the paradox looms that if we live in a post-modern world that purports to value the indigenous perspective, it is difficult to cherry-pick exactly which bits of it we value. A dynamic interaction between worldviews is all that can truly be aimed for when the West seeks out indigenous knowledge.

I. Defining Indigeneity

Representations of Tropical Forests and Tropical Forest-Dwellers in Travel Accounts of *National Geographic*

Anja Nygren

INTRODUCTION

Far-away and fabulously exotic tropical lands and tropical peoples have long inspired curiosity and aroused the imagination, finding expression in both scholarly and journalistic writing (Raffles 2002; Slater 2002; Stepan 2001). Travel writing is one of the most important sources of popular discourses and images of tropical landscapes and tropical peoples. As such, travel writers play a crucial role in shaping public understandings of tropical forests and tropical forest-dwellers, as well as in regulating public opinion in relation to the implications of environmental policies that need to be implemented in the tropics (Arnold 1996: 141–68; Duncan and Gregory 1999; Holland and Huggan 1998: 67–81).

In this essay, I examine the highly selective, essentialist images that have come to represent indigenous and non-indigenous forest-dwellers in accounts of travels in tropical America published in *National Geographic*. This famous U.S. magazine, with an authoritative voice and a total circulation of 9.5 million copies per month, offers an excellent source of discourses and images that have come to represent the neotropics and its peoples in Western popular imagination.[1] My main aim is to show how particular representations and discourses are privileged in the magazine's travel accounts and how they tend to create hierarchical polarities. For the purpose of this examination, I have analysed all the relevant travel writings with a focus on neotropics and neotropical forest-dwellers that were published in *National Geographic* (earlier *The National Geographic Magazine*) during the years 1888–2004. Most of these, thirty-seven accounts in all, were written by British or U.S. travellers.

Anthropologists, geographers, and literary historians have recently called attention to a rich corpus of 'Western' narratives that tend to categorise 'non-Western' peoples as racial, cultural, and gendered others.[2] Torgovnick (1990), Arnold (1996) and Stepan (2001) have analysed the discourses of primitivism and otherness embedded in Western thinking about the tropics, while Pratt (1992, 1994) has examined the imperialist discourses interwoven in the early Western travel accounts of the world outside Europe. Lutz and Collins (1993) have analysed the images of non-Western peoples as exotic others as portrayed in *National Geographic*, while Rothenberg (1994) has called attention to the

Environmental Values **15** (2006): 505–525.

magazine's representations of non-European women as mysterious and naturally erotic others. Ramos (1994, 1998) and Slater (1996, 2002) offer inspiring analyses of the historical trajectories of the images of the Amazonian rainforests and the Amazonian Indians as now infernal, now paradisal others.

Characteristic of many of these analyses is the view that Western narratives of the non-Western world rely on a powerful distinction between 'us' and 'them' (Lutz and Collins 1993: 26, 110–11; Torgovnick 1990). My research on travel accounts by *National Geographic* starts from a somewhat different point of view. As will be shown in the following analysis, the poles of this dichotomy are not simply 'we' and 'the other', or 'the West' and 'the rest'. Instead, the accounts produced by *National Geographic* construct essentialist images of tropical forest-dwellers as either 'good' or 'bad' others. Whereas most of the late nineteenth century and early twentieth century travel writings of *National Geographic* presented rainforest Indians as wild savages and the non-indigenous people as progressive pioneers, this distinction began to reverse in the early 1970s. In accordance with the growing global concern over tropical deforestation and the increasing attention being paid to tropical rainforests as remarkable sites of biodiversity protection, more recent travel accounts tend to produce images of rainforest Indians as 'noble natives', dwelling in nature according to nature, while small-scale settlers and non-indigenous rural poor are portrayed as 'ignoble villains', who are in need of control and order.[3]

This duality in the Western views of the other, as either good and peaceful or bad and violent, is implicitly present in Torgovnick's (1990: 3) statement that primitive peoples 'exist for us in a cherished series of dichotomies; by turns gentle, in tune with nature, paradisal, ideal – or violent, in need of control; what we should emulate or, alternatively, what we should fear'. This challenging idea, also suggested by Slater (2000: 78), has, however, rarely been elaborated further in analyses of the discourses on tropical forest-dwellers. Especially when dealing with representations of Amazonia, the attention has so strongly focused on the 'majestic forests' and 'mysterious Indians' that the non-indigenous residents have largely remained invisible. As remarked by Nugent (1993: 20, 43), non-indigenous Amazonians represent an 'incomplete other', having little culture and little history and thus lacking sufficient status of difference to be included in discussions of Amazonia.

Based on the theoretical argumentations by Lutz and Collins (1993: 1–3), the focus in the following analysis will not be so much on how 'realistic' the images of tropical forest-dwellers presented in the travel writings of *National Geographic* are, but on the imaginative spaces that the tropical peoples occupy in the travel writers' minds. Narratives about foreign places and alien practices, whether scientific or not, are never simple documents or objective mirrors of reality. They also either reinforce or challenge general understandings of cultural similarity and difference, thus reflecting substantially the attitudes of those who are behind the text (Briggs 1996; Graham 2002; Oakdale 2004). As will be demonstrated below, the predominance of particular representations at a particular time depends not so much upon essential differences between the target populations themselves, but upon the prevailing regimes of representation that shape the writers' perceptions and interpretations of the issues under consideration (Duncan and Gregory 1999; Porter 1993; Schwartz 1996).

In this connection, the images presented in the travel narratives of the tropics and tropical peoples can not be simply dismissed as incorrect or false. Instead, they need to be examined in relation to the contexts in which they are generated and the purposes they serve. Certain representations become socially dominant not merely on rhetorical grounds; they are also closely related to issues of authority and power (Conklin 2002; Jackson 1997; Li 2000). In producing feelings of closeness and empathy with some, and distance and discredit towards others, narratives of tropical forests and tropical forest-dwellers build upon social classifications and moral cartographies that construct hierarchical patterns of otherness. At the same time, these narratives have considerable influence on popular understandings of environmental agendas and environmental policies related to the tropics.

My examination of these travel accounts combines qualitative content analysis with textual interpretation in an effort to identify the characteristic representations they construct and the transformations and consistencies that can be observed in them over time. For this purpose, I have utilised the QSR N6 qualitative data analysis programme.[4] Owing to space limitations, I will focus here on what I take to be the dominant representations in the analysed accounts. This does not, however, imply that the perceptions of neotropical forests and neotropical peoples portrayed in these accounts were absolutely reified or monolithic. As remarked by Arnold (1996: 142–57) and McEwan (1996), the images of tropical forests and tropical peoples, although essentialist and stereotypical, can also be ambivalent, containing elements of different sets of imagery side by side.[5] In this respect, it is important to note that the selection of the narratives and representations accepted for publication in such a well-established magazine as *National Geographic* is often a result of negotiation and compromise among various stakeholders and their personal and institutional ambitions. The following analysis aims to understand the significant alteration in the images of indigenous and non-indigenous forest-dwellers in the accounts by *National Geographic* during the 1970s within the changing context of travelling and the changing role of tropical rainforests in the global environmental discourses and policies.

CONQUERING THE 'GREEN HELL'

Characteristic of late-nineteenth century and early-twentieth century travel writings on tropical America in *National Geographic* is the view of tropical forests as an impediment and/or a challenge. At that time rainforests were considered to be demeaning peripheries as well as landscapes of abundant potentiality. These views have much to do with the social and political climate of scientific and economic exploration in which many of these travellers entered the neotropics. Many of them were 'scientist-adventurers' who were searching for new knowledge and new economic opportunities in largely 'unexplored' tropical lands.[6] This spirit of intrepid exploration was indicated already in the headlines of many of these accounts, including 'Across Nicaragua with Transit and Machéte' (Peary 1889), 'Exploring the Valley of the Amazon in a Hydroplane' (Stevens 1926), and 'A New World to Explore: In the Tree-Roof of the British Guiana Forest Flourishes Much Hitherto-Unknown Life' (Hingston 1932).

Typical of these accounts is the representation of tropical forests as lands that are completely unknown. They are behind beyond, terrains that have never been trodden by the foot of a white man. The writers present heroic stories of solitary explorers who survey hitherto unknown rivers and hack their way through tangled forests, carrying heavy packs of supplies and enduring the pervasive isolation of the jungle (Holt 1933; Schurz 1936; Stevens 1926; Sultan 1932). This atmosphere of emptiness and desolation is necessary in order to justify the Western discovery and conquest of the tropical regions, represented as uninhabited peripheries (Lowenthal 1997; Stepan 2001). As a symbol of conquest, some of the explorers carry the flag of their own country during their travels.

Many of these narratives also present allegories, reminiscent of Robinson Crusoe, of men taming the wild forests. Gill's (1934: 139–43) account of pioneering in Ecuador offers an illustrative example: 'It seemed incredible that, out of all this disorder, we could establish a small area of civilisation for ourselves by the primitive means available.' Gill's account also portrays a picture of the giant cinnamon tree which was the first to fall when the 'modern Crusoes' began the difficult task of carving a modern home out of the jungle. As is typical of imperialist travel accounts (Pratt 1992; Spurr 1993), many of these writers use sexual images to portray their masculine conquest of the virgin forest. The term 'man' is used to refer to humans, and the female pronoun represents nature in accounts describing the traveller's eagerness to probe nature's secrets and tap the forest's wealth.

Accounts of the hazards encountered during the expedition merely magnify the achievements of the intrepid explorers. The explorers are beset by pounding rains and insufferable heat; they are covered with insect bites, ravaged by malarial fevers, and frightened by snakes and jaguars. Such hardships abound in the account by Robert Peary (1889), the famous discoverer of the North Pole, of his journey through the untouched forests of Río San Juan, Nicaragua. According to Peary, the days were filled with constant obstacles, and the tropical thicket was so dense that it was impossible for even a strong, active man to penetrate through it without a machete. One had to wade in knee-deep mud and be alert for crocodiles, peccaries, and venomous insects. For Peary, Río San Juan was an awesome jungle, with few links to the outside world, and for this reason its exploration could not be delegated to just anyone. As such, Peary's account repeats the epoch's conventional narrative of the white man conquering the hostile tropics.

In the travel accounts of *National Geographic* at this time, the rainforest is commonly presented as a 'Green Hell' and an 'enemy of mankind'. Romantic views of tropical forests as a source of nostalgia and a cradle of peacefulness are not absent, but they are more uncommon.[7] In several accounts, the rainforest is presented as a gloomy and disease-ridden jungle that represents an untamed savagery. This heart of darkness is considered to be a source of fear and panic that easily engulfs lonely travellers. In his account of an expedition through British Guiana, Hingston (1932: 625–8) describes his feelings 'of being completely shut in' by a jungle with no horizon and his 'immeasurable relief on getting back at last into the open and enjoying the spaciousness and freedom'. Sultan (1932: 593) conjures up visions of the riotous vegetation of Nicaraguan rainforests, 'where the footing is always insecure' and the jungle is 'so thick that you can rarely see ten feet in any direction'.

Given these conceptions, a great number of the late nineteenth-century and early twentieth-century travel writings published in *National Geographic* consider the value of the tropical forest to arise from the possibility that its remarkable potential could be harnessed in the service of human progress. In these accounts, rainforests appear as an obstacle to be overcome and/or as a mysterious 'El Dorado' to be discovered. Such views are closely linked to the epoch's Western-oriented political-economic ambitions, which considered tropical forests as inexhaustible resources to be exploited and dominated by means of Western scientific and technological innovations (Nugent 1994). In accordance with this conception, many of the epoch's travel writers tend to project categorical views of tropical forest-dwellers as either primitives living in the backwoods or as frontier-breakers taming the hostile jungle.

INDOLENT SAVAGES VERSUS PROGRESSIVE PIONEERS

Concerning the people living in the neotropics, the travel accounts of *National Geographic* during the late nineteenth and early twentieth century tend to produce strict distinctions between those who are cultivated and those who are not. Like tropical nature itself, the rainforest Indians are considered dangerously unpredictable until controlled (Rothenberg 1994: 164–5). In the photographs of these accounts, the Indians are portrayed either as powerful hunter-warriors who glare wild-eyed at the camera or as backward savages who are dazzled by the modern devices of the white man. Stevens (1926: 400–2) presents a picture where 'the tallest of the Mayongong Indians came hardly more than shoulder-high to members of the expedition', claiming that this was the first time that these natives 'had any contact with the civilisation'. By emphasising the Indians' primitivism, the travel writers provide moral justifications for 'modernising' them.

At the same time, non-indigenous farmers are presented as backward but hard-working pioneers who are eagerly participating in the development of modern society. Although the accounts of *National Geographic* of this time regret the general nonchalance of the rural people living in the neotropics, several texts laud the tropical settlers, who are rescuing their living spheres from a state of idleness and abandonment. De Pinedo (1928: 283) describes with relief how 'here and there, along rivers and the Matto Grosso's fringe, farms and plantations are cut from the ever-engulfing jungle', while Stevens (1926: 383) describes how the Boa Vista forest frontier 'reminds one of bygone American frontiers', with open ranges and cattle shipped down river. According to Sultan (1932: 609), the few cars on the Nicaraguan roads, carrying some government officials or landed proprietors on business, 'add the touch which shows that you are traveling in a civilised country'.

The differences between indigenous and non-indigenous livelihoods are interpreted as the effects of uneven cultural development. The rainforest Indians are portrayed as eking out a miserable hand-to-mouth existence through hunting and gathering and considered doomed to give way to more progressive ways of exploiting the tropical resources. An illustrative example of this perception is Holt's (1933: 600-1) description of how agriculture among the natives of Brazilian Amazon 'has not progressed beyond the simple stage' and how an Indian, when he travels, 'carries his entire household. The whole family go, and with a pot, a few fishhooks, bow and arrows for baggage, are prepared to live indefinitely off the country'. Correspondingly, Sultan (1932: 608) describes how 'the huts of the Sumo Indians are simple structures, thatched with palm leaves' and 'their worldly possessions are confined to bows, arrows, blowguns, and one or two pots and pans'.

Even though Indians serve as indispensable guides for many of these travellers, the native conceptions of nature rarely become an object of study (cf. Miller 1996: 12). Hingston (1932: 642) describes how the Arawak Indians paddle the travellers' boats, clear the forest paths, climb tree trunks, skin specimens, and accomplish a dozen other tasks associated with life in the jungle. Although the Arawak are depicted as 'born naturalists', their environmental knowledge is still described as primitive. In several narratives the native inhabitants are considered as culturally backward which is reflected in the writers' tendency to use animal metaphors when referring to Indians. Stevens (1926: 412) makes disparaging comments on Parima Indians, who stand on one leg 'like storks' when he tries to photograph them.

The non-indigenous residents of the neotropics are, in contrast, described as assiduous entrepreneurs who are introducing more rational ways of utilising the tropical resources. Holt (1933: 587, 602) portrays a picture of Santa Isabel on Río Negro where three houses, standing in a small grassy clearing, 'greet the eye at this remote outpost of commerce'. He also praises the residents of Pará for their persistent efforts to remedy the backwardness of an environment 'still beleaguered by the jungle'. 'After voyaging up the Amazon for 900 miles between forest walls with hardly a gap', Holt (1933: 599) admires the modernity of Manaós, where 'handsome buildings, electric lights, and boulevards lined with artistically trimmed fig trees' have been created 'as if by magic from the jungle'.

One of the most powerful indicators of Indian primitivism is their nakedness: the Indians are described at going naked, without showing any shame, until civilised. An illustrative example of this conception is Stevens' (1926: 385) description of how the Indian women in Branco learn to 'look askance at their still naked sister' after they have been garbed 'for the first time in their lives'. The non-indigenous residents are, in contrast, portrayed as people dressed in civilised clothes, and their festivals are depicted as beautiful gatherings where gentlemen in neat shirts and trousers keep company with ladies clad in showy print dresses. Peary (1889: 331) tells with delight how, after getting from the jungle to more civilised parts of Nicaragua, he meets 'black-eyed and brown-limbed *señoritas*, instead of wild hogs and turkeys', and at night, 'he hears, not the scream of tigers [jaguars], but the songs of the *lavandera*'s [laundress'] ecru daughters floating across the stream'.

Concerning the 'marvels of modernity', the early travel writings of *National Geographic* depict Indians as innocent primitives who are afraid of radios and flashlights and who examine mirrors with wonder, thus implicitly bolstering the view that the Indians lack any self-consciousness until contacted by modernity (cf. Hyndman 2002: 50). Sultan (1932: 621) frets over their Indian servants, who do not understand why sheets should be tucked in at the foot of the bed or what possible use there could be for two forks at one meal, while Gill (1934: 158, 161–2) makes light of the Ecuadorian 'jungle Indians' among whom 'one's calling card is his own painted face'. She also makes disparaging comments on how these Indians do not comprehend the idea of typing but are astonished 'why anyone should spend so much time poking the keys merely to make rows of black marks on paper'. Correspondingly Stevens (1926: 408) laughs at the Maku Indians, who have difficulties in putting on the clothes received as a gift from the white man. He describes the plight of a Maku chief, who finds difficulty in fastening his first shirt around his neck and 'calls on his squaw for assistance, much as a civilised husband sometimes calls on his wife for help in a similar difficulty'.[8]

In contrast of these portrayals of primitive Indians in the woods, the travel accounts of *National Geographic* of this time praise tropical fields and plantations as islands of opulence and civilisation. In accordance with the epoch's global political economy, many of these accounts express fervent optimism about the future prospects in the neotropics. At the end of his journey in Nicaragua, Peary (1889: 334–5) gazes down upon the valley of Tola and paints an enthusiastic vision of this landscape, when 'fields have replaced forests' and 'the fertile shores

Anja Nygren

of the Tola basin are occupied by cacao plantations'. This picture gives Peary much pleasure, for it means 'the dream of centuries realised' and 'the cry of commerce answered'. Hulse (1927) and Marden (1944) are equally convinced of the success of converting tropical forests into coffee plantations and cattle estates in Central America, while de Pinedo (1926) has ambitious visions of the introduction of steamboats and sawmills in Amazonia. A similar vision of the Western exploitation of the neotropics continues in the magazine's travel writings up till the 1970s, when the representations of neotropical forests and neotropical forest-dwellers portrayed in the magazine start to reflect new attitudes toward global environmental policies and public opinions concerning the rainforests.

PRESERVING THE VERDANT EDEN

Whereas the late nineteenth century and early twentieth century travel writings of *National Geographic* on the neotropics were stylised as heroic accounts by 'modern Crusoes' who were struggling to conquer the hostile Inferno of Nature, many of the magazine's more recent travel writings present the rainforest as a fragile sanctuary that demands assiduous care. It is lauded as an earthly Garden of Eden, where eternal greenery and untamed luxuriance reign. This evergreen realm is seen as a refuge from the ills of civilisation and as a source of tranquillity for the human spirit.

This reversal from the rhetoric of the Green Hell to that of Paradise Threatened has close links to the international concern for tropical deforestation and forest degradation that came into prominence in the early 1970s (Nugent 1994; Slater 2002: 9–16). Under the banner of a new environmental awareness and new environmental politics, the rainforests have since then been regarded as endangered global resources whose destruction presents worldwide risks and whose rescue from obliteration requires worldwide efforts. These rainforests are repositories of incredible richness and sources of scintillating insights awaiting wide-eyed scientists; they are the Earth's green belt, the world's largest reservoir of genetic traits and an irretrievable memory bank that has evolved over billions of years (Melham 1990; Morell 1999; White 1983).

As a consequence, the tropical rainforests are losing their earlier hostile image and becoming places of impassioned protection. Many of today's writings on the neotropics published in *National Geographic* are written by persons who have participated in scientific expeditions to learn about the natural wonders of the rainforests and to understand the complexity of tropical ecosystems. They include detailed descriptions of scientists who devote themselves to studying the tropical hotspots of biodiversity and training the native peoples to preserve their environmental and cultural heritages (Garrett 1989; O'Neill 1993; Kamper 2000; Morell 1999).

Representations of Tropical Forests

The second type of contemporary travel accounts published in *National Geographic* is that which elaborates originality and adventure. Whereas the earlier accounts portrayed images of explorers who went to the tropics to discover 'untouched' territories, modern accounts tell about travellers who go to the tropics to (re)discover their 'authentic' self (cf. Blanton 1997: 1–29; Short 1991: 60–1). These writers emphasise the spiritual regeneration and the mental renewal experienced during their tropical journeys. Typical of these individual odysseys is the portrayal of rainforests as places of recreation instead of work, and as spaces of spectacular feelings instead of sober observation. To distinguish themselves from 'ordinary tourists', the travellers trek into the most pristine forests and visit the most secluded peoples to know the rainforests and their inhabitants profoundly (Risse 1998; Steve 1999). Schreider and Schreider (1970), and Chmielinski (1987) describe their participation in the race to be the first team to navigate the Amazon from the source to sea and to see with their own eyes every foot of this unique waterway.

The third group of contemporary travel accounts are those narratives in which the writer devotes himself/herself to efforts to understand the native culture and to record the Indians' traditional way of life. These writers focus their narratives on descriptions of their extraordinary experiences while living among the 'last' Indians. Their accounts are framed in the language of cultural relativism, and they eagerly position themselves as specialists in other cultures. Webster's (1998) account of his visit among the Yanomami offers an illustrative example. Ignoring the rich anthropological literature on the Yanomami, Webster (1998: 8–13) argues that the Yanomami territory is so little explored that only rarely are anthropologists 'allowed into these forests'. He then describes how the Yanomami greeted him with hoots and screams and how their headman repeatedly slapped him on the chest as a sign of welcome. In these accounts, native people are referred as 'my' Indians, whose acceptance the traveller appears to have won to the point that he/she is treated like kin.

Characteristic of the contemporary travel writings published in *National Geographic* is the writers' tendency to focus on their own feelings and their own reactions to the tropics. The traveller's unique appreciation of the tropical landscape becomes the main focus of inquiry in repeated descriptions of hikes taken in tropical forests that stretch to the horizon, devoid of any sign of human presence. According to Schreider and Schreider (1970: 109), 'cruising for hours past walls of unbroken forest, we sometimes seemed to be the only boat on the river [Amazon], the only people'. Kamper (2000) describes how the first view of the Madidi National Park in Bolivia compensated for all the difficulties of the journey: 'Breathtaking landscapes, abundant birdlife, utter wildness as far as the eye could travel'.

At the same time, today's travel writings on the neotropics convert the earlier images of chaotic jungles into pictures of marvellously complex rainforests. This perception becomes clear already in the titles of recent *National Geographic*

Anja Nygren

accounts, among them 'Nature's Dwindling Treasures: Rain Forests' (White 1983) and 'Wilderness Headcount' (Morell 1999). The image of the rainforest as a threatened Garden of Eden is reinforced by verbal and visual references to paradisal icons. The rainforest is depicted as lush, breathtaking, and primeval, and although there are also descriptions of rainforests as gloomy and tangled, they do not appear in the same frequency as in earlier writings. More commonly, the rainforest is a place that 'feels like paradise', where 'scarlet macaws fly overhead, their wings pounding against the jungle twilight's electric blue sky' (Webster 1998: 11). It is also an immense library whose unbridled burning resembles the burning of the ancient library of Alexandria (White 1983: 24).

The current situation, in which the tropical rainforests are 'rapidly being clear-cut, strip-mined...bulldozed, and burned' (Melham 1990: 113), leads the contemporary travel writings to underscore the urgent need for international intervention on behalf of tropical conservation. Although it is evident that there are severe problems of deforestation and environmental degradation on many tropical forest frontiers, in some of the *National Geographic* accounts the pictures painted of the worldwide disaster that will result from the conversion of this lush greenery into a rusty red desert may be too ominous (Ellis 1988; McIntyre 1988). In contrast to earlier accounts which championed the uncontrolled exploitation of tropical resources, current narratives draw ugly pictures of tropical forest frontiers whose resources have been brutally degraded. To substantiate this view, the rainforests are described as razed, raped or denuded (Garrett 1989: 439; McIntyre 1977: 708; Kamper 2000: 16). Through simplified interpretations of ongoing environmental policies and programmes of forest management in the neotropics, any use of rainforests is condemned as an abuse.

FOREST-FRIENDLY VERSUS FOREST-UNFRIENDLY CULTURES

Parallel to the changing views of the rainforests, the dominant images of tropical forest-dwellers produced in the travel accounts of *National Geographic* have changed considerably in recent decades. Indians, whose impacts on nature were dismissed or ridiculed in earlier accounts, are now depicted as guardians of the forest, blessed with inherent environmental wisdom. They stand in sharp contrast to tropical settlers and non-indigenous rural poor, whose relationship to nature is considered to be based on short-term forest encroachment and who are easily portrayed as forest ravagers, with little awareness of the need to protect nature.

This new perception has much to do with the growing worldwide concern about the survival of the tropical forests, which has promoted the environmentally-founded rediscovery of the rainforest Indians as providers of a human face for the global attempts at rainforest protection (Brysk 2000; Conklin and Graham 1995; Oakdale 2004). Characteristic of this discourse is the portrayal of native peoples as a part of this magnificent nature. The Indians are included with the tropical flora and fauna as part of the overall spectacle of a divine fragility that needs protection (Slater 2000). This view becomes clear in various more recent *National Geographic* accounts of Indians, who blend into the forest background so completely that they seem to be part of the landscape. They either stand silently in the forest like statues – which makes it difficult to visualise one separated from the other – or when speaking for themselves, they present themselves as traditional stewards of nature.

At the same time, current *National Geographic* travel writings project tropical settlers as nomads who flock into the tropical forests to hack out homesteads. While the Indians are described as profoundly territorialised peoples in metaphors that refer to roots and soils, the settlers are depicted as culturally uprooted populations with liquid metaphors of movement in the descriptions which speak of waves of colonists and floods of migrants that stream into the forest frontiers (cf. Malkki 1992). According to Ellis (1998: 788), these settlers 'came on foot and by bicycle. They came clinging to one another as they rode on top of cargo in trucks that undertook the axle-breaking journey along the access road. They came in the rain, and they came during the dry season...And when they came to claim the land, they brought with them all the social ills of a frontier boom.'

This juxtaposition of nature-friendly versus nature-unfriendly forest-dwellers has brought the rainforest Indians into the limelight of global media's admiration. According to recent travel writings, the Indians live in the rhythm of the forest and feel a sense of oneness with nature. The small-scale colonists seldom receive a similar amount of attention, simply because they do not represent such an exotic way of life, nor can they claim the same status as champions of nature (Nugent 1994). When they appear in current travel writings, the settlers are rather portrayed as rootless penetrators who mindlessly destroy nature's precious gifts. Their livelihood strategies are described as a 'drama of man against the jungle' (Schreider and Schreider 1970: 117) or a 'veritable war between man and nature' (White 1983: 38). In the *National Geographic*'s special number on Tropical Forests in 1988, McIntyre's article on the Urueu-Wau-Wau Indians is entitled 'The Last Days of Eden', while the article by Ellis on Amazonian colonists is entitled 'Brazil's Imperiled Rain Forest: Rondônia's Settlers Invade'.

Characteristic of these representations are the portrayals of the breathtakingly beautiful rainforests, with giant trees and cascading rivers. Brilliantly-plumaged macaws and jewel-like orchids form an integral part of this enchanted realm, in which sensitive humans harmoniously participate. This view of Indians living in the forest free from the chains of civilisation characterises, for example, Melham's (1990: 154) description of his trip among the Yanomami: 'Just outside [the Yanomami lodge] stood eight or ten visiting Mucajai Indian women, nude save for their *tangas* (string girdles), and all smiled as they painted stripes, zigzags, and delicate stylised flowers on each other's skin. They were so happy, so full of childlike delight and innocence in the bright sunlight.' Correspondingly, Webster (1998: 13) describes how the Yanomami life goes on 'as it has probably been lived for thousands of years...Families doze in hammocks and huddle around fires.'

According to contemporary travel writings, this transcendent harmony of tropical world is on the verge of being inundated by the massive flow of colonists who are portrayed as harmful elements in the global drive to protect tropical nature. The images of the tropical settlers produced in recent travel accounts of *National Geographic* include portrayals of intractable invaders who are attracted to the jungle by an ardent pioneer spirit. The settlers' parcels are depicted as denuded spaces carved from burn-downed jungles, 'littered with charred logs and smoldering stumps' (McIntyre 1977: 708). They are also described as 'malarial acres' which 'lie blackened and scarred like a battlefield in war' (Ellis 1998: 782). The settlers' life on the forest frontier is portrayed as gloomy and unattractive, where makeshift houses proliferate across the landscape, and 'insects are the only wildlife around cleared areas' (Ellis 1988: 788). The forest frontiers are also depicted as political trouble zones and undercurrents of lawlessness where issues are decided at the point of a gun. According to Ellis, the Amazonian frontier drama is being ratcheted up 'to full throttle' where men 'staring and seldom smiling', fight, guzzle and hang around listening to the night music of debauchery. Many of the writings emphasise the similarity of tropical forest frontiers to the North American 'Wild West' a century ago.

At the same time, the rainforest Indians are represented as colourful performers whose culture is manifested in spectacles rich in myth and mystery. In contemporary travel accounts, the native people are often illustrated in ritual adornment, sometimes even when working. There are also descriptions of Indians with their 'pre-Columbian' dugout canoes, the canoe thus becoming an icon of the Indians' timeless life-style. A revealing example of this kind of iconisation is van Dyk's (1995: 27) description of 'six Indians paddling past just a few feet away, arrow-shaped paddles dipping silently in the river. It seemed an ancient image, almost a dream.' Likewise, Devillers (1983: 66) describes how a Wayana Indian hunter skilfully bends his bow and takes careful aim, motionlessly awaiting the proper moment until his arrow penetrates his prey. This dynamic unity, formed by the Indian body and the tropical landscape, is depicted as a symbol of the fragile beauty of the indigenous world, associated with a timeless attachment to tropical nature.

The settlers' life is, on the contrary, described by accounts of poverty and brutality. While Indians are presented as skilful archers, the settlers are portrayed as miserable colonists, clearing the land more as a 'burden rather than a skill' (Lutz and Collins 1992: 146). In the photographs, the Indians are typically presented in profile against a greenery that gives no evidence of social context, thus showing more interest in the Indians' place in nature than in their links to the larger society (Ramos 1994). The settlers are, instead, illustrated standing in a crowd and staring directly at the lens, all of which conveys a message of their anarchic way of life and their threatening potential for violence. While Indian women are depicted as sexually aesthetic and alluring others, with naked bronze bodies, non-indigenous women are portrayed as indiscreet and excessive in their sexuality. Von Puttkamer (1971: 440) tells about his encounter with three Cinta Larga women: 'who wore necklaces of dyed nutshells and almost nothing else. Though demure, they were unabashed and headed directly for our kitchen.' Impressed by the Indians' poise, von Puttkamer calls them the 'Three Graces'. In contrast, McIntyre (1977: 691) describes his unpleasant experience on a Brazilian frontier, where thirty men 'showed up on Saturday night to share dollar-a-bottle Coke and warm beer by candlelight with forlorn-faced women … The bouncer wore a Smith & Wesson .38.'

Analogous to these images, the journeys to Indian territories are depicted as voyages of discovery, creating the impression that visiting these people is like moving back through the millennia. The travellers launch forth from the modern city to a remote jungle, where the Indians have survived with minimal changes in their 'Pre-Columbian cultures' (Garrett 1989; von Puttkamer 1979; Webster 1998). In this pre-modern realm, modern techniques have little validity and the Western sense of superiority easily looses its power. Schreider and Schreider (1970: 62–3) describe how a Campa Indian 'glided through the jungle like a wraith … moving so silently that the symphonic trill of unseen birds and insects hardly changed its pitch'. As the Schreiders clumped after him, 'feet squishing in the damp sponge of rotting vegetation, it seemed as though the conductor had dropped his baton'. Correspondingly von Puttkamer (1971: 435) tells about how Cinta Larga boys allowed the travellers to accompany them on hunting trips, but expected them to carry the game they killed. When they came to streams deep enough to harbour electric eels, the Indians silently climbed upon the travellers' backs.

At the same time, the settlers' living spheres are presented as an ugly monotony of the curses of modernisation or as a helter-skelter of pre-modern backwardness and post-modern chaos. Schreider and Schreider (1970: 109) describe Amazonian frontier towns where their journey 'began to follow a script that never varied: The same greying wood shacks, the same bleached thatch roofs and dusty red streets, the same stocks of cane alcohol, cigarettes, and canned beef in the same bare-shelved shops'. Sartore, the photographer who accompanied Kamper (2000: 28) in his journey to Bolivia, has a similar impression of the frontier town of Pelechuco: 'Poverty abounds. Everything is worn out or broken. The kids … swim in raw sewage'. Van Dyk (1995) and Webster (1998) both wonder at the architectural disharmony of Amazonian frontier cities, where cardboard shacks and skyscrapers, dugout canoes and high-tech containerships mix abruptly with each other. In contrast to earlier travel accounts, which presented the modernisation of the tropics as a panacea, current travel writings evoke nostalgic sentiments recalling a vanishing pre-modern world, unspoiled by the ills of globalisation.

Anja Nygren

CONCLUSION: TRAVEL WRITINGS AND TROPICAL TROPES

Instead of rehearsing the conventional dichotomy between 'we' and the 'other', this essay has shown that the travel writings of *National Geographic* produce representations of neotropical peoples that rest on essentialist categorisations of 'good' and 'bad' others. In the late nineteenth and early twentieth century accounts, rainforests are typically presented as 'Green Hells' and rainforest Indians as primitive savages, while the non-indigenous settlers are praised as virtuous pioneers who are exploiting the tropical jungles. Since the rise of global concern of tropical deforestation and environmental degradation in the early 1970s, the accounts of *National Geographic* have instead tended to depict rainforest Indians as faithful guardians of marvellous forests, while tropical settlers are seen as mindless destroyers of tropical biodiversity. These historically changing, but equally essentialist images are based on repeated contrasts between the 'virtuous' and the 'vicious'.

The importance of the role of travel writings in formulating popular conceptions of tropical forests and tropical peoples can hardly be overestimated. With the global spread of tourism, travel narratives – like travel itself – have been made available to a large audience, and the genre of travel writing has become one of the most popular and widely read forms of literature today (Holland and Huggan 1998: 1–2). This essay has intended to show the crucial role that travel accounts in such a well-established magazine as *National Geographic* have played in how the neotropics and the neotropical peoples have been envisioned over time. While referring to actual people and places, the *National Geographic* accounts are interspersed with highly selective tales and images of neotropical forests and neotropical forest-dwellers. In spite of these elements, the writers present their narratives as authentic transcriptions of reality with little recognition of the inequality of the encounter and little analysis of how the encounterer's life is related to that of those whom he/she encounters. This is not to say that all the presentations of tropical forest-dwellers in *National Geographic* are categorically reified. Several writers also note the discrepancy of images in tropical resource conflicts, by stating that the conflicts are complex and difficult to understand. Some of them also recognise that the problem of tropical deforestation reaches far beyond settler culture in a situation where colonists suffer from a lack of economic assistance and intimidating bureaucracy.

Many of the travel accounts in *National Geographic*, however, operate within a pre-established semantic field, repeating the same tropes over and over again, while excluding the elements that do not accord with the conventional conceptions. Although the prevailing images portrayed in the accounts have changed over time, the same categorisations such as friendly versus unfriendly and pure versus impure, crop up repeatedly. The majority of the accounts present the Indians' and the settlers' conceptions of their environments monolithically, with little recognition of the existing intracultural diversity in the ways human beings experience nature. At the same time, the differences between the indigenous and the settler landscapes are categorised as cognitive, with limited recognition of the variety of conditions that these people encounter in meeting the daily requirements of their livelihoods. There is also little recognition of these people's differentiated positions in relation to governmental policies and global environmental agendas and advocacy networks.

While the late-nineteenth and early-twentieth century travel writings of *National Geographic* reflect a desire to control tropical nature-society relationships by constructing essentialist images of culturally primitive Indians, the present-day images of rainforests as a threatened Eden that demands global protection rest on a similar kind of desire for control. The overall concern for what is natural overshadows the tropical forests as historical, political, and cultural spaces (Nugent 1994; Slater 2000: 76–7). The contemporary travel writings of *National Geographic* on neotropics are especially thin in history, positing tropical forest-dwellers as ahistorical beings with little knowledge of the world around them. By portraying cultural difference in terms of distance and isolation, they show little interest in analysing the links between the representations and the wider issues of politics and power. The argument that the protection of nature is an inherent aspect of native life portrays the Indians in terms of Western images of 'stewards of nature' (Kirsch 1997). This image does not necessarily coincide with the Indians' own visions, which are based on increasing self-determination and control over their own resources (Conklin and Graham 1995; Oakdale 2004; Ramos 1998).

In the same way, the disregard of non-indigenous settlers for the environment is explained by their lack of forest culture, with limited attention to the structural roots of resource destruction on tropical forest frontiers, including unequal control over resources and settlers' vulnerable position in relation to the global economy. These accounts show little awareness of forest frontiers as places of social injustice and political marginalisation. This reification of people's relationship with nature fails to recognise the diversity of lived environmental relations and the complexity of the power struggles that mediate the ebb and flow of competing environmental images and environmental policies concerning tropical forests and tropical peoples (Graham 2002; Nygren 2004; Raffles 1999). It also ignores the fact that, in reality, the majority of the indigenous and non-indigenous neotropical forest-dwellers have for centuries been the most marginalised members of their national societies. In the earlier travel accounts of *National Geographic*, the social exclusion of Indians is wrapped in the pejorative comments on their cultural primitiveness, while today's narratives package it with picturesque images of Indians living in the peace of mystery. Correspondingly, the social marginalisation of the small-scale settlers is shrouded in their cultural uprootedness and their homesteader mentality.

All this shows how the representations of tropical forests and tropical forest-dwellers produced in travel writings of *National Geographic* create categorical

Anja Nygren

narratives with limited interest in allowing these people to speak with their own voices. Despite the writers' professed intentions to approach tropical peoples with open minds, essentialist distinctions prevail. In this context, there is a need for more sensitive narrative approaches that would permit more diversified views and more plural voices to be heard, without diminishing the popularity of this magazine. Such perspectives could reflect more complex portrayals of tropical peoples and reveal the multifaceted patterns of exchange and interaction that take places when different realities mingle together. They could also show the artificiality of making categorical distinctions between what is 'authentic' and what is 'spurious'. In today's world of globalisation and hybridisation, one can no longer predict who will put on a loincloth and lift up a blowgun and who will slip into jeans and pick up a mobile telephone.

NOTES

This essay draws on research financed by the Academy of Finland and the University of Helsinki.

[1] See www.nationalgeographic.com (visited on 10 June 2005).

[2] I am well aware of the problems involved in terms 'Western' and 'non-Western', or 'First World' and the 'Third World'. My use of these terms includes a critical notion that none of these categories is a monolith nor can they be considered to exist in sharp distinction with each other.

[3] A similar dichotomy is found in much advocacy literature, which distinguishes ecologically benevolent indigenous peoples from settlers as enemies of sustainability. For criticism of such dichotomies, see Li (2000), Nugent (1997), Nygren (1998, 1999) and Slater (2000).

[4] Non-Numerical Unstructured Data Indexing Searching & Theorizing Qualitative Data Analysis Program, version 6.0, 2002. QSR International, Melbourne, Australia.

[5] In this respect, see also the study by Conklin (2001) on the European colonisers' attitudes toward Wari Indian cannibalism, which offers an inspiring analysis how the discourses of horror and disgust became interwoven with the discourses of humanism and cultural relativism in the colonisers' accounts.

[6] Although some of these 'scientist-travellers' had no direct links to Western political and commercial ambitions in the tropics, their intensive investigations nevertheless contributed to Western economic exploitation of the tropics (Slater 2000: 12; Stepan 2001: 31).

[7] For more on the historical ambiguities in the images of tropical forests, see Arnold (1996), Putz and Holbrook (1988), Raffles (2002), Slater (2002) and Stepan (2001).

[8] In this respect, see also von Puttkamer (1971: 433).

REFERENCES

Arnold, David. 1996. *The Problem of Nature: Environment, Culture and European Expansion*. Oxford: Blackwell Publishers.

Blanton, Casey. 1997. *Travel Writing; The Self and the World*. New York: Twayne Publishers.

Briggs, C. 1996. 'The Politics of Discursive Authority in Research on the Invention of Tradition', *Cultural Anthropology* **11**(4): 435–69.

Brysk, Alison. 2000. *From Tribal Village to Global Village: Indian Rights and International Relations in Latin America*. Stanford: Stanford University Press.

Chmielinski, P. 1987. 'Through Wild Andes Rapids: Kayaking the Amazon', *National Geographic* **171**(4): 461–72.

Conklin, Beth A. 2001. *Consuming Grief: Compassionate Cannibalism in an Amazonian Society*. Austin: University of Texas Press.

Conklin, B.A. 2002. 'Shamans Versus Pirates in the Amazonian Treasure Chest', *American Anthropologist* **104**(4): 1050–61.

Conklin, B.A. and L.R. Graham 1995. 'The Shifting Middle Ground: Amazonian Indians and Eco-Politics', *American Anthropologist* **97**(4): 695–710.

de Pinedo, F. 1928. 'By Seaplane to Six Continents', *The National Geographic Magazine* **LIV**(3): 247–302.

Devillers, C. 1983. 'What Future for the Wayana Indians?', *National Geographic* **163**(1): 66–83.

Duncan, James and Derek Gregory. 1999. 'Introduction', in J. Duncan and D. Gregory (eds), *Writes of Passage: Reading Travel Writing* (London: Routledge), pp. 1–13.

Ellis, W. 1988. 'Brazil's Imperiled Rain Forest: Rondônia's Settlers Invade', *National Geographic* **174**(6): 772–99.

Garrett, W. E. 1989. 'La Ruta Maya', *National Geographic* **176**(4): 424–79.

Gill, R.C. 1934. 'Mrs. Robinson Crusoe in Ecuador', *National Geographic Magazine* **LXV**(2): 133–72.

Graham, Laura 2002. 'How Should an Indian Speak?', in K.B. Warren and J.E. Jackson (eds), *Indigenous Movements, Self-Representation, and the State in Latin America* (Austin: University of Texas Press), pp. 181–228.

Hingston, M.R. 1932. 'A New World to Explore: In the Tree-Roof of the British Guiana Forest Flourishes much Hitherto-Unknown Life', *National Geographic Magazine* **LXII**(5): 617–42.

Holland, Patrick and Graham Huggan. 1998. *Tourists as Typewriters: Critical Reflections on Contemporary Travel Writing*. Ann Arbor: University of Michigan Press.

Holt, E.G. 1933. 'A Journey by Jungle Rivers to the Home of the Cock-of-the-Rock', *The National Geographic Magazine* **LXIV**(5): 585–632.

Hulse, S. 1927. 'Nicaragua, Largest of Central American Republics', *The National Geographic Magazine* **LI**(3): 370–8.

Hyndman, D. 2002. 'Indigenous Representation of the T'boli and the Tasaday Lost Tribe Controversy in Postcolonial Philippines: Interpreting the Eroticised, Effeminising Gaze in *National Geographic*', *Social Identities* **8**(1): 45–66.

Jackson, J. 1997. 'The Politics of Ethnographic Practice in the Colombian Vaupés', *Identities* **6**(2–3): 281–317.

Kamper, S. 2000. 'Madidi: Will Bolivia Drown its New National Park?', *National Geographic* **197**(3): 1–29.

Kirsch, S. 1997. 'Lost Tribes: Indigenous People and the Social Imaginary', *Anthropological Quarterly* **70**(2): 58–67.

Li, T. 2000. 'Articulating Indigenous Identity in Indonesia: Resource Politics and the Tribal Slot', *Comparative Studies in Society and History* **42**(1): 149–79.

Lowenthal, David. 1997. 'Empires and Ecologies: Reflections on Environmental History. In T. Griffiths and L. Robin (eds), *Ecology and Empire: Environmental History of Settler Societies* (Edinburgh: Keele University Press), pp. 229–36.

Lutz, Catherine A. and Jane L. Collins. 1993. *Reading National Geographic*. Chicago: University of Chicago Press.

Malkki, L. 1992. 'National Geographic: The Rooting of Peoples and the Territorialization of National Identity among Scholars and Refugees', *Cultural Anthropology* **7**(1): 24–44.

Marden, L. 1944. 'A Land of Lakes and Volcanoes', *The National Geographic Magazine* **LXXXVI**(2): 161–92.

McEwan, C. 1996. 'Paradise or Pandemonium? West African Landscapes in the Travel Accounts of Victorian Women', *Journal of Historical Geography* **22**(1): 68–83.

McIntyre, L. 1977. 'Treasure Chest or Pandora's Box? Brazil's Wild Frontier', *National Geographic* **152**(5): 684–719.

McIntyre, L. 1988. 'The Last Days of Eden: Rondônia's Urueu-Wau-Wau Indians', *National Geographic* **174**(6): 800–17.

Melham, Tom. 1990. 'Man and the Forest', In *The Emerald Realm: Earth's Precious Rain Forests*. (The National Geographic Society), pp. 112–63.

Miller, David P. 1996. 'Introduction', in D. Miller and P.H. Reill (eds), *Visions of Empire: Voyages, Botany, and Representations of Nature* (New York: Cambridge University Press), pp. 1–18.

Morell, V. 1999. 'Wilderness Headcount', *National Geographic* **195**(2): 32–41.

Nugent, Stephen. 1993. *Amazonian Caboclo Society: An Essay of Invisibility and Peasant Economy*. Oxford: Berg.

Nugent, S. 1994. 'Invisible Amazonia and the Aftermath of Conquest: A Coda to the Quincentenary Celebrations', *Journal of Historical Sociology* **7**(2): 24–41.

Nugent, S. 1997. 'The Coordinates of Identity in Amazonia: At Play in the Fields of Culture', *Critique of Anthropology* **17**(1): 33–51.

Nygren, A. 1998. 'Environment as Discourse: Searching for Sustainable Development in Costa Rica', *Environmental Values* **7**(2): 201–22.

Nygren, A. 1999. 'Local Knowledge in the Environment-Development Discourse: From Dichotomies to Situated Knowledges', *Critique of Anthropology* **19**(3): 267–88.

Nygren, A. 2004. 'Contested Lands and Incompatible Images: The Political Ecology of Struggles over Resources in Nicaragua's Indio-Maíz Reserve', *Society & Natural Resources* **17**(3): 189–205.

Oakdale, S. 2004. 'The Culture-Conscious Brazilian Indian: Representing and Reworking Indianness in Kayabi Political Discourse', *American Ethnologist* **31**(1): 60–76.

O'Neill, T. 1993. 'New Sensors Eye the Rain Forest', *National Geographic* **189**(9): 118–30.

Peary, R. 1889. 'Across Nicaragua with Transit and Machéte', *National Geographic Magazine* I(4): 315–35.

Porter, Dennis. 1993. *Haunted Journeys: Desire and Transgression in European Travel Writing*. Princeton: Princeton University Press.

Pratt, Mary Louise. 1992. *Imperial Eyes: Travel Writing and Transculturation*. London: Routledge.

Pratt, Mary Louise. 1994. 'Travel Narrative and Imperialist Vision'. In J. Phelan and P.J. Rabinowitz (eds), *Understanding Narrative* (Columbus: Ohio State University Press), pp. 199–221.

Putz, Francis E. and N. Michele Holbrook. 1988. 'Tropical Rain-Forest Images', in J.S. Denslow and C. Padoch (eds), *People of the Tropical Rain Forest* (Berkeley: University of California Press), pp. 37–52.

Raffles, H. 1999. 'Local Theory: Nature and the Making of an Amazonian Place', *Cultural Anthropology* 14(3): 323–60.

Raffles, Hugh 2002. *In Amazonia: A Natural History*. Princeton: Princeton University Press.

Ramos, Alcilda. 1994. 'From Eden to Limbo: The Construction of Indigenism in Brazil', in G.C. Bond and A. Gilliam (eds), *Social Construction of the Past: Representation as Power* (London: Routledge), pp. 1–22.

Ramos, Alcilda. 1998. *Indigenism: Ethnic Politics in Brazil*. Madison: University of Wisconsin-Madison.

Risse, Marielle. 1998. 'White Knee Socks Versus Photojournalist Vests: Distinguishing Between Travelers and Tourists', in C.T. Williams (ed.), *Travel Culture: Essays on What Makes Us Go* (Westport: Praeger), pp. 40–50.

Rothenberg, Tamar Y. 1994. 'Voyeurs of Imperialism: *The National Geographic Magazine* before World War II', in A. Godlewska and N. Smith (eds), *Geography and Empire* (Oxford: Blackwell), pp. 155–72.

Schreider, Helen and Frank Schreider. 1970. *Exploring the Amazon*. National Geographic Society.

Schurz, W.L. 1936. 'The Amazon: Father of Waters', *National Geographic Magazine* LXIX(4): 445–63.

Schwartz, J.M. 1996. 'The Geography Lesson: Photographs and the Construction of Imaginative Geographies', *Journal of Historical Geography* 22(1): 16–45.

Short, John R. 1991. *Imagined Country: Environment, Culture and Society*. London: Routledge.

Slater, Candace. 1996. 'Amazonia as Edenic Narrative', in W. Cronon (ed.), *Uncommon Ground: Rethinking the Human Place in Nature* (New York: W. W. Norton & Company), pp. 114–31.

Slater, Candace. 2000. 'Justice for Whom? Contemporary Images of Amazonia', in C. Zerner (ed.), *People, Plants and Justice: The Politics of Nature Conservation* (New York: Columbia University Press), pp. 67–82.

Slater, Candace. 2002. *Entangled Edens: Visions of the Amazon*. Berkeley: University of California Press.

Spurr, David. 1993. *The Rhetoric of Empire: Colonial Discourse in Journalism, Travel Writing, and Imperial Administration*. Durham: Duke University Press.

Stepan, Nancy Leys. 2001. *Picturing Tropical Nature*. New York: Cornell University Press.

Anja Nygren

Steve, Clark. 1999. 'Introduction', in S. Clark (ed.), *Travel Writing and Empire: Post-colonial Theory in Transit* (London: Zed Books), pp. 1–28.

Stevens, A.W. 1926. 'Exploring the Valley of the Amazon in a Hydroplane', *National Geographic Magazine* **XLIX**(4): 353–420.

Sultan, D. 1932. 'An Army Engineer Explores Nicaragua', *National Geographic Magazine* **LXI**(5): 593–627.

Torgovnick, Marianne. 1990. *Gone Primitive: Savage Intellects, Modern Lives.* Chicago: University of Chicago Press.

van Dyk, J. 1995. 'Amazon: South America's River Road', *National Geographic* **187**(2): 2–39.

von Puttkamer, J. 1971. 'Brazil Protects Her Cinta Largas', *National Geographic* **140**(3): 420–44.

von Puttkamer, J. 1979. 'Man in the Amazon: Stone Age Present Meets Stone Age Past', *National Geographic* **155**(1): 60–82.

Webster, D. 1998. 'The Orinoco: Into the Heart of Venezuela', *National Geographic* **193**(4): 1–31.

White, P. 1983. 'Nature's Dwindling Treasures: Rain Forests', *National Geographic* **163**(1): 2–47.

When 'The Environment' Comes to Visit: Local Environmental Knowledge in the Far North of Russia

Timo Pauli Karjalainen and Joachim Otto Habeck

INTRODUCTION

On the basis of field research in the Komi Republic (North Russia), we discuss the interrelations of environmental perception and environmental knowledge. From this interplay arises local environmental knowledge, a term that we specify and characterise towards the end of this paper. Thereby we wish to contribute to the debate about the role of local knowledge in environmental policies.

The need to address the multiple ways of perceiving environmental change in different sectors of society is very topical in arenas of environmental policy-making (Bickerstaff and Walker 2001). It is also argued that a diversity of knowledge – including local knowledge and expertise – is necessary for the regulation of environmental risks (Fischer 2000: 200). Some scholars, however, have stated that scientists and policy-makers often fail to acknowledge local forms of knowledge as meaningful sources of information (Kroll-Smith et al. 1997) or do not 'hear' certain aspects of what is being said by local people. While they are interested in local knowledge for its possible factual content, they do not always realise that such knowledge 'includes statements that are not simply descriptive, but moral and performative. They inform and direct those with whom the knowledge is shared to act on that information' (Feit 2001: 34).

The initial point of our argument is the distinction between environmental perception and environmental knowledge. Environmental perception refers to the environment as 'that which surrounds'; in this sense, environmental changes are perceived in a framework of everyday action, and through direct experience of other people and the non-human world (Ingold 2000). In contrast, environmental knowledge is part of environmental concern (Dunlap and Jones 2002) or environmentalism, and stems mainly from media, science, education and other forms of communication. If perception situates the individual *within* the environment (environment with lower-case 'e'), viewing it as a life-world; environmentalism often applies the environment in the global sense ('from the street corner to the stratosphere', Cooper 1992: 167) and as if from *outside* ('The Environment' with upper-case 'E'), as if the viewer were detached from it (Ingold 2000: 209–210, 218). In a local context, these two interrelate and intermingle.

Environmental Values **13** (2004): 167–186.
© 2004 The White Horse Press. doi: 10.3197/ 0963271041159877

Timo Pauli Karjalainen and Joachim Otto Habeck

More detailed empirical research on the interrelation between environmental perception and environmental knowledge may help to fill practical lacunae such as failings in the 'information deficit' model (Irwin et al. 1996) of environmental policies. This model can be found in the background of many global change projects and national environmental policies, and in this science-centred view of 'the environment', environmental issues are seen as rather static in their nature and meaning, even when a specific issue is compared in different national, cultural and social contexts.

Global surveys (e.g. Dunlap 1998; DeBardeleben and Heuckroth 2001) indicate that levels of expressed environmental concern are high in Russia but that levels of knowledge are comparatively low. We argue that low levels of abstract and cognitive knowledge found in survey studies do not entirely embrace the whole issue of the knowledge base of environmental concern, especially in the Russian context where environmentalism ('The Environment' as an interpretative category) has had a different history and political significance than in Western countries (see Pickvance 1998; Weiner 1999). Thus, context-dependent knowledge needs to be examined, and our intention here is to provide a close-up picture of 'the environment' in a relatively unstudied regional context: the Komi Republic in the northeastern part of European Russia. We shall examine how local people perceive and experience their surroundings and environmental changes, and how their knowledge of environmental issues is formed.

This study is part of the interdisciplinary research project TUNDRA[1], which set out to examine environmental pollution and climate change in the catchment area of the Usa River, a tributary of the Pechora in the Komi Republic. We are concerned with both scientific knowledge about regional environmental problems and the (re-)interpretations of this knowledge among local inhabitants. Some of the environmental concerns of local people are not corroborated by any scientific studies, but nonetheless people feel highly worried about them. If local knowledge is more closely related to the experience of the environment that surrounds the individual, science appears to yield a knowledge about that environment that is at once more abstract and more authoritative. The relation between local and science-based knowledge becomes particularly apparent and salient in cases of major environmental disasters. This relation may be further highlighted through the visits of environmental activists, as happened in one of the areas of our case study. Here we examine the reasons behind the congruencies and the incompatibilities between the two kinds of environmental knowledge.

ENVIRONMENTAL PERCEPTION, KNOWLEDGE AND CONCERN

It is often claimed that nowadays science and scientific research are the source from which people receive their environmental knowledge. In the background of many global change research projects and national environmental policies stands a science-centred 'information deficit' model (see e.g. Blake 1999). This model presumes that in order to promote environmentally benign behaviour and action, individuals and local communities have to be given more scientific facts. Environmental policy makers should have more appropriate methods of social engineering (the domain of social scientists) in order to contribute to a change of attitudes via scientific knowledge (the domain of natural scientists).

Many findings indicate, however, that in the formation of environmental concern, scientific knowledge and discourse do not necessarily play a major role (Irwin et al. 1996; Brand 1997). When people receive scientific knowledge about environmental problems, they restate it in their everyday life contexts. Thus even 'global concepts of the environment and environmental change are always localised in particular socio-political and cultural contexts' (Burningham and O'Brien 1994: 914).

Similarly, some recent studies of the public understanding of environmental issues stress the importance of everyday direct experience and the local context in the formation of environmental concern (e.g. Blake 1999; Bickerstaff and Walker 2001). Bush et al. (2002: 130) claim 'direct sensory perception and commonsense understandings continue to be important in framing environmental understandings and concerns'. The local socio-economic and cultural context strongly influences the way concerns are expressed about the global as well as the local environment (Darier and Schüle 1999). Environmental changes and issues are not separated from other changes and issues in the society–environment milieu, such as crime and economic insecurity. Consequently, environmental policy should be 'sensitive to the everyday contexts in which individual intentions and actions are constrained by socio-economic and political institutions' (Blake 1999: 274).

We argue, as Tim Ingold (2000) does, that the individual discovers the meanings of his/her environment in action and interaction and through direct experience of other people and the non-human world. Ingold takes much of his theoretical framework from Gibsonian ecological psychology, stating that

> [I]f perception is a mode of action, then what we perceive must be a direct function of how we act. Depending on the kind of activity in which we are engaged, we will be attuned to picking up particular kinds of information. The knowledge obtained through direct perception is thus *practical*, it is knowledge about what an environment offers for the pursuance of the action in which the perceiver is currently engaged. In other words, to perceive an object or event is to perceive what it *affords*. (…) [O]ne learns to perceive in the manner appropriate to a culture, not by acquiring programmes or conceptual schemata for organising sensory data into higher-order representations, but by 'hands-on' training in everyday tasks… (Ingold 2000: 166–167, original emphasis)

Perception is the key to local expertise, which is often non-verbal and bounded with context and practice. It refers to the way one obtains knowledge within one's environment. If this direct perception applies to the individual's engagement with the surrounding 'home' environment, knowledge is then grounded in experience in a particular local context. We want to stress that this is not only the case with hunting and gathering societies but also with town-dwellers.

In a recent article, Dunlap and Jones (2002: 485) define environmental concern as 'the degree to which people are aware of problems regarding the environment and support efforts to solve them and/or indicate a willingness to contribute personally to their solution'. In attitude theory studies, environmental concern is taken to be equivalent to 'environmental attitude', comprising cognitive, affective and conative dimensions. Cognitive expressions of environmental concern usually have to do with the individual's knowledge and beliefs about the nature of an environmental problem, its causes and possible solutions. For Dunlap and Jones (2002: 490) the cognitive dimension is 'a multidimensional construct (environmental cognition) that can be inferred from people's expressed knowledge and beliefs about environmental issues'.

Surveys examine mostly the latter, i.e. quite abstract knowledge (environmental cognition) and concern about environmental issues, and can hardly grasp the actual commitment for 'The Environment' embedded in a local context. This deficiency is usually discussed as a gap between environmental awareness and behaviour. We consider that the role of perceptual knowledge grounded in everyday life is essential in the formation of environmental concern, and that relationships between forms of perceptual and cognitive knowledge need to be better understood.

Specific levels of environmental knowledge do not predict environmental action or behaviour. As Karl-Werner Brand (1997: 204) notes, '(a) pronounced environmental consciousness in one field of behaviour combines with an astonishing indifference in others'. The common survey setting does not acquire local or cultural meaning of the environment and environmental issues, and this is why global survey data 'have to be interpreted with caution. Results are highly dependent on the wordings of questions and the cultural context of interviewees' (Brand 1997: 205).

Nature conservation has a long history in Russia, but in the Soviet Union discussions about it were at times suppressed and in any case tolerated only in a scientific and non-ideological form. Consequently, environmental discourse has been different from that in the West. It has arisen from the nature protection debate in the decades before 1960 (e.g. the protection of Lake Baikal) and became heated by the end of Perestroyka, when environmental critique was an essential part of the critique directed against the Soviet regime (Mirovitskaja 1998; Weiner 1999). In comparison to the Perestroyka period and the subsequent years, nowadays in Russia environmental concern is of much less political relevance. The public debate about environmental issues and official environmental policies had faded away by the time of Putin's presidency (Peterson and Bielke 2001). Nevertheless, the majority of citizens of the Russian Federation do feel concerned about the state of their immediate environment and surroundings. Studies show, in particular, that local sources of pollution (mainly from industry and transportation) are seen as health threats by the citizens in many towns and cities of Russia (Karjalainen et al., submitted). However, according to survey results, citizens of Russia appear to be less concerned about global environmental issues than do those of Western countries (Dunlap 1998). In 'the all-encompassing risk society of Russia', as Oleg Yanitsky (2000) depicts present-day society in Russia, citizens' environmental concerns are pronounced as personal concerns about health and well-being.

It is clear that a given political and socio-economic context influences public discourses, agenda setting, the strategies of environmental organisations and environmentalism in all its shapes and forms (Brosius 1999). Klaus Eder (1996: 163) contends that 'rather than an evolution of environmentalism toward some kind of universal ethics, there is an evolution of different worlds of environmentalism which are cultural responses to specific social conditions. The project of environmentalism is a series of 'green particularisms' rather than a collective project.' As we are interested in these green particularisms in different cultural and social contexts, there is a need to know the role of perception in the formation of environmental concern. Hence the question: What is the meaning of the 'The Environment' in the milieu-specific life-worlds of the local people in the Komi Republic?

EMPIRICAL SETTING

Here we shall briefly introduce the fieldwork region: the Usa Basin in the northeast of European Russia. In terms of landscape, the region is characterised by forests and tundra. In administrative terms, most of the territory belongs to the Komi Republic, the name of which is derived from the predominant indigenous group living in this area: the Komi. Russians have lived in adjacent regions for many centuries, but it was only after 1930 that large numbers of Russians and representatives of other ethnic groups (Ukrainians, Tatars, Germans etc.) settled in the northern part of the Komi Republic. First they came there as a consequence of deportations and forced labour, but later many more arrived voluntarily. The question of how, why and when newcomers arrived in the northern towns has significant implications for the perception of the environment, local discourses and the formulation of environmental concerns, as shall be discussed below. Nowadays the so-called newcomers (*priezzhie*) outnumber the indigenous (*korennye*) inhabitants in this as well as most other parts of the Komi Republic.

However, it is not possible to make a clear-cut distinction between these two groups, nor would this distinction coincide with the one between rural and urban communities (Komi do not live only in villages, nor do newcomers only live in towns). We rather want to emphasise differences in livelihoods and land-use strategies characteristic of the various communities. In the villages of the research region, inhabitants live by fishing, small-scale agriculture, berry-picking, hunting and partly reindeer herding. The towns in the north of the Komi Republic can be defined as *resource communities*, which were built to operate coal mines or oil and gas fields, and in which the main industry constitutes the unifying social bond that sustains the community. But although the two towns under study in this article, Usinsk and Vorkuta, have much in common, they have divergent histories, are located in different landscapes and vegetation zones, and have contrasting industrial functions. These functions give their inhabitants different resources, leading to distinct 'local logics' concerning environmental, but also other socio-political, issues.

In comparison to Vorkuta (and most other towns of Russia), Usinsk is an affluent place because the oil industry brings in revenues. The sense of a local urban identity is weaker in Usinsk (founded c. 1970) than in Vorkuta (founded in the 1930s), where a considerable proportion of the inhabitants were born, grew up and have always lived, thus acquiring a kind of 'home-town' feeling. The Usinskians feel more like newcomers than the Vorkutinians. The fact that Vorkuta has existed for a longer time also means that nowadays there are only a few indigenous inhabitants who can remember how things were before the town existed, so Vorkuta is already embedded in its environment to a greater extent than Usinsk.

Despite the fact that the Komi Republic has coped fairly well economically, by Russian standards, during the crises of the 1990s, economic development within the Komi Republic has been unequal in the various districts and fields of production. The adjustment of the coal mining industry to the new market conditions has been rather difficult. The lack of processing, the low level of technology and the rise of transportation charges have made many of Vorkuta's coal mines unprofitable[2]. During the 1990s, the city-dwellers of Vorkuta underwent a crisis that affected the local economy as well as self-identification[3].

Against the backcloth of a comparatively pristine natural environment, the oil production zone of Usinsk and the coal mining area around Vorkuta stand out, both in scientific observations – by the peaks in emissions of methane, carbon oxides and sulphur oxides – and in the visual perception of local people – by the oil derricks, pit-heads and smoke stacks. Around Vorkuta as well as around Usinsk, indications of environmental pollution can be seen with the naked eye: layers of soot in the accumulated snow, dead forest around old oil wells, the bus that brings workers to the oil spill sites that still need cleaning.

It was the oil spill in autumn 1994 close to the River Kolva, north of Usinsk, which caused 'The Environment' to visit Usinsk and the adjacent villages. This event caught the attention of the international media, although it was neither unique nor the largest oil spill that has ever taken place in the north of Russia. After a number of smaller leakages in the regional trunk pipeline, in autumn 1994 the situation culminated in its temporary closure, while an estimated 110,000 tonnes of oil poured out into the bogs of the forest tundra (Sagers 1994; Poklad 1995; Vil'chek and Tishkov 1997; Lodewijkx and Hirsch 2000). Some of the oil reached the River Kolva during the same autumn, but the main charge came down the river after the snow melt in spring 1995, at a time when the clean-up was still in its initial stage. The lower course of the Kolva, the mouth of the Usa and therefrom the Pechora underwent heavy pollution. The 1994 oil spill is vividly recollected by all our informants in Usinsk and surroundings. It has led to very prominent manifestations of environmental concern, which is markedly different from the concerns we found prevalent in the other fieldwork region, Vorkuta.

METHODS

Our case study is based on a combination of fieldwork among rural (mainly indigenous) and urban (mainly newcomer) inhabitants. In the sociological component of the research, environmental perception and knowledge were examined among town-dwellers in Usinsk and Vorkuta. This study was carried out from August to October 1998 by means of face-to-face thematic interviews. Semi-structured, open-ended questions focused on the interviewee's life-history, hobbies and views of the surrounding environment, perceptions and acuteness of socio-economic problems and changes in the state of environment; and likewise knowledge of, responsibilities for and solutions to environmental issues. Interviews ranged in duration from 25 to 105 minutes. All interviews were recorded and transcribed. For analysing and evaluating this material, we used a 'grounded theory' approach based on the constant comparison of emergent themes and an exploration of deviant cases (Glaser and Strauss 1967).

The interviews among town-dwellers focused on three occupational groups: 114 industrial workers, 30 teachers, and 33 managers and administrators. The data set comprises 89 interviews in Usinsk and 86 in Vorkuta[4]. The occupational groups studied were chosen because of their 'strategically important' position. The administration, particularly in Russia, has great influence in environmental decisions (see Yanitsky 2000); teachers are largely responsible for environmental education; and industrial workers, as a large and organised group, have shown their ability to gain public attention all over Russia through their protests for better living conditions (Burawoy and Krotov 1994).

In the anthropological component of the research, fieldwork was conducted in six villages[5] and in reindeer herders' camps in the adjacent forest-tundra and tundra areas from July 1998 to July 1999. Semi-structured questionnaires with open-ended questions, initially designed for interviewing, did not prove useful as interviewees generally found this way of communicating too formalised. Instead, a couple of initial questions could help to start open discussions whereby both the interviewer and the interviewees had opportunities to elaborate on subjects that were deemed most topical in the given situation. Part of the

Timo Pauli Karjalainen and Joachim Otto Habeck

data was acquired in 'informal talks', which proved to be valuable for gaining a detailed understanding of the local inhabitants' main concerns and attitudes. Participant observation, predominantly in reindeer husbandry, constituted another key element of the anthropological research. Migrating with Komi reindeer herders and their families for three months provided a better grasp of how they perceive their environment, and helped to assess the question of whether their perceptions differ from those of oil workers. In particular, encounters between reindeer herders and oil workers in the tundra shed light on this question. Altogether, the anthropological field notes record conversations with approximately 180 different individuals, ranging from occasional ten-minute talks to repeated encounters and long-term company with some 30 key informants.

LIVELIHOODS AND THE PERCEPTION OF 'NATURE'

The local identity of Vorkuta and Usinsk as *resource communities* becomes apparent in their officially promoted heroic image. Travellers who come by train to Vorkuta are greeted by a large inscription on the platform: 'Vorkuta – the outpost for the opening-up of the North' (*Vorkuta – forpost osvoyeniya Severa*). Vorkuta is a true frontier town, it seems, where man struggles against a harsh and hostile natural environment in order to secure coal for the sake of the national economy. A similarly zealous inscription in Usinsk encourages its inhabitants to produce 'more oil from Usinsk for the motherland' (*Bol'she Usinskoy nefti rodine*). The local museum in Usinsk shows a photograph of pioneers who had come in tank-like vehicles to build a whole town amidst bogs and forests. Such imagery of the two towns and their inhabitants is in stark contrast with the officially promoted traditionalist image of the indigenous population, and newcomers as well as indigenous inhabitants appear to buy into this discourse and tend to see each other along these lines (Habeck 2003). For example, many newcomers speak about Komi reindeer herders as 'children of nature', allegedly too honest and naïve for living in urban places, but feeling 'at home' in the tundra[6].

However, such ethnic stereotypes do not help to elucidate the question of whether there are differences in the perception of the environment between various groups. Differences of this kind cannot be explained simply by ethnicity but rather by different livelihoods and everyday activities, as we shall illustrate in what follows. The Komi reindeer herders – to return to the above example – do not see themselves as 'children of nature', but rather as people who 'know the tundra' (compare Anderson 2000) with sufficient experience and endurance to make a living there. It is in this sense that practical environmental knowledge and skills provide for a livelihood in the tundra and forest.

Many of the villagers work in, or at least depend on, fishing, reindeer husbandry and, to a smaller extent, hunting. Like the townspeople, specific occupational groups of the rural population have distinct spatial spheres of activity. A reindeer herder cannot pursue his work without having a thorough knowledge of his environment. Migrating between summer and winter pastures, herders travel up to 1,000 km in the course of one year, the whole way on reindeer-drawn sledges. Fishermen have a smaller radius of activity but some may travel more than 100 km to get to their preferred places. The biographies of fishermen, hunters and herders are closely connected with specific places; many place names testify to the deeds of their ancestors. For the villagers, it is beyond any doubt that humans are capable of staying in this region for a lifetime.

Most of the urban informants regard the northern 'natural' environments – forests, bogs, the tundra etc. – mainly as settings for leisure activity. Mushrooming and berry-picking, fishing, hunting and hiking are forms of relaxation and give the opportunity 'to breathe fresh air'. These environments are held to be important for health and well-being, as a counterbalance to industrial or other work and town life. At the same time, these activities also have significance for subsistence economy and urban inhabitants make use of them to various degrees. Several townspeople spend their holidays fishing in a way similar to the activities of rural fishermen: they travel to a river or lake in the tundra and stay there for a couple of weeks. While townspeople use the tundra for leisure and subsistence, villagers use the residency of their children or relatives in town to buy cheaper products. Hence, both townspeople and villagers have some degree of experience of each other's everyday environments; but these are not, as we would say with Brand (1997), their milieu-specific life-worlds, within which environmental perception is constituted.

Overall, newcomers to the North consider living conditions to be very harsh. This harshness derives partly from the uncertain 'transitional' circumstances. Yet what is more, there is a *narrative of the North* – a kind of frontier discourse (Keskitalo 2002: 59), which depicts northern nature as very austere. The narrative of the North used to be part of the rhetoric needed to justify the special status of these towns with regard to subsidies and benefits within the Soviet economy, and remains significant within the Russian economy which is still dependent on the export of natural resources (Karjalainen 2001). This narrative is particularly discernible in Vorkuta, and it seems to confirm the presupposition that newcomers tend to see nature as an alien domain whose appropriation involves struggle, suffering and conquest. However, it is their environment as a whole that involves suffering, rather than the encounter with a 'nature' that is separated from the human world. Urban newcomers feel confronted with the adversities of the natural environment as well as those of the built surroundings and the social milieu (notably institutions of the official sphere). Our urban interviewees usually discussed 'the environment' without separating the natural from the social[7]. Similarly, for rural informants changes in the natural environment are closely connected to the socio-economic sphere (compare Berglund 1998: 54).

To sum up this section, both townspeople and villagers make no clear distinction between the natural and the social environment. We also argue that for

Timo Pauli Karjalainen and Joachim Otto Habeck

individuals in both groups the temporal and spatial range of everyday life activities are constituent for their perception of the environment. Not the individual's ethnic identity but his/her engagement and interaction with the surroundings give rise to environmental perception. This finding can also account for the specific differences in environmental perception between the inhabitants of Vorkuta and Usinsk, the two towns under study. In the next section, we shall explore in more detail how environmental perception interrelates with sudden environmental changes as well as local discourse and mediated knowledge about 'The Environment'.

PERCEPTION AND KNOWLEDGE OF POLLUTION

In general, environmental issues hold a low profile compared with other social problems among townspeople of Usinsk and Vorkuta. Under current circumstances income level and employment are clearly more important to people than environmental issues. However, interviewees from the town of Usinsk were far more oriented towards environmental problems than people in Vorkuta (Karjalainen 2000).

Environmental concern strongly increases following any major disaster (compare Berkes 2002: 337). It was not only the obvious changes in the environment that leapt to the eyes of the Usinskians, but also the presence of correspondents from abroad, the feeling that all of a sudden their concerns were receiving attention in many other parts of the world. The whole topic of environmental pollution and the protection of the environment was more open to public debate in Usinsk, and the public became more alert to related questions, than in Vorkuta. In the Usinskians' experiences of the major oil spill in 1994, its wide media profile and the visits of environmentalists created a special discursive space for environmental affairs. This generated a greater eagerness to discuss environmental issues, thereby also laying the grounds for environmental concern. In this way, 'The Environment' – environmental concern from outside the local context – came to visit Usinsk and merged with the inhabitants' environmental perception. In Usinsk almost all discussions about the environment derive from the major oil spill in 1994.

Question: 'Do you have any problems with the state of the environment in the area where you live?'

'Yes, they are connected to the oil production. I don't think that everything is fixed up after the oil spill in 1994. (...) People start to think about the environment only when something extreme happens, like the disaster in Chernobyl or the oil spill in 1994. (...) After the oil spill there was a lot of discussion, but whether or not something was done, I don't know.' [Ut11, teacher, 20s, female, Russian]

'Yes, we have. But only in recent times have people started to pay attention to them, although they have existed all the time.' [Uw53, oil worker, 40s, male, Russian]

Air pollution is environmental threat number one for the city-dwellers of Vorkuta. The sources are two heating plants, the district heating centre for water, coal mines, the cement factory and (less significantly) automobiles. Some estimates claim that people in Vorkuta 'eat 1.4 tonnes of dust' per year (Vt07). The lay public evaluates air pollution by direct observations and experiences, such as cleanness of clothes ('a shirt is clean for a day', Vw06; Va10) and snow ('around Vorkuta, in a circle of about 5 km, snow is black because of coal dust', Vw37), and also by breathing difficulties. Scientific knowledge plays only a minor role here, although we noticed that administrators speak about levels and effects of pollution in more technical and scientific terms.

Although Vorkuta is classified as the most polluted region in the Komi Republic (Taskayev 1999: 117), environmental issues are not really hot topics of discussion in the city. Based on his study of the suburbs of Manchester, Irwin wrote that '[e]nvironmental pollution is one important characteristic of life..., but it is not the *sole* characteristic' (Irwin 1995: 94). The same can be said about Vorkuta. Moreover, in Vorkuta, public discourse often refers implicitly to the economic and historical role of the city. Official and informal talks are frequently connected to subsidies from the central government, income levels and privileges. This is due to the fact that Vorkuta was 'a true Soviet city' with coal mines and miners ('élite workers' in one of the key industries), and received major subsidies from the central government. On the other hand, it has anticommunist, radical political traditions of underground activity, which have remained from its former role in the GULAG (the system of forced-labour camps in the Stalin era), although the political radicalism of the miners derives primarily from the fact that their economic conditions have persistently been much worse than those of miners elsewhere (Burawoy and Krotov 1994).

The effects of water pollution are assessed quite similarly in both towns. Townspeople are worried about health risks connected to the quality of their drinking water, and their recreational activities related to the diminishing fishing potentials and swimming possibilities. The town-dwellers of Usinsk have responded to water pollution by boiling water and purchasing filters, and some buy their drinking water in bottles. The poor quality and quantity of fish are very evident results of water pollution in Usinsk (and also in Vorkuta in connection with the Vorkuta River). According to Usinskians, fish have an oily smell and taste. 'Previously there were plenty of salmon in the Usa River, but now there are only few left and those smell oily' (Ut09). The impacts of water pollution are 'real' for the town-dwellers, even if they are already accustomed to the situation.

For the inhabitants of the villages around Usinsk, the impacts of the 1994 oil spill are even more salient. Previously they used to drink water from the rivers Pechora and Usa, but nowadays they deem it too risky. They too say that the quality of the fish has deteriorated and the quantity has diminished[8]. More than the town-dwellers, the rural inhabitants rely on fish as a staple diet because their monetary income is very low. What is more, the river meadows, which are used for making hay, were all polluted by oil during the subsequent spring flooding. It was reported that in 1995 cattle died by the hundreds in this region and that since then the milk yield of the surviving cows has almost halved. The reduced milk output might stem from the lack of artificial feed but the interviewees see it also as being connected with the oil spills.

We saw that in some situations the local inhabitants are not willing or able to reconcile their personal experiences with scientifically established knowledge. Once an ecological disaster has occurred and environmental change becomes visible within a short period of time, the local inhabitants seem not only more interested in receiving information about the state of the environment, but also more critical of its validity. The example of the Usinsk oil spill illustrates this. Experts from both the Ministry of Agriculture of the Komi Republic and a Moscow-based consulting company, independently of each other, conducted studies of the extent and the results of oil pollution in the area. The studies concluded that, although large-scale pollution had happened, human diseases and the illness or death of animals could not be directly connected with the oil spills. Such conclusions disappoint the rural inhabitants, for they feel that the causal connection is all too obvious and even acknowledged by oil companies themselves. Many people do not trust these environmental reports and call for additional studies, hoping that these would prove their point.

Some inhabitants have started to collect data themselves, for example statistics on diagnoses in the local hospital. This is where the validity of scientific knowledge comes into play. Although everybody in these villages is entirely sure about the causal connections, their local environmental knowledge, mainly based on personal experience, does not suffice to argue their case; instead, they have to tangle with a realm of knowledge which, albeit not alien to them, in its abstraction lacks the sensual perceptibility that their knowledge affords (Berglund 1998: 152–74; Grove-White 1993: 21–2).

In Brown's terms (1997), the response to the Usinsk oil spill exemplifies 'popular epidemiology', a form of citizen science in which people engage in 'lay' ways of collecting knowledge about environmental and technological hazards. Public-health officials and scientists work with abstractions, but their knowledge does not link up with local people's reality; it does not fit into practical 'lived' experience. All local concerns are expressed by referring to personal experiences and observations. The fact that interviewees are comparing how things were before and after the oil spills implies that not only their perception of the environment has changed, but also their environmental cognition; in other words, their concern about 'The Environment'. During our research, we noticed that local citizens have also begun to assess and compare the various oil companies operating in the region by their environmental standards, production methods and social commitment within the district (Habeck 2002; Karjalainen 2001).

For the newcomers, environmental threats are clearly connected with questions about the future of the local industry, employment, incomes and housing, and attached to anxiety over the whole 'disarray' in Russia. Perceived environmental changes are experienced as part of the unstable societal situation and social change processes. In many people's thinking, the deterioration of the environment reflects the prevailing 'disorder' of Russia: dirt, litter in the streets and other forms of pollution are seen as signs of disorder (for similar findings from St. Petersburg, see Simpura and Eremitcheva 1997: 468). The collapse of the old practical and symbolic world has caused a strong feeling of insecurity, especially in the heavily subsidised northern cities.

Question: Do you think that the state of the environment has changed during your living time in Usinsk/Vorkuta?

'Everything has changed, not only the environment. Nothing has changed for the better. In the year 1981, the gradual worsening started and the 1990s are very hard to understand.' (Uw51: electrician, 50s, male, Usinsk)

'Worsened. Some kind of abandonment of the North has happened. Previously we felt Vorkuta was needed... Now we feel no one needs us.' (Vt13: teacher, 40s, female, Vorkuta.)

CONCLUSIONS AND IMPLICATIONS

During the past three decades, several international survey studies (e.g. Dunlap et al. 1993) have revealed a substantial growth in public concern over environmental issues. Environmental awareness and concern are now seen as global phenomena (Dunlap 2002). However, some scholars have asked why the expression of concern is not translated into environmentally conscious behaviour in people's everyday life (Brand 1997). One attempt to explain this gap between concern and action is the 'information deficit' model, which claims that people do not have enough knowledge to act, or have misconceptions about environmental issues, and first need to be 'enlightened' through scientific knowledge. Dunlap (2002: 168) identifies this 'as the cognitive (or knowledge) fix, which assumes that information and persuasion will suffice to produce the necessary changes in behaviour'.

Yet the 'information deficit' model cannot fully explain the gap between environmental awareness and environmental action. Rather, as we have sought to show, we need to study local contexts of everyday life, and people's experiences of environmental issues *in their own environments*. Using the approach of Ingold (2000), we situate people in the context of an active engagement with the constituents of their environment. This means seeing the individual as placed *within* the environment, rather than in the position of having to reconstruct it from the *outside*. Thus, we see 'environment' as 'life-world'. To 'dispense' information in local communities, or to 'harvest' so-called traditional environmental knowledge from them without taking into consideration the embeddedness of different kinds of knowledge in certain practices, would only create new gaps, between local people, scientists and policy-makers.

Our findings from the north of the Komi Republic support the view that the environmental problems reported by local inhabitants are by no means imaginary. They force themselves on people's attention through the constraints they place on practical activities of livelihood. For the people of Usinsk, Vorkuta and surroundings, air and water pollution are perceived first and foremost through their own senses and experiences in the context of everyday life. It is this direct perception of environmental changes that accounts for the 'highly localised nature of environmental concerns and interests' (Bush et al. 2002: 129), which has been discerned in many studies. Hence we also agree with Bickerstaff and Walker that '[t]he importance of primary experience is evident in the widespread public recognition of pollutants that could be distinguished through physical senses' (2001: 143). External sources of information play a minor role in this process.

In several respects, our research corroborates the results of qualitative case studies in Western countries. First, environmental issues are not separated from other issues and changes in a local context; and they are not accorded highest priority in the communities that we studied, for social and economic issues are at least as salient as environmental ones (Bickerstaff and Walker 2001; Bush et al. 2002; Irwin et al. 1996). Second, historical and economic peculiarities in the local discourse frame the way in which environmental concerns are balanced (Bush et al. 2002). This is evident in the differences between Usinsk and Vorkuta. Third, science and scientific knowledge have no major significance for the identification or formulation of environmental concerns in everyday life. Public discussion is not about the 'facts' of pollution or global environmental issues, but is based on qualitative relationships, experiences and people's own observations. And fourth, global environmental issues and concepts are 'contextualised in terms of the routines and everyday problems in which individuals are embroiled' (Burningham and O'Brien 1994: 917). We have to add that global environmental issues (e.g. global climate change) are less widely discussed in the north of the Komi Republic, and in Russia as a whole, than in Western countries (Darier and Schüle 1999; Dunlap 1998; Karjalainen et al., submitted).

Both groups, villagers and town-dwellers in the north of the Komi Republic, perceive pollution as a threat and have experienced some impacts of pollution on their daily lives and everyday surroundings. However, there are differences in the perception and experiences of environmental impacts among these groups. The rural individuals and communities experience biophysical impacts of environmental change (oil spills) more directly in their livelihoods, as fish are inedible and cattle cannot drink water from the rivers. Rural inhabitants are more directly dependent on the local ecosystems than newcomers working in hydrocarbon extraction and other industries. Although the townspeople, too, engage to some extent in subsistence activities, they are less affected by the consequences of environmental pollution because their livelihood is based on a different combination of income sources.

A fair proportion of the newcomers had come to the north in order to earn money in hydrocarbon extraction and were planning to move back to the south after a couple of years. In this respect, not only the function but also the value of places and 'natural resources' (such as rivers and lakes) are quite different for newcomers from what they are for indigenous dwellers. Hence, we can say that social and occupational groups differ in their perception of environmental changes (e.g. with regard to 'reading signs' of changes in plant species valuable for them) because they engage in different tasks and use different skills; they have different functional relationships to the 'local space', and only in this respect one might say that indigenous inhabitants and newcomers inhabit different life-worlds.

To conclude, environmental changes are perceived in a framework of everyday life, whereas environmental knowledge originates mainly from outside the immediate context or 'real-life' experience and is imported along various forms of communication. *Local environmental knowledge* is the result of the transactions and interactions, within a local context, between environmental perception and environmental knowledge. Local environmental knowledge is knowledge through engagement at two levels. At the first level, the individual engages with his/her surroundings giving rise to environmental perception, and at the second level, environmental perception is engaged with externally derived cognition giving rise to local environmental knowledge. This knowledge cannot be treated solely as factual information since it has its own moral and symbolic dimension within a social, cultural and political context.

NOTES

We are grateful to Aleksander Maksimov for assistance in conducting the interviews. Prior drafts have been improved thanks to extensive comments by Riley E. Dunlap and Tim Ingold. We would also like to thank Timo Järvikoski, Kalle Reinikainen and the journal's reviewers for their comments. This research was supported by the Finnish Graduate School for Russian and East European Studies, the Kone Foundation, Finland and the Daimler Benz Foundation, Ladenburg, Germany.

[1] TUNDRA (Tundra Degradation in the Russian Arctic) is supported by the EC Environment and Climate Research Programme (contract nr. ENV4-CT97-0522, climate and natural hazards). For a general description of this project, see Kuhry and Holm (1999).
[2] The fact that the town is located in the treeless tundra with its harsh temperatures makes the plight of the city-dwellers of Vorkuta more severe; whereas Usinsk is located

in the northern forest zone, where small-scale food production at the cottage (*dacha*) is to some extent possible.

[3] On a practical level, the mines (their work) could no longer provide the same kind of social safety net they used to do. Savings were devastated by inflation and the government abandoned many privileges that were meant to compensate for the 'hard living conditions of the North'. On the symbolic level, Vorkuta and its workers lost their prestige and the dispensations they had enjoyed as 'élite workers' in one of the key Soviet industries.

[4] The number of interviews was limited by the 'saturation point' at which we found that further interviews added virtually nothing to what we had already been told. In this article, when referring to the urban interviews, we shall use the following abbreviations: U = Usinsk; V = Vorkuta; a = administration staff; t = teacher; w = worker.

[5] These six villages are: Mutnyy Materik, Novikbozh and Ust'-Usa (in the Usinsk District); Petrun' and Abez' (in the Inta District) and Kharuta (administratively belonging to the Nenets Autonomous Okrug, but geographically located on the territory of the Komi Republic, Inta District).

[6] While the forest is clearly a traditional habitat for the Komi, the tundra is not. Their arrival in the tundra zone was concomitant upon the adoption of reindeer-herding practices from their northern neighbours, the Nenets. In the process of learning these skills, the northern Komi have also adopted some Nenets concepts about plants and animals, but it is not clear how far this has influenced ideas and notions of the environment in general.

[7] The Russian term for 'environment' can comprise both the natural and the social environment. The colloquial term for 'environment' is *okruzhayushchaya sreda* (literally, 'surrounding milieu' or 'surrounding environment'). The more formal (scientific, juridical) term is *okruzhayushchaya prirodnaya sreda* (literally, 'surrounding natural environment').

[8] This affects the entire catchment area of the Pechora, because many of the most valuable fish species are migratory or semi-migratory.

REFERENCES

Anderson, David 2000. *Identity and Ecology in Arctic Siberia: The Number One Reindeer Brigade*. Oxford: Oxford University Press.

Berglund, Eeva 1998. *Knowing Nature, Knowing Science: An Ethnography of Environmental Activism*. Cambridge: The White Horse Press.

Berkes, Fikret 2002. 'Epilogue: making sense of Arctic environmental change?', in I. Krupnik and D. Jolly (eds), *The Earth is Faster Now – Indigenous Observations of Arctic Environmental Change*. Fairbanks, Alaska: Arctic Research Consortium of the United States, pp. 334–49.

Bickerstaff, Karen and Gordon Walker 2001. 'Public understandings of air pollution: the 'localisation' of environmental risk', *Global Environmental Change*, 11 (2): 133–45.

Blake, James 1999. 'Overcoming the 'Value-Action Gap' in environmental policy: tensions between national policy and local experience', *Local Environment*, 4 (3): 257–78.

Brand, Karl-Werner 1997. 'Environmental consciousness and behaviour: the greening of lifestyles', in M. Redclift and G. Woodgate (eds): *The International Handbook of Environmental Sociology*. Cheltenham: Edward Elgar, pp. 204–17.

Brown, Phil 1997. 'Popular epidemiology revisited', *Current Sociology*, 45 (3): 137–56.

Brosius, J. Peter 1999. 'Analyses and interventions: anthropological engagements with environmentalism', *Current Anthropology*, 40 (3): 277–309.

Burawoy, Michael and Pavel Krotov 1994. 'Class struggle in the tundra: the fate of Russia's workers' movement', *Antipode*, 27 (2): 115–37.

Burningham, Kate and Martin O'Brien 1994. 'Global environmental values and local contexts of action', *Sociology*, 28 (4): 913–32.

Bush, J., S. Moffatt, and C. E. Dunn 2002. 'Contextualisation of local and global Environmental issues in north-east England: implications for debates on globalisation and the "risk society"', Local *Environment* 7 (2): 99–133.

Cooper, David E. 1992. 'The idea of environment', in David E. Cooper and Joy A. Palmer (eds): *The Environment in Question: Ethics and Global Issues*. London, New York: Routledge, pp. 165–80.

Darier, Éric and Ralf Schüle 1999. '"Think globally, act locally"? Climate change and public participation in Manchester and Frankfurt', *Local Environment*, 4 (3): 317–29.

DeBardeleben, Joan and Kimberly Heuckroth 2001. 'Public attitudes and ecological modernization in Russia', in I. Massa and V.-P. Tynkkynen (eds), *The Struggle for Russian Environmental Policy*. Helsinki: Kikimora Publications, Series B: 17, pp. 49–76.

Dunlap, Riley E. 1998. 'Lay perceptions of global risk: public views of global warming in cross-national Context', *International Sociology* 13 (4): 473–98.

Dunlap, Riley E. 2002. 'Environmental Sociology', in Robert Bechtel and Arza Churchman (eds), *Handbook of Environmental Psychology*. New York: John Wiley & Sons, Inc, pp. 160–71.

Dunlap, Riley E., George H. Gallup and Alec M. Gallup 1993. 'Of global concern: results of the Health of the Planet Survey', *Environment*, 35 (9): 7–39.

Dunlap, Riley E. and Robert Emmet Jones 2002. 'Environmental concern: conceptual and measurement issues', in Riley E. Dunlap and William Michelson (eds), *Handbook of Environmental Sociology*. Westport: Greenwood Press, pp. 482–524.

Eder, Klaus 1996. *The Social Construction of Nature: A Sociology of Ecological Enlightenment*. London: Sage.

Feit, Harvey 2001. 'Long-term local knowledge, historically-bound science and the politics of environmental policy changes: new roles for northern peoples and scientists'. *Arctic Feedbacks to Global Change*. International Symposium, Arctic Centre, Rovaniemi, Finland, October 25–27, 2001. Rovaniemi: Arctic Centre, pp. 34–6.

Fischer, Frank 2000. *Citizens, Experts and the Environment: The Politics of Local Knowledge*. Durham and London: Duke University Press.

Glaser, Barney and Anselm Strauss 1967. *The Discovery of Grounded Theory*. Chicago: Aldine.

Grove-White, Robin 1993. 'Environmentalism: a new moral discourse for technological society?', in Kay Milton (ed.), *Environmentalism: The View from Anthropology*. London, New York: Routledge, pp. 18–30.

Habeck, Joachim Otto 2002. 'How to turn a reindeer pasture into an oil well, and vice versa: transfer of land, compensation and reclamation in the Komi Republic', in Kasten, Erich (ed.): *People and the Land: Pathways to Reform in Post-Soviet Siberia*. Berlin: Reimer, pp. 125–47.

Habeck, Joachim Otto 2003. 'What it means to be a herdsman: the practice and image of reindeer husbandry among the Komi of northern Russia'. Unpublished PhD dissertation, University of Cambridge.

Ingold, Tim 2000. *The Perception of the Environment: Essays in Livelihood, Dwelling and Skill*. London: Routledge.

Irwin, Alan 1995. *Citizen Science: A Study of People, Expertise and Sustainable Development*. London: Routledge.

Irwin, Alan, Alison Dale and Hilary Rose 1996. 'Science and hell's kitchen: the local understanding of hazard issues', in Irwin, Alan and Brian Wynne (eds), *Misunderstanding Science? The Public Reconstruction of Science and Technology*. Cambridge: Cambridge University Press, pp. 19–46.

Karjalainen, Timo P. 2000. 'Environmental awareness among the city dwellers of Northern Komi: local environmental changes as a part of societal and global transformations'. Paper held at the sixth ICCEES World Congress, Tampere, 29 July–3 August 2000 (Panel XII-27). Abstract in: Sinisalo-Katajisto, Petra and Paul Fryer (eds): *VI World Congress for Central and East European Studies: Abstracts*. International Council for Central and East European Studies (ICCEES); Finnish Institute for Russian and East European Studies (FIREES), p. 192.

Karjalainen, Timo P. 2001. 'Institutional framing of environmental issues in the Komi Republic', in I. Massa and V.-P. Tynkkynen (eds), *The Struggle for Russian Environmental Policy*. Helsinki: Kikimora Publications, Series B: 17, pp. 77–106.

Karjalainen, Timo P., Timo Järvikoski and Pentti Luoma (submitted): 'Climate change and citizens in the Komi Republic (Russia)'. Manuscript submitted to *Global Environmental Change* (Part A).

Keskitalo, E. C. H. 2002. *Constructing 'The Arctic': Discourses of International Regionbuilding*. Acta Universitatis Lapponiensis 47. Rovaniemi: University of Lapland.

Kroll-Smith, Steve, Stephen R. Couch and Brent K. Marshall 1997. 'Sociology, extreme environments and social change', *Current Sociology* 45 (3): 1–18.

Kuhry, Peter and Tuija Holm 1999. 'Arctic feedbacks to global warming: Tundra Degradation in the Russian Arctic (TUNDRA)'. http://www.urova.fi/home/arktinen/tundra/tundra.htm [accessed in August 2001].

Lodewijkx, M. and H. Hirsch 2000. *Komi: Der Preis des Erdöls: die sozialen und ökologischen Kosten der Ölförderung in der Komi-Republik und Nordwest-Russland* [The price of oil: The social and ecological costs of the oil production in the Komi Republic and northwestern Russia]. Hamburg: Greenpeace.

Mirovitskaja, N., 1998. 'The environmental movement in the former Soviet Union'. In A. Tickle and I. Welsh (eds), *Environment and Society in Eastern Europe*. Longman, Edinburgh, pp. 30–66.

Peterson, D. J. and E. K. Bielke 2001. 'The reorganization of Russia's environmental bureaucracy: implications and prospects'. *Post-Soviet Geography and Economics* 42 (1): 65–76.

Pickvance, Katy 1998. *Democracy and Environmental Movements in Eastern Europe: A Comparative Study of Hungary and Russia*. Boulder: Westview Press.

Poklad, Yuriy 1995. 'Ekologiya sushchestvovaniya [The ecology of existence]', *Severnyye prostory*, 10 (4–5): 27–8.

Sagers, M. J. 1994. 'Oil spill in the Russian Arctic', *Polar Geography and Geology*, 18 (2): 95–102.

Simpura, Jussi and Galina Eremitcheva 1997. 'Dirt: symbolic and practical dimensions of social problems in St. Petersburg', *International Journal of Urban and Regional Research* 21 (3): 476–80.

Taskayev, A. I. (ed.) 1999. *Gosudarstvennyy doklad o sostoyanii okruzhayushchey prirodnoy sredy Respubliki Komi v 1998 godu*. Syktyvkar: Ministerstvo Prirodnykh Resursov i Okhrane Okruzhayushchey Sredy Respubliki Komi; Departament po okhrane okruzhaiushchei sredy Respubliki Komi; Institut biologii Komi Nauchnogo Tsentra Ural'skogo Otdeleniya Rossiyskoy Akademii Nauk.

Vil'chek, G. Ye. and A. A. Tishkov 1997. 'Usinsk oil spill: Environmental catastrophe or routine event?' in R. M. M. Crawford (ed.), *Disturbance and Recovery in Arctic Lands: An Ecological Perspective*. Dordrecht, Boston, London: Kluwer Academic Publishers, pp. 411–20.

Weiner, Douglas. R., 1999. *A Little Corner of Freedom: Russian Nature Protection from Stalin to Gorbachev*. Berkeley: University of California Press.

Yanitsky, Oleg 2000. *Russian Greens within a Risk Society: A Structural Analysis*. Helsinki: Kikimora Publications. Series B:11.

Riding the Tide: Indigenous Knowledge, History and Water in a Changing Australia[1]

Heather Goodall

Indigenous knowledge and water are at the centre of the conflicts in Australia today over land ownership. Federal Court Judge Olney used metaphors of water to naturalise his 1998 rejection of the Native Title claim by the Yorta Yorta people of the Murray River to be recognised as the continuing custodians of their land and the river running through it: 'The tide of history has indeed washed away any real acknowledgement of their traditional laws and any real observance of their traditional customs'.[2] This was the Yorta Yorta's seventeenth attempt since the 1860s to reclaim secure title over their land from the colonising British who had taken control of the Australian continent in 1788. Central to the Yorta Yorta's increasingly bitter demands, just as it was to the Olney judgement, is the question of history. What does the passage of time and the effect of dramatically changing conditions mean to the complex of beliefs, understandings and practices which are 'indigenous knowledge'? Is such knowledge a fixed archive which can be eroded and 'washed away' over time as Justice Olney claimed?

The questions around the continued presence and value of 'indigenous knowledge' are of high interest in environmental politics both internationally and locally. Since the World Parks Congress in Durban in 2003, the recognition of indigenous people's rights to and knowledge of environmentally sensitive and endangered lands has been escalating.[3] In Australia, one of the most progressive non-government environmental advocates, the Wilderness Society, has recently launched a national program engaging actively with Aboriginal people in planning and implementing its Wild Country campaigns across the continent. At all levels of government, conservation agencies have recognised the importance of indigenous knowledge in various ways. Yet in each of these initiatives, the meaning of 'indigenous knowledge' is uncertain and undefined. So while the importance of indigenous knowledge rises on the agenda of government and non-government conservationists, so too do the continuing unresolved questions about how to understand indigenous knowledge in contemporary circumstances.

This essay will argue that 'indigenous knowledge' can be more effectively understood as a process rather than as an archive, both before and after colonisation. Water and rivers have played a key role in the continuing practice of such cultural processes by Australian Aboriginal peoples, not only in the recently colonised 'remote' areas but throughout the turbulent centuries of intensive colonisation in the country's south east. This means putting the recognition of historical change back into the analysis of indigenous knowledge both before

Environment and History **14** (2008): 355–84.

and after the invasion by the British. It contradicts the more usual 'watershed' view of colonial impact which suggests that there is an unbridgeable difference between indigenous life, or indeed ecologies, before and after the invasion. The cost of an argument which reintroduces history like this into the post-invasion period is that it destabilises the concept of 'indigenous knowledge', opening it up to questions about its loss or dilution. The result of recognising historical change is to offer a more fruitful way to recognise the high value of indigenous people's understanding of changing places and environments both past and present.

The focus for this discussion is the floodplain of the upper Darling River in rural north western New South Wales, an inland delta crossed by several interlaced rivers which flow into the Darling. Australia is the driest continent on earth and so water everywhere is a key resource. The upper Darling is more fertile than other areas but its waters are unpredictable: it faces severe droughts and expansive floods, so Aboriginal harvesting demanded extensive knowledge of its extreme conditions.[4] Water is a constant presence in the early collections of the legends of the Yuwalaraay and the Ngiyampaa, suggesting its central role in the symbolic as well as the material life of pre-invasion Aboriginal societies along the river.[5] The region was first invaded violently by the British in the 1830s and penetrated by the settler grazing economy by the mid-1840s. Settler management has since then been aimed in essence at controlling its water: storing it in weirs and dams, modifying flows to contain the rivers strictly within surveyed banks and locking up the land in between as private property. My work has involved a series of projects investigating the relationships between Aboriginal people, settlers and environmental change in the Darling River region and in Central Australia.[6]

The Darling River in rural New South Wales is not the type of area usually discussed in relation to indigenous environmental knowledge in Australia. Popular accounts of conservation movements and their interaction with indigenous knowledge concentrate on northern and Central Australia as does the new initiative of the Wilderness Society, focusing on the remote Aboriginal communities living a most recognisably 'traditional' lifestyle and whose lands have only recently been drawn into the western economy.[7] The vast majority of identified Indigenous Protected Areas advertised enthusiastically by the Federal government as heralding a new era of recognition of indigenous knowledge, are also all in northern and north-western Australia. Yet the state with the greatest number of Aboriginal people is the longest settled, intensively farmed and densely populated New South Wales in the south east, which holds 30 per cent of the overall Aboriginal population of 500,000.[8] It is followed closely by the adjoining south-eastern areas of Queensland. In the upper Darling itself, which straddles these two states, between 40 per cent and 50 per cent of the region's rural population of 50,000 are Aboriginal.[9] Does this mean that the majority of the Aboriginal population, located in this south-eastern quadrant of the continent, has no 'indigenous knowledge' of interest in conservation matters? If so, how are the nine 'co-managed' protected areas in NSW to be managed?[10] What role will Aboriginal people play in them? While there are no simple answers, the themes of water and history are central to understanding how indigenous knowledge has been sustained and is mobilised in these south-eastern states today.

RELATIONAL INDIGENEITY

Perhaps the first unresolved question concerns the meaning of the word 'indigenous' (or 'aboriginal', meaning 'original'), which continues to be widely used in Australia and in much of the west, as if it is a simple concept which has a global meaning. The concept of 'indigeneity' is a complex one which invariably involves an interaction between the self-representation of the individuals and groups asserting their indigeneity on the one hand and, on the other, the pressures and goals of allies and enemies, whether within the nation state or internationally.[11] Aboriginal analysts in Australia have been cautious in their use of the term, reserving it for the context of international comparison and preferring to use local language names for groups of Aboriginal people within Australia.[12] An unproblematised definition of 'indigenous peoples' tends to be used by researchers and activists working in 'first world' and 'settler colonial' situations, where Europeans became the majority population after displacing small-scale societies practising economic forms labelled 'hunter gatherer' or 'shifting cultivator'.[13] For analysts like Baviskar and Li, working in India and Indonesia respectively, the definition of indigeneity is relational and unstable and needs to be considered cautiously. Nor can cultures and economic practices be regarded as congruent, because societies alter their economic strategies in conditions of pressure. People regarded as shifting cultivators in India could move to dependence on harvesting (hunting/gathering) if circumstances changed and at other times might chose cultivation over harvesting, regardless of their categorisation by others as 'tribals' or as 'farmers'.[14] This continuing complexity is evident in the Durban and later IUCN documents, which by 2006 had recognised the shared interests and at times shared identities across groups identified as 'indigenous peoples, mobile peoples and local communities'.[15]

THE MYTH OF TIMELESSNESS AND PRESSURES FOR STATIC INDIGENOUS KNOWLEDGE

Unresolved questions also exist around whether 'indigenous knowledge' was a fixed body of information before and after colonisation and then after 'modern' development. The confusion around this question is reflected in the variety of terms used to identify indigenous environmental knowledge. Some authors have referred to it as Traditional Ecological Knowledge[16] and others discuss it as 'pre-colonial' or 'non-western'.[17] The implication of each of these terms has been that this body of knowledge was static in time and was opposed to 'western' systems of scientific knowledge of environments and their changing ecologies.

There have been strong pressures which have led to a focus on pre-colonial 'tradition' as the model for all 'indigenous knowledge' and which have defined this as if it were unchanging even in pre-colonial cultures. One pressure has arisen from settler interests in the European dominant colonies. Lands which had been shaped by centuries of harvesting or swidden agriculture were mis-read by settlers as previously untouched and stable 'wilderness'. These myths of a 'pristine wilderness' were used to justify undisputed settler possession and have continued to shape relationships between indigenous colonised peoples and dominant populations in countries with settler colonial backgrounds like Australia, Canada and the United States.[18] Another pressure to see indigenous knowledge as static has arisen from the western environmental movements which emerged in the 1960s and which rejected 'modern' commercial exploitation of environments but retained the mythology of pre-modern 'wilderness' where indigenous people were depicted as exotic 'noble environmentalists' living 'in harmony' with the non-human environment. This movement continued the as-sumption that indigenous societies had taken no role in shaping and managing a 'wild' environment.[19] The mythology of 'wilderness' held by early conservation advocacy groups was used to exclude Aboriginal people from a role in manage-ment and this continues to be a pervasive attitude among the more conservative wings of the movement, as Aboriginal environmentalist Fabienne Bayet-Charlton has described.[20] The unrealistic yardstick of 'noble environmentalist' is used to criticise contemporary Aboriginal people who do not live a recognisably 'traditional' lifestyle, and who use guns and four-wheel-drives to hunt game or who seek an economic return on community owned land.[21]

Yet the pressure to consider indigenous knowledge as a static repository of pre-colonial knowledge has not arisen only from colonial settlers and non-indigenous conservationists. The victories of long-fought Aboriginal campaigns to have their rights of prior ownership to land recognised in Land Rights and Native Title legislation have ironically locked inflexibilities into the small gains made from those achievements. Both the bureaucratic nature of land registra-tion under these acts and the intensely adversarial court cases necessary to 'prove' title have shaped the outcomes to fit entirely into a model of western property rights based on a slice of time frozen at the point of colonisation. The tests of evidence rely on biological inheritance and settler-authored historical documentary records. The flexibility of traditional cultural land responsibilities and the complexity afforded by oral accounting of land affiliation are ignored.

One of the few Aboriginal people to have written about indigenous environ-mental knowledge in the long-settled south east is Tex Skuthorpe, a Yuwalaraay man from the Nhunggabarra clan on the Darling River floodplain, whose long history of creating visual art and storytelling about the river will be discussed below. His recent writing in collaboration with a western researcher in Business Knowledge Management has been directed towards environmental management but depicts indigenous knowledge as a timeless, static and 'intact' pre-invasion knowledge system which can be viewed whole and in opposition to western land management. Nhunggabarra society, according to Skuthorpe and Sveiby, ended in 1828 with the first appearance of British invaders.[22] While few other Aboriginal analysts would agree with the depiction of an abrupt end to indigenous society or culture, there has still been a focus on considering indigenous knowledge of the environment as it is exists in remote areas. Marcia Langton is the most widely published Aboriginal analyst of land and environmental knowledge and she has addressed indigenous knowledge largely in terms of its maintenance and resilience in conditions of high retention of traditional languages and of relative ecological stability and biodiversity maintenance.[23] Neither Skuthorpe nor Langton answer questions about how to understand indigenous knowledge under conditions of long colonisation and intensive cultural interaction.

INDIGENOUS KNOWLEDGE AND ENVIRONMENTAL MANAGEMENT

The attempt to integrate the knowledge of indigenous people into environmental management has largely been enacted within this paradigm of a static repository which was complete prior to colonisation. Its fragments now need to be 'captured' in order to use it to restore health to ecologies disrupted by globalising commercial management. The result has usually been to present 'indigenous knowledge' as if it were a list or a database because these are the forms in which such information is recognisable to scientifically trained professionals and it is the most readily searchable for use in planning resource management.[24]

Yet 'indigenous knowledge' is not held or transmitted within indigenous communities in the form of a list or a database. It may be passed on during practical activities but it might also be remembered and orally performed as narrative in very different genres to the catalogued arrangements of data familiar to the cultures of literacy. Several theorists have drawn cautionary attention to the idea of straightforward 'information transfers'. Bruno Latour's work has demonstrated how 'field work' and the necessity to catalogue specimens of everything from soil samples to 'knowledge', changes the meanings we can make from that material.[25] Virginia Nazarea has asked whether the cultural production of environmental knowledge is reducible to the Linnean taxonomic systems of western science.[26] Oral narratives are dismembered in the same damaging way for legal or historical research.[27] Roy Ellen argues that rather than static, permanent structural relations, classifications should be seen as situational and dynamic.[28]

The Dene people of Canada have asserted that indigenous knowledge must be seen in a holistic sense to include both everyday knowledge and the more formal narrative 'stories' which are recognised as oral tradition. They hosted an international symposium in 1990 which suggested both the strengths and the limitations of the concept of Traditional Environmental Knowledge.[29] The course of the discussions between indigenous people from very different areas demonstrated the continuing questions around the actual use of such knowledge and the difficulties of taking the outcomes beyond the static database approach.

Work which does recognise historical change is Firket Berkes' extensive research with Aboriginal people in Canada and elsewhere.[30] Rejecting romantic notions of essentialised indigenous knowledge, Berkes explores the responsive capacity of indigenous and local knowledge systems as environments change. Trained in natural resource management rather than cultural analysis, he distinguishes everyday environmental information gathered in hunting and gathering from the formal narrative conventions of 'stories', ceremonies and mythology. Berkes is only able to trace processes of flexibility and historical change in the elements of indigenous knowledge which comprise everyday environmental understanding, which is transformed by feedback in isolated communities in which the 'resources', like caribou, remain under the sole control of the indigenous people. Then 'social learning' occurs when, for example, ecological feedback demonstrates over-harvesting, thus allowing adjustments to occur over time.

The most recent work on indigenous knowledge engages anthropological approaches with natural resource management but it largely returns to considering remote rather than long-settled societies. Benjamin R. Smith's 2007 account of the development of hybridised knowledge systems in the mid Cape York area of sub-tropical northern Queensland points out the fragmented nature of western science, rather than just the local indigenous system.[31] Change is discussed in Smith's account as being the active engagement of a relatively stable pre-invasion indigenous knowledge system with a localised variant of western science, producing a hybridised and responsive body of environmentally specific new approaches to land management. It still does not allow us to understand the long and heavy impact of colonial economies and social controls on indigenous knowledge in the south east of Australia, or indeed in any long settled area.

THINKING THROUGH ORAL TRADITIONS

Historians may have something to contribute to this work because they have tried to make sense not only of what may have happened in the past, but of how the past has been represented. This has included the oral traditions of societies which did not use writing as well as the historiographies of societies which rely on written accounts of the past. Even literate societies, like those of Europe, have oral traditions maintained by marginalised groups such as the Roma or women midwives. There was a great deal of interrogation of oral tradition by historians in the 1960s, as western trained historians like the Belgian Jan Vansina tried to fit the oral narratives of African and Pacific societies into the rigid templates then demanded of written sources in order to justify their use.[32]

Vansina revised his earlier simplistic approach in 1985 and made a major contribution to the better understanding of the flexible creation and reception of oral tradition.[33] This in turn allowed rich insights into the social processes of memory and historical change in cultures which did not use writing. Written sources have themselves since been opened up for intensive critique, first on the basis of their frequent origin within colonial processes and later as discourse analysis has effectively undermined claims for unquestioned 'authenticity'. It is clear, as recent African historians have demonstrated, that every medium, whether written, visual or oral, has its own qualities but that none can be drawn on as a source without careful interrogation.[34] However, the question has now reemerged in the very different forum of conservation politics as indigenous knowledge is celebrated but at the same time called on to carry the burden of finding solutions to major environmental crises, without allowing such reflection on how such knowledge might be constructed and transmitted.

Yet while oral tradition is open to creative interventions in the socially mediated and interactive performances of any oral culture, this is not at all how oral traditions present themselves within indigenous societies, including Australia's.[35] Instead oral traditions contain a rhetoric of enduring permanence built structurally into their narratives which asserts an unchanging quality to their forms and content.[36] The words used to describe oral tradition in Pitjant-jatjara country in central Australia, for example, is *Tjukurpa* or Law, suggesting unchanging permanence, while the identification of the narrative participants as ancestors locates the stories far in the past. Such narrative strategies assert authority by claiming trans-human creation of both stories and their forms, by ancestral or divine figures whose power is said to be far greater than that of today's human population.

Certainly some types of knowledge are transmitted unchanged over many generations, entrusted to skilled experts in verbatim memorisation and faultless recall. These are generally those few relating to survival, which no society can afford to lose, like the skills of over-the-horizon navigation in Pacific Island cultures or those of inland desert navigation in Australia.[37] Most oral knowledge is passed on in the far more flexible conditions of performance, often, as in much Australian ceremony, in participatory and interactive settings. Here there are opportunities to engage apparently unalterable narratives with the historical changes in both environment and social life. The important observation from historians working on oral tradition is that this process was occurring in 'pre-colonial' times. It is how such oral performances have always been created and how they are able to negotiate the continuing dynamic of lived change with the cultural imperative of appearing to be enduring and authoritative. There was therefore no 'colonial watershed' in the way that indigenous oral societies recorded, transmitted and enacted cultural learning. Oral traditions have always been a dynamic form, which engaged with and reflected changing social and environmental circumstances however much they then presented themselves as fixed, received truth. This continued after settlement began just as it had beforehand.

As stories about historical events moved into oral traditions, whether this happened before or after colonisation, they lost their chronological markers and took up the thematic, narrative and locality-related markers which allowed them to be fitted seamlessly into the existing oral performance. Only in situations of sudden cultural change can we see this process occurring. A striking example is the development by the Yanyuwa people in the Northern Territory of a whole ceremonial performance known as 'Aeroplane Dance' which tells the story of the rescue of a World War 2 American bomber pilot whose plane had crashed nearby in 1942. The traditional narrative form of the dance and song cycle was able completely to dramatise the sequence of events, and only the unusual subject matter demonstrated that this was not a 'traditional' event, but instead a recent 'historical' event which had been woven seamlessly into a traditional genre.[38] Examples from the western inland desert, but also many other areas, show how the key symbols of western imperialism in Australia, like Captain Cook's voyage of 1770 in which he claimed the country for the British Crown, have been appropriated into the very traditional narrative and performative genres of oral traditions to offer a powerful counter analysis of colonialism.[39]

Not only does oral tradition allow the recording and analysis of recent, historical events. The flexibility of oral tradition and traditional knowledge also allows societies to have some mechanisms to cope with enormous, sudden changes like displacement and distant resettlement. Francesca Merlan has described this process as it occurred in Katharine, a town on the edge of the tropical wetlands in the Northern Territory to which the Jaywon people were moved for resettlement. They were then at some distance from their traditional country and while continuing to maintain interest in that original country, they paid close attention to their surroundings in Katherine, expecting and seeking a meaningful connection to their new inescapable home. As Merlan has written:

> …there is always the possibility of the 'discovery' of existing but newly revealed and interpreted significances, whether or not these be clearly attributed a mythic dimension[40]

One such site was 'Catfish', an area near a long established Aboriginal camp in the town which over many years came to be seen as a place of significance which offered a link to more distant ceremonial stories in the areas from which people had migrated.[41] The concepts of revelation and discovery allow communities to feel that close attention to the new site might be rewarded with the affirmation of traditional legitimacy. The many genres of oral traditions which may carry environmental knowledge are often transmitted in this participatory performance mode, which offers the capacity to be responsive to the recording of changes in the environment within which humans were participating. The possibility of discovering newly revealed episodes to story cycles, particularly in unfamiliar places, offers a powerful stimulus to close observation of environments. This dimension of pre-invasion cultural process developed even more importance with the increasing experiences of displacement which occurred after British settlement.

MAKING PLACES, MAKING PEOPLE

Seeing how apparently unchanging oral tradition actually develops as a flexible and interactive engagement with the past and the present leads us to consider the broader questions of how societies relate to places. Arjun Appadurai has sketched out an ethnography of modernity which might encompass both small scale and large scale societies. He argues convincingly that the link between small scale societies and place, which is so often presented as if it were just as unchanging and enduring as oral tradition, is in fact a work in progress. He argues that 'locality', (the ways humans know and understand material places), is an 'inherently fragile' social creation, reached and sustained only because societies work at it.[42] Rather than 'local knowledge' being the enduring record of a revealed truth about an ideal and stable environment, Appadurai focuses on the ceremonies which are seen to be a record of the connection between people and place. He argues that they are the means to continuously create and then regenerate that bond. He discusses the way these processes intersect with the ordinary, everyday conditions of life, making what is actually uncertain and precarious look ordinary and taken for granted. [43]

Appadurai describes the production of 'local subjects', that is people who are confident of their links to and ownership of the places they live in because they know them, and the networks of social relations between 'local subjects', people who feel they are secure because they *have* a place. It is this which Appadurai argues is the central role of much of the performative ceremony in any society. His argument is helpful in considering societies undergoing substantial change and in states of displacement, such as the present case study on the Darling River floodplain and in other research in which I am involved with Aboriginal people living on a river in suburban Sydney. Many are recent migrants from rural areas and struggle with producing locality in drawing on their conceptions of themselves as Aboriginal.

Kingsley Palmer's discussion of dramatic change in remote desert societies of Western Australia offers other insights into indigenous knowledge and place making. Palmer argues that the concept of a responsible adult in traditional, pre-invasion societies was one who had and was exercising custodial rights over country. Land custodianship developed in a flexible way over a person's life, and the social processes of marriage and alliance linkages reshaped responsibilities to and power over land, which all meant that attention to places was a necessary part of daily life. Palmer documents this flexible means by which extension of traditional social processes could generate affiliations to new places when western desert peoples were forcibly moved into the iron ore mining areas of the Pilbara.[44] The possibility of such flexibility in creating *locally*-affiliated people must have existed throughout the two centuries of colonised land and social relations in the south east, offering a means to understand how the Aboriginal communities devastated by invasion violence and either displaced themselves or taking in people displaced from elsewhere, might have been able to make some form of cultural recovery. The expectation that such a process could occur placed demands on newcomers that they accumulate the knowledge about the new homeland which would allow them to fulfil appropriately the roles of owner and custodian. So both customary social arrangements and resulting custodial roles could contribute to a means to cope with disruption and dislocation in the turbulent conditions of colonial life.

HISTORY AND THE DARLING FLOODPLAIN

By drawing history back into the analysis, we can consider how the changes caused by colonial economies and technology intersected with indigenous peoples' continued interactions with their environments. The upper Darling floodplain is an area of relatively fertile grasslands which was subject to intense, violent invasion in the 1830s. Rivers, creeks and water holes were invariably the places over which Aboriginal owners and British settlers fought because the water sources were vital to both for the survival of people and livestock. For each, these waters held a symbolic value far beyond their essential biological and economic role. For Aborigines, water forms a key structural role in traditional narratives, as the local stories collected on the Darling floodplain in the 1890s by Katie Langloh Parker demonstrate, where many of the stories are about the creation of rivers and springs.[45] They tell about ancestral heroes battling over water or creating river beds in their travels or burrowing the invisible, underground water channels which are said to connect one river or spring with another, which the ancestors used to travel secretly across country to outwit their enemies, rescue their loved ones or revenge their deaths. So it is unsurprising that water might be a significant element in contemporary narratives. But it has played a more complex role.

Settler pastoralism became the dominant economic land use by 1860 and Aboriginal workers were recruited into the pastoral companies as seasonal and casual workers. In that role, and for most of the twentieth century, Aboriginal people would not be regarded as living a 'traditional life'. Yet today, despite the dramatic changes which have occurred, most of the Aboriginal population in the upper Darling area know where their family's traditional country, in the broadest sense, and their language area lies and they live in reasonable proximity to it. Most of these Aboriginal people regard themselves as being traditional owners of land in the region in which they are living and they exercise an active role in land campaigns or management processes. In these rapidly settled areas of the south east, at least until the 1920s, stock densities on large properties were low enough to allow some compatibility of economies, and Aboriginal workers combined subsistence harvesting with stock work and droving. This meant they were effectively subsidising the settler economy but it allowed Aborigines to maintain both ceremonial and kinship obligations across wide distances. But from the 1920s onwards the big pastoral runs shifted to mechanised pastoral management or were cut up into smaller, family-run grazing businesses using less labour or were turned into more intensively farmed wheat and horticultural farms, making further compatibility with Aboriginal subsistence harvesting virtually impossible. The fencelines around properties had been of little significance when Aborigines were widely employed on the properties and they had continued to move freely across land they still regarded as their own country. But with the widespread loss of employment, the fencelines became closed borders. Most recently, rising hostility by white property holders to Aboriginal claims for land and native title have meant that the gates into the few remaining hospitable properties have been locked and real access to country had been choked off.

Water had always been essential to the pastoralists and Aboriginal knowledge of where to find water and how to move between water sources was an invaluable resource for the stockowners who employed Aboriginal drovers, shepherds and stock workers. Periods of high employment in the pastoral industry had meant learning a whole new range of uses for water knowledge as Aborigines developed skills in managing large numbers of sheep and cattle in relation to the rivers, soaks and springs they had known as far more fragile watering points for people and native stock like kangaroos. The developing settler infrastructure involved expanding the watering points. First, settlers dug earth tanks, in technologies for rainwater harvesting learnt from India via the British. Then, in 1878, the ground water resources from the Great Artesian Basin were tapped by the first deep bores at Bourke in north western NSW and then in south western Queensland, increasing the number of off-river water supplies not only for domesticated stock but for native marsupials and birds, allowing kangaroo and emu to multiply rapidly.

But water remained scarce and the legal structure of access to it reflected its high value for life rather than profit. In NSW the rights to flowing water had been retained in the public hands, in a careful set of decisions in the mid-nineteenth century, which were made after inquiries in all colonies into the riparian property models available in British and United States. Beyond public rights in flowing water, the access to water was retained as a public right.[46] Both water itself and, in theory, the routes across land to gain access to it remained open to the general public, including Aboriginal people, even as their real access to the lands of pastoral properties began to close down with the loss of employment. The most reliable access routes to water were the Travelling Stock Routes (TSRs), long strips of land also reserved for public ownership for drovers moving stock long distances to markets. The TSRs included access to watering points at regular intervals along each route, following the natural above ground water courses and so showing the way water flowed.

REMEMBERING COUNTRY THROUGH WATER

The ways in which Aboriginal people in rural NSW today are documenting their environmental knowledge reflects this history. Earlier general research in anthropology[47] or history[48] was framed in a search for the sites of cultural significance or social history, like work sites, camp sites and conflict sites. Later historical and environmental studies[49] have been focused on water because the severe impact of water scarcity has been felt during the last 25 years of low rainfall or drought, and government agency catchment management strategies, such as Streamwatch, emerging in this situation in the 1990s tried to learn more about alternative approaches. Most recently, rather than imposing a priority theme, studies have asked Aboriginal people to map out the places of significance to them, seeking to chart an alternative geography defined by Aboriginal people rather than by the infrastructure of settler fences and surveys, and to identify those places where Aborigines are aware of the presence of high environmental knowledge among members of their community.[50] The results for all of these methodological approaches are strikingly similar: water, rivers and springs appear frequently and are of high significance in all these studies as Aboriginal people recount important places and tell the stories which carry environmental knowledge. Such accounts are fragmentary. There are many stories which appear no longer to circulate and there are only segments of others which are known. More notable is the geographic unevenness of the information: it largely focuses on places along or close to rivers, springs or water sources.

(i) *Lists/ecologies/networks*

The types of information which can be derived from these documentations in collaboration with Aboriginal people tend to occur in three forms. Firstly there is the sort that readily translates as catalogued items into databases and encyclopedia entries of 'traditional' knowledge. This offers a rich body of information on the biology and hydrology of water. There are many forms of plants and fish, water creatures, birds and land animals both in and around rivers, lagoons, estuaries and springs which have been recorded in this format according to their distributions and uses for nutrition, medicine or crafts like weaving and fishing, as well as for their cultural meanings and presence in various stories and performances.[51] What is evident from these studies is the prevalence of knowledge about water-related biota throughout the Aboriginal community. Cotter points out that although water sites are most commonly the location of high concentrations of environmental knowledge among the Gamilaraay, people speak also about travelling stock routes along which they travelled between water points and the higher stoney ridges which have not been intensively developed. She argues that sustained access and relatively lower levels of damage from the incoming settler industries have each contributed to this higher transmission of knowledge about native species.[52] The Gamilaraay and Pikampul people working with Thompson on Boobera Lagoon and the Wiradjuri working with English and Gay on the Macquarie Marshes have all explained that they were very conscious of the loss of their access to other places on their country and that these water sites have become increasingly important to them for this reason.

These lists of plants and animals are different from those which tend to be generated in the 'local knowledge' of white grazing and cotton farming residents in the Darling floodplain, because the purposes brought to activities by farmers have been different from those of most Aborigines, despite often sharing a productivist dimension to their interest.[53] Graziers have been looking for sloping banks down which they can safely lead stock to drink, whereas Aborigines have been interested in steep or high banks as valuable sites for yabby fishing and other forms of harvesting. Cotton farmers want empty water, with no fish or reeds which will clog up the pumps so they can fill their storage tanks, and they want predictable even flows to water their crops. Aborigines want variable flows, to make the fish run and to refresh the river for the many other species of river creatures which they use.

Although it is older people who are most often the contributors of such information in this study, younger people were involved too and were active in learning, particularly in relation to frequent activities like fishing. Thompson, Cotter and English and Gay each argue that cultural knowledge, meaning both the stories within which such biological information is entwined and the context in which these stories are retold and discussed are essential to understanding the full meaning of the animal, fish or plant to the Aboriginal people involved. Rather than a classificatory database of individual species, the stories suggest the ecologies of interaction within which such lifeforms are actively sustained. The contexts for transmission allow an insight into the distribution of species, for example, are they found below the waterline or above, in drought or flood, what season are they present. Context offers information about the practical enactment of the knowledge about particular species: whether it is eaten or avoided, for example, or how it might be found. Perhaps most importantly, it suggests the conditions necessary for this form of knowledge transmission to continue. Continued fishing, for example, means continuing conversations about bait, habits of fish, troublesome or interesting insects on the bank, the state of the river and of course the stories about them all. Contexts also indicate the anomalies which signal change. Phrases like 'we used to get ...' or 'you don't see them now …' are common in discussions about species and about behaviours of the river water. People involved in the above studies and in my own research in the north west talk frequently about the river water being more or less turbid than it was in the past, having more or less of any species of reeds or mussels or the invasive carp and of the water itself moving in a different way. They grieve particularly about the loss of 'the freshes', the unpredictable small changes in the flow pattern as water entered the system in the some distant northern tributary and flowed suddenly past.

(ii) *'Water shows us country'*

The second form of indigenous knowledge documented in the upper Darling area is much harder to dismember into a taxonomy. It might be thought of more usefully as an approach to land and water management embedded in narrative, rather than as an item of data. One example is the awareness commonly expressed among Aboriginal people on the floodplain that the river system cannot be thought of as being 'naturally' confined within banks. This approach is evident in the landscape paintings of Tex Skuthorpe, who as an artist has been teaching young Aboriginal people for many years, in work which tends to contradict his recent published work arguing that the circulation of Nunggabarra knowledge had ceased. A painting he did in the early 1990s of the Yuwalaraay region showed three rivers flowing south west into the Darling and a fourth which ended just to the north of the main river, in the Narran Lake. Official maps of the area show the rivers neatly confined to their banks, flowing past the towns and paddocks on down to the south. Tex used concentric lines, a feature of traditional Yuwlaraay graphic design previously incised on skins and wood, to show the flow of water beyond the river banks, onto the floodplain and through the areas identified as townships. In the ebbs and flows of the concentric designs he has drawn young fishlings, mussels and other animals which breed on the plains when the river is in flood. His painting depicts a 'flood dependent' ecosystem, which needs flooding to regenerate. This painting, like so many of the ground designs and earth sculptures of the region, is a medium intended to be one element in complex performative oral genres which are interactive and participatory. So Tex talks about his painting and as he does so he explains the traditional Yuwalaraay stories of the area. 'The water *shows* us the country' is a phrase Tex repeats often in his explanation, stressing the need to see not just one but many floods to gain a deep understanding of the landforms beyond the river banks, made up of subtle variations of low black soil and higher stony ridges. The water not only creates the land of the floodplain by depositing its black silt. More than the shape of the country, the flow is important for the meanings it reveals. Tex explains that an important site in his country is a series of rocks within a river bed. Only when the level of the river reaches a certain depth does the water flowing over the rocks make visible the shape of the ancestral being whose spirit is embodied within the rock, allowing the story not only to be told but to be seen. Again, Tex repeats, 'the water *shows* us'.[54]

Another example is suggested in the cautionary approach to the environment embodied in the narrative of the Kurriya at Boobera Lagoon. Thompson has documented the extensive oral tradition about this site, actively passed on to many younger people in frequent visits to the area over many generations under colonial conditions.[55] The Kurriya is a powerful and frightening ancestral figure with creative powers. Through these powers, it created much of the region's landforms and watercourses, above and beneath ground. It is understood by Aboriginal owners to rest in the deep recesses of Boobera Lagoon, a large body of water understood to be permanent because it was fed from a mysterious and very deep underground water source which never ran dry, a fact witnessed by many Aboriginal people who relate how they saw the waters rising in the middle of dry periods with no explanation.[56] Hydrologists now believe that there are strong indications that the Lagoon is fed by a deep recharge spring from the Great Artesian Basin, but there is as yet no conclusive evidence.[57] While the narrative refers to the water source, the real issue of concern for Aboriginal communities is the terrible power of the Kurriya, the need to respect and protect it and particularly to avoid swimming in or making noise near the Lagoon, due to the spirit's ability to consume anyone who goes into the water. The recounting of this story makes it clear that this power to destroy has continued since the invasion and is just as effective against white settlers and their stock as it is against Aboriginal people.[58]

The Lagoon has over the last fifteen years been the site of a new conflict as Aboriginal people tried to gain protection over the Lagoon not only from the stock of graziers and the cotton farmers who were seeking to irrigate, but also the region's boating recreation body, whose high power waterskiing activities was not only damaging the Lagoon's banks but desecrating the cultural meaning of the site with their noisy and intrusive presence on the water. The central effect of the Kurriya narrative was to protect the water body by denying entry to it. The impact of the settler activities is now clear: siltation from stock and power boat bank erosion as well as clearing for grazing and irrigated farming has silted up the floor of the lagoon and appears to have obstructed the underground water recharge inlet to the lagoon from the Great Artesian Basin.[59] This precise outcome is not explicit in the Boobera Lagoon narrative, but if the general precautionary principle had been honoured in this case, impacts on the water body would be have been minimised and siltation would not have occurred. Aboriginal people are arguing that their knowledge contained an approach which would have protected an important resource, of value to both settlers and Aborigines, which has now been harmed and perhaps irreparably damaged by ignoring the warnings inherent in the traditional narrative.

(iii) *Water and colonialism narratives*

The third form in which indigenous knowledge can be identified is in emerging narratives and performances about the ways Aboriginal societies in south eastern Australia have engaged with and remembered colonial life. This process is normally discussed in terms of loss, considering the decimation of population, the disruption of ceremonies and the denial of access to country have all made it harder to perform and transmit the fullest versions of any oral tradition. But the conditions of colonialism have intensified Aboriginal people's experiences with water and this has been reflected in the ways indigenous knowledges about water are expressed. One of the narratives of colonialism relates to the way Aboriginal people's knowledge of water sources was used by settlers when they hired Aboriginal stockworkers and drovers. Aboriginal people based their new employment on established traditional knowledge, but they had to learn innovative ways to manage the limited water sources they knew because they now had to water large flocks of sheep or herds of cattle. Once artesian water was discovered, in 1878, the new bores became additional watering points on the long routes for droving stock across the arid areas down to metropolitan markets. There was some congruence with the previously Aboriginal-known mound springs, the naturally occurring outlets from the deep artesian sources, but many of the bores were in country which before had been entirely unwatered. Aboriginal drovers became confident authorities in navigating from water to water, building the new water knowledge into the traditional frameworks. Many, like George Dutton in far western NSW, were able to incorporate their fulfilment of custodial and ceremonial obligations into their droving routes, maintaining an active ceremonial life by taking part in long ceremonial routes across long distances in the central desert areas of South Australia, Queensland and the Northern Territory, all adjoining to the NSW border and accessible to Dutton because he was a respected drover.[60]

Such interweaving of traditional water knowledge and European pastoral skills was not all the drovers did. As they travelled, they taught young male relatives new trades, and they also taught them the invasion histories of the country over which they travelled. Wilpi, an old Wangkumara man I interviewed in Bourke, recalled how as a young droving apprentice, he was taught by his elders as they moved from water hole to water hole:

> Old fellas used to tell us, 'you want to come out, learn to work' and we was pleased to too, didn't know what horses was like. So we went down onto the Cooper then, onto the flood water country then, they took us out there. And the old fellas used to show us sandhills here and sandhills there, all different islands, y'see. And they had names for these waterholes, see, where all the Abos got shot down there when the troopers came in to shoot them. They was killin' cattle, see, at the waterhole. So anyway, they told us all these names, showin' us where they were shot and all..... So we went out, we were workin' with'em there, oh

for a good while, riding' about with'em, mustering cattle and they used to say, 'well, you go to a waterhole', you know they name'em there. Like they call'im *Watuwara*, that's 'water where the birds live', then next, where they shot the Murris[61], they call that *Thuliula*, that's a mussel see, *Thuliu*, and the next one, about a mile away, they call that 'little *Thuliula*'....[62]

This was an oral transmission of the memory of invasion violence across generations, and into the present, not only conserving but situating historical knowledge. It allows Aboriginal people to pose a counter narrative to the colonisers' history of 'peaceful settlement', which continues to be retold in school history texts of the twenty-first century. Where there are some European authored accounts of these incidents of invasion violence, the differences between indigenous oral accounts and the non-indigenous written accounts can offer important insights into the way indigenous people have understood invasion and colonisation.[63] What researchers have not yet done is identification of the environmental knowledge, and changing ecologies under the impact of settler land management, which may be entangled in these new narratives.

Virtually all of these stories of massacre violence occurred at water places either because the conflict was over a contested watering site or because Aboriginal people were camped beside water when they were attacked. It is the role of water places as both resource and as a central element in the human use of the landscape which structured the patterns of violence. Just in the area of the upper Darling there is Hospital Creek, Boobera Lagoon and Myall Creek, where massacres occurred which were partially documented by Europeans. Others remain known only in the oral record, but no less powerful for that. Such emplaced oral accounts were experienced by young Aboriginal people growing up in the 1910s and 1920s. The stories continue to be retold in the same manner today, tangled up with language learning and family histories, taught to young Aboriginal people as their families travel. But they are also of high importance in the ways rural Aboriginal communities induct and orient newly arrived non-Aboriginal lawyers, teachers and other staff in Aboriginal-controlled organisations. I was one of those people, taken out to see Hospital Creek by local Aboriginal spokespeople Kevin Williams and Tombo Winters in the 1970s. I was shown the creek side location of this disturbing story, was introduced there to bush foods and traditional medicines, and was shown the landscape conditions around the creek. Nick McClean, a current graduate student and environmental activist, has recorded similar experiences with Ted Fields, a senior Yuwalaraay man from Walgett.[64] There are deep analytical and symbolic dimensions to these stories, offering political analyses and histories which are embedded in the land and which demonstrate continuing Aboriginal knowledge to both younger Aboriginal people and to non-Aborigines, testing newcomers, challenging their complacency and demanding their allegiance.[65] This has become very much a ritual occasion – and certainly an important example of the 'place-making' which Appardurai has discussed as *making* local subjects, in which political, social and cultural knowledge is imbricated with environmental knowledge.

Finally, there are the narratives of family life which circulate actively. They are located in the intersection of life story and oral tradition, but again environmental knowledge is threaded throughout the narratives as they anchor episodes to places of work, camping and water. Working life under colonialism involved movement for Aboriginal families, as the jobs available on the Darling River pastoral properties were seasonal. Aboriginal people had a 'beat' of stations they regularly worked on, living in the camps on the station, and travelling across country, often on the TSRs, from station to station for the next job. Children grew up familiar with camping out next to creeks and waterholes, gathered round campfires listening to stories under the stars at night, and navigating more by the water courses than fencelines. Many people working around Boobera Lagoon, for example, camped on the lagoon when they were travelling between jobs, and so children learnt the stories about the Kurriya and how they must not swim in the lagoon.[66] But many people were forced to live more sedentary lives, particularly after 1912 when the state government began systematically to remove any Aboriginal children it could argue were 'neglected', in order to incorporate them into an indentured labour scheme which it hoped would 'cure' them of their desire to return to their families.[67]

Rivers again played a critical role. Rental accommodation was invariably segregated, and many Aboriginal families lived on vacant land near the river banks. The river was a necessary economic resource. While families lived near towns, they often had to do without paid work and the fish, yabbies, mussels and birds to be found around the rivers became their only source of nutrition. When parents were working on properties out of town, or mothers were employed cleaning in the hotels or hospitals or private white town homes, grandmothers particularly would take children out along the river to fish and catch yabbies. Long days on the river bank became opportunities for teaching and learning about country. As access to the wider countryside began to close down because employment was falling, the only remaining safe places for Aboriginal people to live and travel along became the rivers. Whether going fishing for food (or for the love of it), to escape the pressures of the hostile white town or the increasingly crowded camps, many Aboriginal people found their main access to their country was now along the river banks.[68]

The rivers clearly reflected the harsh politics of country racism. White townships frequently planned their development so the rivers functioned as a border and a barrier to Aboriginal access. Aboriginal people were allowed to camp but only on the 'other' side of the river or out of town – always the floodprone side. There were unofficial curfews in most towns in which Aboriginal people could not be seen on the 'white' side of the river after dark and times when Aboriginal people remember swimming the river towards the camp to escape arrest from police for breaking the curfew, while in other situations men who had been drinking in the camps tried to swim the rivers drunk to avoid arrest, and sometimes didn't reach the safe side at all.

The continuous struggle to protect children from removal was intimately linked to the river as well. Women recall swimming in the rivers away from the camp with children on their backs to escape the authorities who had come to take children away. Even if children were enrolled in schools (from which they were often excluded on racial grounds) they would be vulnerable if they were noticed for not being clean enough or for having pediculosis or scabies, the perennial minor contagious infestations faced by all children in poor schools. But for Aboriginal families, it could mean the intervention of the state to take away their children, so faces had to be shining and nails scrubbed. Even so, children still faced the humiliating line up each day to check their heads and nails. Such daily attention meant many buckets of water hauled up the steep river banks by women to boil in the coppers so there would be hot water to wash kids and clothes. If school children did develop scabies or head lice, there were traditional remedies involving infusions from local plants. But as mothers recall: 'that meant *another* bucket of water!'[69]

The river banks were important for other reasons. The Darling and its tributaries on the flood plain have banks with deep gullies and tangled gum tree roots in black silty soil, which forms a sucking, impassable bog when wet. On the riverbanks Aboriginal people were also safe, at least to some extent, from the pursuit of police who came to regulate their lives, control who they associated with and sometimes to take their children. Transgressive meetings for drinking, gambling and sex were all possible, for whites as well as Aborigines, and at night the river banks were sites where daytime colour bars were sabotaged. Some of the most powerful political campaigns of the 1970s were assertions of the collective energy regained from having river banks as safe places in which to conserve a sense of identity and counter solidarity. The demands to restore rights to land in NSW were generated by the urgent need to reclaim rights to water as in Brewarrina in 1974, and in later years when the cultural identity of the Aboriginal community was reasserted to demand control over local cultural festivals which had appropriated Aboriginal river symbols.[70] Water sites have been sites of segregation but also of resistance, sites of massacres and exclusion but also of learning and social regeneration.

IMPLICATIONS: HISTORY, WATER AND INDIGENOUS KNOWLEDGE

Indigenous knowledge has been sustained since the invasion although in substantially altered forms, at some times reflecting pre-invasion conditions and at others reflecting newly emerging content arising from traditional bases but in engagement with very changed conditions. The capacity of indigenous people in conditions of historical change to identify and reflect on environmental changes is an important dimension of the broader value of indigenous knowledge. It is not a dimension which is welcomed by the contemporary Australian state or its legal structure. In 2002 the High Court decision on the Yorta Yorta appeal against Olney's 'tide of history' judgement was one of a cluster that year which narrowed the already limited rights available to Aboriginal traditional owners. It confirmed Olney's approach that no indigenous knowledge, however directly based on continuing oral tradition, but which had been generated after the invasion began, could be considered as 'authoritative' or 'legitimate' tradition. The decision effectively excised history from any consideration of what indigenous knowledge might be or of the high value it might hold.

This discussion in this paper demands the question be posed: to what extent, if at all, can what has been described in this paper be regarded as 'indigenous knowledge'? It has been explicitly dismissed as such by Justice Olney in the Yorta Yorta native title case. It has been largely ignored by the conservation movement to date.

There are a number of reasons to consider this as indigenous knowledge. First, it is based on and sourced in pre-invasion knowledge of oral traditions, formal and informal. Secondly, its production and circulation occurred because it is motivated by the desire to fulfil traditional social and cultural goals of achieving responsible adult roles by becoming knowledgeable land custodians. It is expected by Aboriginal people of themselves and each other that they will notice and comment on the state of the land and waters around them, and that they will care about what happens to them. This reflects a continuing social and cultural process of engaging with the material environment to generate locality and from there, to relate to people as neighbourhoods, even if far flung. Thirdly, this knowledge of the state of the rivers has been acquired and to some extent intensified because of the historical conditions of colonisation in the repression, dispossession and impoverishment of indigenous peoples. This has forced them into an even more continuous and intimate relationship with rivers and river banks than would ever have been the case under the conditions of mobility of pre-invasion life. Their knowledge about rivers, creeks and waterholes now records the events of the invasion and the exercise of colonial power. Finally, this new knowledge has been recorded in stylised forms and retold in conventionalised performances which echo the processes of pre-invasion indigenous knowledge. The memories of massacres, conflicts and a life working in the grazing industry are now inscribed onto the landscape through being incorporated into stories which are themselves embedded in places. Such stories are retold, across generations, in a similar way although no longer in the same forms as those transmitted in pre-invasion oral traditions. So the stories record the events in a traditional way, but the content of the stories, is a dramatised and analysed account of colonial interactions.

What are the implications then for the practice of research in environmental history and conservation to recognise historical change and to see indigenous knowledge as a process rather than an archive? Once rivers and water sources are understood to have played a critical role, not only in sustaining life or the pre-invasion oral tradition, but in the historical, social, spatial and political life of Aboriginal people, it is no surprise that such places will have concentrations of meaning and significance for Aborigines including much knowledge about pre-invasion conditions. Waterways are the places which will offer a partial glimpse of the ecological relationships in pre-invasion times, in very different environmental conditions of active Aboriginal management, more riverine flooding and less artesian water. The extent to which oral traditions have been retained is the extent to which these narratives which thread human dramas with environmental details and embed them in places are available. And so research in collaboration with indigenous communities to record and sustain such knowledge associated with rivers is a priority.

There are insights too into approaches to land management which are outside western development paradigms, although still productivist. They may differ from the goals of some environmental movements which seek to reduce production of any sort from protected areas. The conception of a river which assumes that water will be present across the floodplain, rather than being 'normally' confined in a river bed is a significantly different approach to living with variable environments than is found in the British-Australian water management strategies. This parallels approaches that Rohan d'Souza has discussed, relating to eastern Indian deltaic systems, between a managed landscape which is 'flood dependent' and the British strategy to control rivers which generated a 'flood vulnerable' landscape.[71] Benjamin Weil has identified similar contrasts in relation to the western Indus river.[72] So collaborative work with indigenous communities should be seeking to learn the broadest forms of narrative and performative expressions of community knowledge, in order to understand approaches to and interpretations of environmental relationships, rather than expecting to reduce indigenous knowledge to a taxonomy.

The body of knowledge held by indigenous people in western NSW today offers an account of changes in the environment under colonial economies of the last 200 years. While not systematic or blanket coverage, it is unique and invaluable for identifying the types and pace of change. It is geographically focused as well, as Aboriginal people have increasingly found that only the rivers and their banks remained accessible to them. Other special places, about which environmental knowledge might have been retold and learnt have not been visited so often or recently because the access to them has been closed off. Nevertheless, given the central role water plays in both pre-invasion and settler post invasion economies, working towards gathering perceptions of change in rivers, springs and water systems will continue to be of high importance.

Aboriginal people in north-western NSW continue to be deeply concerned about the ongoing changes. The interest in fulfilling custodial responsibilities continues to be relevant and enacted by Aboriginal people, perhaps the most important continuation of the social processes of indigenous tradition. The most detailed oral traditions about important places away from the rivers have become harder to maintain in active circulation as access has been cut off, but Aboriginal interest in re-engaging with off-river land management and regeneration has been rising. This is most evident where Aboriginal people have real security of tenure over significant areas of land, a possibility which has been rare until recently. Only now, with some land acquisitions directly in Aboriginal hands and tentative steps towards co-management of some protected areas, have communities begun to reacquaint themselves with the country from which they had been excluded for many years. With their communities still living in impoverishment, they have often had to make hard decisions between managing the few acres they have for short term profit or giving up hopes of profits in order to develop regeneration strategies.

Water remains an urgent priority. One elderly Yuwalaraay Walgett man explained to me in 2000 his worries about the large amounts of water being sucked out of the river by cotton irrigation pumps on one of the properties he knew well, his traditional country and land he had worked as a stockman on horseback for most of his life. He decided to show me the damage so we drove across the black soil plains towards the river. We entered the property and crossed ungrazed and heavily wooded paddocks to where we should have been able to see the water, but found our way blocked by a massive water storage, with bulldozed earthen sides rising 15 or 20 metres and stretching far into the distance on either side. This was where the river water was going. More deeply disturbing for this knowledgeable senior man was that he had lost his way on country he had known intimately. The huge scale of the water storage meant that all his landmarks had been wiped out. He eventually admitted that he was defeated, humiliatingly lost on his own country. But he was beaten only in the short term. Soon after, he embarked on the process of recording his knowledge of the complex watercourses, tracing out the water and the stories with young researchers, black and white, in tow.[73] He sustained his recordings until his death in 2006, drawing together his memories of traditional stories and performance, his historical knowledge and awareness of change. Most importantly he was teaching: his stories, overflowing again, continue to dissolve the symbolic walls of that massive water storage. In a way that doesn't look at all like a traditional ceremony, this Yuwalaraay man was producing *locality*, making indigenous knowledge live on.

ACKNOWLEDGEMENTS

My thanks to the friends and colleagues who have read drafts of this paper and offered valuable critiques and, even more important, asked difficult questions. These people are in no way responsible for any of the essay's conclusions. I am grateful to Mahesh Rangarajan, Gunnel Cederloff, Ranjan Chakrabarti, Libby Robin, Tom Griffiths, Denis

Heather Goodall

Byrne, Peter Thompson, Peter Read, Michael Adams, Heidi Norman, Damian Lucas and Nick McClean, as well as to the two anonymous referees for this Journal. Allison Cadzow and Lindi Todd were careful and insightful editors who allowed this final version to take shape.

NOTES

[1] This paper was first discussed in a presentation to The History of Waters conference at Jadavpur University, Kolkata, India, March 3, 2006. An early version will be published in the conference proceedings, to be edited by Professor Ranjan Chakrabarti, for the Association for South Asian Environmental History. When used in relation to human beings, Australian Aboriginal people sometimes refer to themselves or other *people* as 'Indigenous people' but this usage and spelling is not universal. The use of capital 'I' to spell the word 'indigenous' is varied around the world. In this essay, 'indigenous' is used as an adjective without capitalisation.

[2] Atkinson, 1995; Commissioner for Social Justice, 2002.

[3] Borrini-Feyerabend et al., 2004; IUCN, 2003; World Council on Protected Areas/ IUCN, 2003.

[4] Australian Aboriginal people usually identified as 'hunters and gatherers', are more usefully described as 'harvesters' in acknowledgement of the high degree of environmental knowledge, planning and active intervention in the landscape which allowed reliable food gathering. Aboriginal responses to British agriculture from 1860 included a range of strategic adoptions of farming in independent blocks across the south eastern coastal and central districts at precisely the same time as the settler government was pronouncing them irretrievably primitive and unable ever to learn the rudiments of farming. This paper uses the term 'harvesters' to describe the Aboriginal economy and society.

[5] Langloh Parker, 1953 [1897]; Robinson, 1965, pp.126, 131.

[6] Goodall, 1996; Flick and Goodall, 2004;. Goodall, 2002; 1994; 1999; 2001; Goodall and Lucas,1997.

[7] Mulligan and Hill, 2001.

[8] ABS, 2004.

[9] Chief Health Officer, 2004.

[10] NSW National Parks and Wildlife Service: http://www.nationalparks.nsw.gov.au/npws. nsf/Content/Which+parks+are+co-managed+in+NSW

[11] Baviskar, 2005.

[12] Bayet-Charlton, 2003; Langton and Rhea Zane, 2003.

[13] As an example, consider the difference between the work of Tania Murray Li, researching in Indonesia, with that of Ronald Niezen, discussing First Nations societies in Canada. Li, 2000; Niezen, 2000.

[14] Cederlof, 2005; Morrison, 2005.

[15] IUCN, 2003; Kothari, 2006.

[16] Johnson, 1992.

[17] Sillitoe, 2007, passim.

[18] Cronon, 1992; Dove et al., 2007; Griffiths and Robin, 1997.

[19] Adams, 2004; Cronon, 1996; Dove et al., 2007.

[20] Bayet-Charlton, 2003; Langton, 1996.

[21] Adams and English, 2005); Head, 2000; Head et al., 2005.

[22] Sveiby and Skuthorpe, 2006, p. 164.

[23] Langton, 1998; 1996; Langton and Rhea Zane, 2003.

[24] See, as one example, the itemised list of plants and their uses known to the Kamilarai and Pikampul peoples around Boobera Lagoon in northwestern NSW. Hawes, 1993.

[25] Adams, 2004; Adams and English 2005; Latour, 1999; 1987.

[26] Nazarea, 1999.

[27] Goodall, 1992.

[28] Ellen, 1993.

[29] Johnson, 1992.

[30] Berkes, 1999.

[31] Smith, 2007.

[32] Vansina, 1965.

[33] Vansina, 1985.

[34] White et al., 2001.

[35] Merlan, 1998; Myers, 1986; Vansina, 1985.

[36] Magowan, 2001.

[37] Vansina, 1985.

[38] Graham and Wositsky, 1994.

[39] Mackinolty and Wainburranga, 1988; Rose, 1984.

[40] Merlan, 1988.

[41] Ibid.; Myers, 1986; Kolig, 1980.

[42] Appadurai, 1996.

[43] Ibid., p. 181.

[44] Palmer, 1983.

[45] Langloh Parker, 1905.

[46] Powell, 1976; 1991.

[47] Beckett, 1978.

[48] Goodall, 1996.

[49] Goodall, 2001; Lucas, 2004.

[50] Cotter, 2006; English, 2002; English and Gay, 2005; Flick and Goodall, 2004.

[51] Byrne and Nugent, 2004; Cotter, 2006; English, 2002; English and Gay, 2005.

[52] Cotter, 2006.

[53] Goodall and Lucas, 1997.

[54] Goodall, 2001; Herman, 1996.

[55] Thompson, 1993.

[56] Interviews with Ted Fields, 2000, conducted by author for work in progress.

[57] Eigeland, 1993.

[58] Goodall, 1995.

[59] Eigeland, 1993.

[60] Beckett, 1978.

[61] Local language word for 'Aboriginal people' or 'our people'.

64

Heather Goodall

[62] Goodall, 1996, p. 34.

[63] Goodall, 2003; Rose, 1991.

[64] Nick McClean, 'Narran Lakes Oral History Project', unpublished Honours Thesis, UTS, 2007.

[65] Reece, 1982.

[66] Thompson, 1993.

[67] There is now a large literature on this policy, which existed in different forms in each state. The children so 'apprenticed' or otherwise removed are now often referred to as the 'Stolen Generations'. Haebich, 2000.

[68] Flick and Goodall, 2004.

[69] Goodall, 2006a.

[70] Goodall, 2006b.

[71] D'Souza, 2002.

[72] Weil, 2006.

[73] Research being undertaken collaboratively with Nick McClean, conservation activist, for 'Rivers Lakes And Plains: Stories from Yuwalaraay Country – The Narran Lakes', unpublished Honours thesis, UTS, 2007.

BIBLIOGRAPHY

ABS. 2004 'Aboriginal and Torres Strait Islander Population'. In *Year Book Australia 2004*. Canberra: Commonwealth of Australia.

Adams, Michael, and Anthony English. 2005. '"Biodiversity is a Whitefella Word": Changing Relationships Between Aboriginal People and the NSW National Parks and Wildlife Service'. In *Power of Knowledge and the Resonance of Tradition*.

Adams, Michael. 2004. 'Negotiating Nature: Collaboration and Conflict between Aboriginal and Conservation Interests in New South Wales, Australia'. *Australian Journal of Environmental Education* 20: 15–23.

Appadurai, Arjun. 1996. *Modernity at Large: Cultural Dimensions of Globalisation, Public Worlds Vol 1*. Minneapolis, London: University of Minnesota Press.

Atkinson, Wayne. 1995. 'Yorta Yorta Struggle For Justice Continues: a background statement to the Yorta Yorta native title claim, 1993'. LaTrobe University.

Baviskar, Amita. 2005. 'Adivasi Encounters with Hindu Nationalism in MP'. *Economic and Political Weekly,* November 26, 2005: 5105–5113.

Bayet-Charlton, Fabienne. 2003. 'Overturning the Doctrine: Indigenous People and Wilderness – Being Aboriginal in the Environment Movement'. In *Blacklines: Contemporary Critical Writing by Indigenous Australians*, edited by Michele Grossman. Melbourne: Melbourne University Press.

Beckett, Jeremy. 1978. 'George Dutton's Country'. *Aboriginal History* 2: 2–31.

Berkes, Firket. 1999. *Sacred Ecology: Traditional Ecological Knowledge and Resource Management*. Philadelphia: Taylor and Francis.

Borrini-Feyerabend, Grazia, Ashish Kothari and Gonzalo Oviedo. 2004. *Indigenous and Local Communities and Protected Areas: Towards Equity and Enhanced Conservation: Guidance on Policy and Practice for Co-managed Protected Areas and Community Conserved Areas*. Edited by Adrian Phillips, *Best Practice Protected*

Area Guidelines Series No 11: World Commission on Protected Areas [WCPA] – The World Conservation Union [IUCN].

Byrne, Denis, and Maria Nugent. 2004. *Mapping Attachment: A Spatial Approach to Aboriginal Post Contact Heritage*. Sydney: Department of Environment and Conservation.

Cederlof, Gunnel. 2005. 'the Toda Tiger: Debates on Custom, Utility and Rights in Nature, South India, 1820–1843'. In *Ecological Nationalisms: Nature, Livelihoods and Identities in South Asia*, edited by Gunnel Cederlof and K. Sivaramakrishnan (Delhi: Permanent Black) pp. 65–89.

Chief Health Officer. 2004. 'The Health of the People of New South Wales'. NSW Health Dept.

Commissioner for Social Justice. 2002. *Native Title Report*. Canberra: Human Rights and Equal Opportunity Commission.

Cotter, Maria. 2006. 'The Gamilaraay Resource Use Project'. Pers. comm, May 2006.

Cronon, William. 1992. 'A Place for Stories: Nature, History, and Narrative'. *Journal of American History* 78, no. 4: 1347–1376, doi:10.2307/2079346.

Cronon, William. 1996. 'The Trouble with Wilderness, or, Getting Back to the Wrong Nature'. *Environmental History* 1: 7–55, doi:10.2307/3985059.

Dove, Michael R., Daniel S. Smith, Marina T. Campos, Andrew S. Mathews, Ane Rademacher, Steve Rhee and Laura M. Yoder. 2007. 'Globalisation and the Construction of Western and non-Western Knowledge'. In *Local Science vs Global Science: Approaches to Indigenous Knowledge in International Development*, ed. Paul Sillitoe (New York & Oxford: Berghahn Books), pp. 129–154.

D'Souza, Rohan. 2002. 'Colonialism, Capitalism and Nature: Debating the Origins of Mahanadi Delta's Hydraulic Crisis (1803–1928)'. *Economic and Political Weekly* March 30, 2002: 20.

Eigeland, Neil. 1993. 'Hydrology'. In *Boobera Lagoon: Environmental Audit*, edited by Inverell Research Service Centre (Inverell: NSW Department of Land and Water Conservation), pp. 87–130.

Ellen, R.F. 1993. *The Cultural Relations of Classification: An Analysis of Nuaulu Animal Categories from Central Seram*. Cambridge: Cambridge University Press.

English, Anthony, and Louise Gay. 2005. *Living Land Living Culture: Aboriginal Heritage And Salinity*. Sydney: Department of Environment and Conservation (NSW).

English, Anthony. 2002. *The Sea And The Rock Gives Us A Feed: Mapping and Managing Gumbaingirr Wild Resource Use Places*. Sydney: NSW National Parks and Wildlife Service.

Flick, Isabel and Heather Goodall. 2004. *Isabel Flick: The Many Lives of an Extraordinary Aboriginal Woman*. Sydney: Allen and Unwin.

Goodall, Heather. 1992. 'The Whole Truth and Nothing But … Reflections of a Field Worker on the Intersections of Western Law, Aboriginal History and Community Memory'. *Power, Knowledge and Aborigines: A special issue of the Journal of Australian Studies* 35: 104–119.

Goodall, Heather. 1994. 'Colonialism and Catastrophe: Contested Remembrance of Measles and Bombs in a Pitjantjatjara Community'. In *Memory and History in Twentieth Century Australia*, ed. Kate Darian-Smith and Paula Hamilton (Melbourne: Oxford University Press), pp. 55–76.

Goodall, Heather. 1995. 'The significance of Boobera Lagoon to Aboriginal People re a request for protection of the Lagoon under Section 10 of the Commonwealth Aboriginal and Torres Strait Islanders Heritage Act'. Prepared at the request of the Toomelah Community.

Goodall, Heather. 1996. *Invasion to Embassy: Land in Aboriginal Politics in NSW 1770 to 1972*. Sydney: Allen and Unwin.

Goodall, Heather. 1999. 'Telling Country: Memory, Modernity and Narratives in Rural Australia ' *History Workshop Journal*, 47: 161–190.

Goodall, Heather. 2001. 'The River Runs Backwards: The Language of Order and Disorder on the Darling's Northern Flood Plain'. In *Words for Country: Landscape and Language in Australia*, ed. Tim Bonyhady and Tom Griffiths (Sydney: UNSW Press), pp. 30–51.

Goodall, Heather. 2002. 'Mourning, Remembrance and the Politics of Place: A Study in the Significance of Collarenebri Aboriginal Cemetery'. *Public History Review* 9: 72–97.

Goodall, Heather. 2003. 'Evans Head and Area, in Respect of the Application for a Determination of Native Title, Number NG 6034 of 1998: Lawrence Wilson v Minister for Lands and Conservation'. Prepared for the Federal Court of Australia.

Goodall, Heather. 2006a. 'Gender, Race and Rivers: Women and Water in Northwestern New South Wales'. In *Fluid Bonds: Views on Gender and Water*, ed. Kuntala Lahiri-Dutt (Kolkata: Stree Books), pp. 287–304.

Goodall, Heather. 2006b. 'Main Streets and Riverbanks: The Politics of Place in an Australian River Town'. In *Echoes from the Poisoned Well*, ed. Sylvia Hood-Washington, Paul Rosier and Heather Goodall (Lanham, MD: Lexington), pp. 255–270.

Goodall, Heather, and Damian Lucas. 1997. '"Country Stories": Oral History and Sustainability Research'. Paper presented at the Sustainability and Social Research, Centre for Rural Social Research, Charles Sturt University, Wagga Wagga.

Graham, T. (Producer/Director), and J. Wositsky (Writer/Researcher). 1994. *Ka-waya-wayama: Aeroplane Dance* [Motion picture]. Lindfield, Australia:. Film Australia.

Griffiths, Tom, and Libby Robin, eds. 1997. *Ecology and Empire: Environmental History of Settler Societies*. Edinburgh: Keele University Press.

Haebich, Anna. 2000. *Broken Circles: Fragmenting Indigenous Families 1800–2000*. Fremantle: Fremantle Arts Centre Press.

Hawes, Wendy. 1993. 'Flora and Fauna'. In *Boobera Lagoon: Environmental Audit*, ed. Inverell Research Service Centre (Inverell: NSW Department of Soil and Water Conservation), pp. 18–74.

Head, Lesley , David Trigger and Jane Mulcock. 2005. 'Culture as Concept and Influence in Environmental Research and Management'. *Conservation and Society* 3.

Head, Lesley. 2000. 'Conservation and Aboriginal Land Rights: When Green Is Not Black'. *Australian Natural History* 23: 448–454.

Herman, Russ. 1996. *Tex Scuthorpe Interview at Tranby, digital video*. Film.

IUCN. 2003. *The Durban Accord*. Durban, South Africa: Vth IUCN World Parks Congress.

Johnson, Martha, ed. 1992. *Lore: Capturing Traditional Ecological Knowledge*. Hay River, CA: Dene Cultural Institute, International Development Research Centre.

Kolig, Ernst. 1980. 'Noah's Ark Revisited: On the Myth-Land Connection in Early. Aboriginal Thought'. *Oceania* 51: 118–32.

Kothari, Ashish. 2006. 'Theme on Indigenous and Local Communities, Equity and Protected Areas'. Curitiba, Brazil: 8th Conference of Parties of the Convention on Biological Diversity.

Langloh Parker, K. 1905. *The Euahlayi Tribe: A Study of Aboriginal life in Australia*. London: Archibald Constable.

Langloh Parker, K. 1953 [1897]. *Australian Legendary Tales*, ed. H Drake-Brockman. Angus and Robertson.

Langton, Marcia, and Ma Rhea Zane. 2003. 'Traditional Lifestyles and Biodiversity Use: Composite Report on the Status and Trends Regarding the Knowledge, Innovations and Practices of Indigenous and Local Communities Relevant to the Conservation and Sustainable Use of Biodiversity. Regional Report: Australia, Asia and the Middle East'. United Nations Environment Program, edited by Secretariat of the Convention on Biological Diversity.

Langton, Marcia. 1996. 'What do we mean by wilderness? Wilderness and Terra Nullius in Australian Art'. *The Sydney Papers* 8, no. 1.

Langton, Marcia. 1998. *Burning Questions: Emerging Environmental Issues for Indigenous Peoples in Northern Australia*. Darwin: Centre for Indigenous Natural and Cutural Resource Management.

Latour, Bruno. 1987. *Science in Action: How to Follow Scientists and Engineers Through Society*. Cambridge, Mass: Harvard University Press.

Latour, Bruno. 1999. 'Circulating Reference: Sampling the Soil in the Amazon Forest'. In *Pandora's Hope: Essays on the Reality of Science Studies* (Cambridge Mass.: Harvard University Press), pp. 24–79.

Li, Tania Murray. 2000. 'Constituting Tribal Space: Indigenous Identity and Resource Politics in Indonesia'. *Comparative Studies in Society and History* 42: 149–179, doi:10.1017/S0010417500002632.

Lucas, Damian. 2004. 'Shifting Currents: A History of Rivers, Control and Change'. PhD, University of Technology Sydney, 2004.

Mackinolty, C, and Paddy Wainburranga. 1988. 'Too Many Captain Cooks'. In *Aboriginal Australians and Christian Missions*, ed. T. Swain and D.B. Rose (Adelaide: Australian Association for the Study of Religions), pp. 355–360.

Magowan, Fiona. 2001. 'Crying to Remember: Reproducing Personhood and Community'. In *Telling Stories: Indigenous History and Memory in Australia and New Zealand*, ed. Bain Attwood and Fiona Magowan (Sydney: Allen and Unwin), pp. 41–60.

Merlan, Francesca. 1998. *Caging the Rainbow: Places, Politics and Aborigines in a Northern Australian Town*. Honolulu: University of Hawai'i Press.

Morrison, Kathleen. 2005. 'Environmental History, the Spice Trade and the State in South India'. In *Ecololgical Nationalisms: Nature, Livelihoods and Identities in South Asia*, ed. Gunnel Cederlof and K. Sivaramakrishnan (Delhi: Permanent Black), pp. 43–64.

Mulligan, Martin and Stuart Hill. 2001. *Ecological Pioneers: A Social History of Australian Ecological Thought and Action*. Cambridge: Cambridge University Press.

Myers, Fred. 1986. *Pintupi Country, Pintupi Self*. Washington and Canberra: Smithsonian Institute and AIAS.

Nazarea, Virginia D. 1999. 'A View From a Point: Ethnoecology as Situated Knowledge'. In *Ethnoecology: Situated Knowledge/Located Lives*, ed. Virginia D. Nazarea (Tucson: University of Arizona Press), pp. 3–20..

Niezen, Ronald. 2000. 'Recognising Indigenism: Canadian Unity and the International Movement of Indigenous Peoples'. *Comparative Studies in Society and History* 42: 119–148, doi:10.1017/S0010417500002620.

Palmer, Kingsley. 1983. 'Migration and Rights to Land in the Pilbara'. In *Aborigines, Land and Land Rights*, ed. N. Peterson and M. Langton (Canberra: AIAS), pp. 172–179.

Powell, J.M. 1976. *Environmental Management in Australia, 1788–1914*. Melbourne: Oxford University Press.

Powell, J.M. 1991. *Plains of Promise, Rivers of Destiny: Water Management and the Development of Queensland, 1824–1990*. Bowen Hills, Qld: Boolarong Publications.

Reece, Robert. 1982. 'Aboriginal Community History: A Cautionary Tale'. *Australian Historical Association*.

Robinson, Roland. 1965. *The Man Who Sold His Dreaming: Verbatim Narratives by Contemporary Aborigines*. Sydney: Currawong Publishing, 1965.

Rose, Debbie Byrd. 1984. 'The Captain Cook Saga'. *Australian Aboriginal Studies* no. 2.

Rose, Debbie Byrd. 1991. *Hidden Histories*. Canberra: Aboriginal Studies Press, 1991.

Sillitoe, Paul, ed. 2007. *Local Science vs Global Science*. New York/Oxford: Berghahn Books.

Smith, Benjamin R. 2007 '"Indigenous" and "Scientific" Knowledge in Central Cape York Peninsula'. In *Local Science vs Global Science*, ed. Paul Sillitoe (New York: Berghahn Books), pp. 75–90.

Sveiby, Karl-Erik and Tex Skuthorpe. 2006. *Treading Lightly: The Hidden Wisdom of the World's Oldest People*. Sydney: Allen and Unwin.

Thompson, Peter. 1993. 'Aboriginal Cultural Heritage'. In *Boobera Lagoon: Environmental Audit*, ed. Inverell Research Service Centre (Inverell: NSW Department of Conservation and Land Management), pp. 206–282..

Tribunal, National Native Title. 2002. 'Background Information on Native Title Proceedings re Members of the Yorta Yorta Aboriginal Community v. State of Victoria and Others'. Commonwealth Government.

Vansina, Jan. 1965. *Oral Tradition: A Study in Historical Methodology*, translated by H.M. Wright. Chicago: Aldine Publishing Co. First published in French 1961.

Vansina, Jan. 1985. *Oral Tradition as History*. London: James Currey.

Weil, Benjamin. 2006. 'The Rivers Come: Colonial Flood Control and Knowledge Systems in the Indus Basin, 1840s–1930s'. *Environment and History* 12: 3–29, doi:10.3197/096734006776026818.

White, Luise, Stephan F. Miescher and David William Cohen, eds. 2001. *African Words, African Voices: Critical Practices in Oral History*. Bloomington and Indiannapolis: Indiannapolis University Press.

World Council on Protected Areas/IUCN. 2003. *Recommendations, Durban, Vth WCPA/ IUCN World Parks Congress*. Durban, South Africa: IUCN.

II. Indigenous Conservation: Beliefs and Practices

Sacred Groves and Conservation:
The Comparative History of Traditional Reserves in
the Mediterranean Area and in South India

M. D. Subash Chandran and J. Donald Hughes

INTRODUCTION

The practice of protection of patches of woods as sacred is ancient. Groves of trees dedicated to the worship of the gods are mentioned by Greek and Latin authors. Ovid said, 'Here stands a silent grove black with the shade of oaks; at the sight of it, anyone could say, "There is a god in here!"'[1] One might think that such a grove, and such an idea, are things that passed away with the ancient world. But scores of sacred groves still persist in many parts of India.[2] Among those the authors studied was one near Mattigar, a village in Uttara Kannada. In an area largely cleared for agriculture stood a fragment of the original rainforest, tall, cool, dark in colour, about three acres in size. We entered with respect; offerings had been placed, but we could see no temple and no idol. As we left, we met an old man who explained, 'There is no image. The gods there live among the trees.'

A Comparative Study

This study compares sacred groves in the ancient Mediterranean with surviving groves of South India, especially Uttara Kannada, to evaluate the roles of these refugia in maintaining a balance between human groups and the ecosystems of which they are part. In the Mediterranean written records survive, and archaeological investigation provides some information. In South India, we observe a living if declining tradition which has persisted for millennia. There data come from observations, folk traditions, history and literature.

Sacre Groves in the Ancient Mediterranean Basin

The Mediterranean zone, mountainous and maritime, has a warm dry summer and a cool winter with rainfall totaling 380 to 900 mm per year. The original vegetation was, at low elevations, mainly forests of pines and evergreen oaks, and brushy maquis in dryer locales. Higher, to about 1,400 metres, deciduous forests occurred, and above them a coniferous forest belt up to the treeline at

Environment and History **6** (2000): 169–186.

about 2,200 metres. Deforestation was widespread even in ancient times. Sacred groves, 'the first temples of the gods',[3] were created by reserving sections of the original forests. Within them the environment was preserved, as a rule, in its natural state.

Sacred Groves in Uttara Kannada

We have chosen Uttara Kannada (North Kanara) district towards the centre of South India's west coast for comparative study. The hills of the Western Ghats, seldom rising above 700 metres, cover most of the land. The high humidity and annual rainfall of 2,000 to 6,000 mm, and the conservation ethics of the local communities, favoured luxuriant tropical forests which, despite nearly two centuries of commercial forestry, cover about 60 per cent of the district.

Coinciding with the decline of the Indus Civilisation, a major vegetational change occurred. Palynological studies indicate that, beginning about 1500 BC, there was an increase in savanna and a decrease in forest, perhaps due to colonisation by agri-pastoral people rather than climatic change.[4] All the forest species represented by pollen earlier are present today in Uttara Kannada forests, but some occur mainly in sacred groves. Sacred groves probably became more evident with the arrival of shifting agriculture.[5] As population was thin and the fallow period long, ecological succession on the abandoned slash and burn lands tended to restore the forest vegetation.

Despite the dwindling importance of sacred groves in the practice of today's Hinduism, the Western Ghats have scores of them. This may be due to the difficult terrain and consequently minor influence of Brahminic Hinduism among indigenous communities. These groves, known as *kans* in Uttara Kannada, are distinct patches of evergreen or semi-evergreen forests with lofty trees, in contrast to nearby forests of fire-prone deciduous trees and bamboo. The links of kans with the gods of villages were mentioned by a British traveller in 1801: 'The forests are the property of the gods of the villages in which they are situated, and the trees ought not to be cut without having leave from the Gauda or headman of the village ... who here is also the priest to the temple of the village god.'[6] We may infer that the kans were under the control of pre-Brahmin communities. The survival of evergreens in the kans reflects the high degree of protection which they enjoyed. These are cult centres for village communities comprising a spectrum of Hindu caste groups.

POINTS OF COMPARISON

In the Mediterranean and South India, earlier practices were similar, and some changes that occurred are parallel.

Sacred Groves and Conservation

FIGURE 1. Karivokkaliga peasant man worshipping in Devaravattikan. Note the metal tridents planted in the grove as offerings. There is no image or cult figure in the grove, where a mother goddess and father god are honoured.

FIGURE 2. Hill of Kronos, Olympia. This small mountain adjoins the site where the Olympic Games were held in honour of Zeus, the chief Greek god, and his sister/wife; Hera. It was dedicated in antiquity to the father of Zeus, and is a typical natural sacred site. There was undoubtedly a grove here in ancient times, but the pines that presently grace the hill were planted by German archaeologists in the late nineteenth century.

M. D. Subash Chandran and J. Donald Hughes

The Genius Loci

The original dedication in both cases is to local deities perceived as dwelling among the trees. In Uttara Kannada, the deities of the groves were not major Hindu gods such as Shiva, Vishnu, Parvati, etc., but indistinct beings represented by vacant spots, crude stones or termite mounds. For instance, in the grove of Mattigar, the female deity Choudamma and the male deity Jatakappa are represented by vacant cult spots. Their presence is perceived in the entire grove by the Karivokkaliga peasants, for whom this grove is the main temple to this day. That nature itself within the grove is sacred is the world-view of pre-Brahminic societies. Some groves are dedicated to animal deities such as serpent and tiger, survivals of an early hunter-gatherer period. Some deities here received animal and human figurines as offerings.

Aniconic worship of indistinct beings well describes early Roman religion. Deities (*numines*) were felt to be inherent in aspects of the landscape. Every notable spot had its *genius loci*, or spirit of the place. An image was not necessary. One famous grove was dedicated to Bona Dea, or 'good goddess', not otherwise named. Diana Nemorensis, or 'Goddess of the Grove' was a local spirit of Nemi.

FIGURE 3. Forest canopy in Devaravattikan sacred grove, Mattigar, North Canara, Karnataka state, India. The mosaic-like pattern is caused by 'crown avoidance', a phenomenon typical of South and Southeast Asian rainforests. Here the sacred grove is a surviving fragment of the original monsoon evergreen forest, and contains species that have disappeared from the surroundings.

Classical Greece reflects a late stage in evolution toward representation of gods by artistic images, but there are survivals of earlier stages. It appears that a sacred grove was first dedicated, and a cult figure added afterwards. Images in many places, such as that of Artemis of Icaria, were uncarved blocks of wood.[7] Regulations concerning hunting indicate an origin in hunter-gatherer society. Of groves dedicated to major gods, most were of Artemis, goddess of hunting, and her twin brother Apollo, who also carried a bow and was titled 'hunter'.[8] There were many to Pan, hunter and herder, especially in forested Arcadia. Archaeology shows that deities received animal figurines as offerings.[9]

Protective Traditions and Regulations

Practices controlled in the groves were strikingly similar, the idea being to honour the gods by keeping the original forest as undisturbed as possible. Around Mediterranean and Indian groves, borders were distinct. Tree felling, collection of biomass, removal of earth, hunting, fishing, farming, grazing of domestic animals, and use for residences or other buildings were forbidden. Specific rules varied from grove to grove, but in general the biota was protected. Exceptions might be allowed in times of need.

An inscription from the city of Magnesia says, 'By the sacred laws, prohibitions and censures, it is forbidden for anyone to pasture or stable or saw or cut wood in the sanctuary of ... Zeus.'[10] Groves were often walled to mark the boundary between holy and ordinary space.

In the Western Ghats, smaller groves function entirely as the abodes of gods with a taboo on biomass removal, but larger groves served as safety forests offering sustenance and ecological security. In the kans villagers could gather fruits and tap toddy from palms. They cared for wild pepper within the kans, until the British takeover of the forests.

Whereas domestic animals were excluded from Mediterranean groves, in the Western Ghats they do enter the groves. Brandis and Grant, however, noted the remains of barrier trenches bordering kans in Shimoga.[11] In Greece and Rome there were penalties for trespassers. In India, large groves were open to the men and women of their villages for gathering non-wood produce, but only to men for worship. The gathering of fallen deadwood in the evergreen kans seems to be the consequence of state monopoly of the timber-rich deciduous secondary forest.

Administration of the Groves

Responsibility for protecting them and enforcing rules was almost always assumed by the local community. In the Mediterranean, the responsible entity was the city; the *polis* in Greece or municipality in Roman dominions assigned jurisdiction over groves to officers, usually religious magistrates.[12] Penalties for miscreants could be severe fines. The land of the community was delineated in

M. D. Subash Chandran and J. Donald Hughes

part by the location of its groves; a recent study of the Greek polis maintains that its borders were demarcated by sacred enclosures.

In Uttara Kannada the grove was an integral part of village life, and decisions regarding it were made by the community.[13] In theory the state owned the kans but local committees fined the offenders, a system prevailing to this day in Halkar. The village headman supervised the safety of the grove and obtained 'permission' from the god before cutting down any tree.

Both in the Mediterranean and India, protection was believed to be enforced by the gods. On a Greek mountain, if a hunter saw his quarry go into the precinct of Zeus, he waited outside, believing that if he entered he would die within the year.[14] Hunting was community-regulated and season-bound in pre-colonial Uttara Kannada. The first hunting is carried out to this day ritually by the peasants before the rice harvest, and the quarry is sacrificed to the grove gods. Any appeal to the gods for favours goes with sacrifice of a fowl or goat. No hunting, normally, is conducted in the groves. In the Mediterranean too, animal sacrifice was customary for any appeal to the gods of groves.

FIGURE 4. Samos, *temenos* wall in the island city. The wall probably dates from the seventh century BC, under the tyrant Polycrates. The nearby town is now called Pithagoreio, in honour of the mathematician Pythagoras.

Refugia

A 'mosaic landscape' is noted in both cases; the groves formed 'islands' of variable size within a pattern of other land uses. Greek and Roman landscapes included categories such as city, cultivated land, pasture, and woodland.[15] Beyond these lay wilderness, with resources such as wildlife or metallic ores. Another land use was sacred space (*temenos* or *templum*), including groves. Aristotle says the land was 'divided ... into three parts: one sacred, one public, the third private: the first was set apart to maintain ... worship of the gods, the second was to support the warriors, the third was the property of the husbandmen'.[16]

In pre-colonial Uttara Kannada people gathered biomass routinely from ordinary forest (*kadu* or *adavi*).[17] Shifting cultivation (*hakkalu* or *kumri*) was widely practised. Sacred groves were prominent in the village landscape. Havik Brahmans raised spice gardens close to evergreen forests including the kans, which assured water supply, shade and leaf manure for their crops of pepper, cardamom, cinnamon, ginger, arecanut, and recently, also clove, cocoa and nutmeg. Villages, groves, other forests, cultivation fallows in varied stages of forest succession, pastures, fields and gardens, in totality, form a mosaic. Landscape heterogeneity and biodiversity are positively correlated.[18] Groves formed part of a landscape of well-connected natural elements and functioned as refugia for many species of plants and animals.[19] The connectivity of the landscape elements, allowing the mobility of species, would make the groves free from some limitations of small islands.[20] Groves are 'specimens' of the original ecosystems of the areas where they exist.

The Mediterranean groves gave refuge to many species. Some sheltered wildlife from hunting, except rarely for sacrifice. Fish survived in their waters.[21] Species were becoming extinct elsewhere due to hunting, demands of the Roman arenas, and habitat destruction, but survived longer in the sanctuaries.[22]

Sacred groves belong to a variety of cultural practices that helped Indian society maintain an ecologically steady state.[23] The kans of Uttara Kannada, though on the wane, remain centres of biodiversity. From a distance, these fragments of climax forests appear distinguished from surrounding woods by their darker foliage and emergent trees over thirty metres tall. The least disturbed kans have greater biomass. The trees in a one-hectare sample of one of the finest kans comprise a basal area of about 63 square metres, comparable to the best of tropical forests. The kans are centres of plant endemism of the Western Ghats, considered as one of the eighteen biodiversity 'hot spots' on earth. One hectare in Kallabbe kan has 83 per cent endemic trees, but endemism is just 15 per cent in secondary forest near the kan. The kans, mostly evergreen, are richer in species than the deciduous forests: 30 to 50 in a hectare in the former and less than 30 in the latter. The sacred grove at Mattigar has about 60 tree species in one hectare.[24]

Groves shelter rare habitats and endangered species. For example, the mighty *Dipterocarpus indicus* has the isolated north end of its range in some of the fine

M. D. Subash Chandran and J. Donald Hughes

kans of southern Uttara Kannada. The latter is also the location of a threatened ecosystem, a *Myristica* swamp. A notable species of wild nutmeg tree[25] and a rare palm[26] are sheltered here. Groves could be valuable gene banks for restoration of natural ecosystems around them. As they diminish in size, habitat types and rare species in them are in jeopardy.

Commercial forestry which did not spare the kans caused a general decline in the rich wildlife of Uttara Kannada; tigers, leopards, elephants, gaur, and other large mammals are seldom or never seen. However, the endangered lion-tailed macaques still occur in Katlekan.[27] Many interior forest birds occur in the groves in the midst of populated villages. A survey showed that half the bird species in Siddapur taluk occur in sacred groves.[28]

Water

The characteristic image of a sacred grove includes a source of water quite as much as a stand of trees. In the Mediterranean and in India, groves protected watersheds and springs. Ancient writers knew that dense forests regulate runoff of precipitation. Like a sponge, the plants and soil hold water, preventing floods and releasing a year-round supply to streams. Plato said that when mountains

FIGURE 5. Grove of evergreen plane trees in a churchyard at Gortyna, Crete. The existence of this unusual variety of a tree that is ordinarily deciduous may well have caused it to be regarded as sacred, and it is not impossible that this Christian sacred grove may be the descendant of an earlier grove dedicated to one of the ancient gods or goddesses.

were deforested, springs dried up, and the sad bits of evidence were shrines at spots where they formerly existed.[29]

The Bombay Government in 1923 highlighted the watershed value of the kans: 'Heavy evergreen forests hold up several feet of monsoon rain ... If evergreen forest is felled in the dry season the flow of water from any spring it feeds increases rapidly though no rainwater may have fallen for some months.'[30] It is, of course, the last such; a case of killing the goose to get the golden eggs. The government made this a reason for reservation of the kans, ordering that they never be cut for timber. Unfortunately this wise prescription was later forgotten.

Peasants believe, with some justification, that forests bring rain. Watershed forests in every village have sanctity attributed to them. Sometimes the only source of water in a village is the grove's spring. Groves can supply fresh water in regions where other water is saline.[31] Indian tradition considers rivers and springs as divine, and many are places of tirthayatra, meaning 'pilgrimage to sacred water'.

The Process of Religious Change

Cultural change presents interesting parallels. In both cases, local deities of the groves were identified with the great gods of the pantheon, due to the influence of a literary tradition dominated by epics and priestly rituals. These resulted in the erection of temples and the use of materials from the groves for construction and sacrifices. The rules protecting the groves got relaxed as the centre of ritual moved away from the trees toward the temple building.

Greek and Latin literature abound in evidence that local deities were subsumed into the state pantheon. Early temples in Greece and Rome were of wood, later of stone but requiring timber beams for roofs. Temple builders looked to large trees which survived in the groves, as elsewhere they had been cut. A tall cypress was felled in the precinct of Apollo on Carpathos and used in rebuilding a temple in Athens.[32] Priests were political officers, members of urban elites desirous of enhancing their power and the prestige of their communities.

The decline of groves in India was a gradual process linked to the absorption of local cults by text-based Brahminism. In the cultural complex of Hinduism two major coalescing traditions are discernible. The 'Little Tradition' is the unwritten lore of people who remain outside the fold of organised religion. Its origins are lost in prehistory. The other, known as the 'Great Tradition', can be traced to the Vedas and other scriptures. The religion of the followers of the textual tradition is more institutionalised than that of the folk cults. Through millennia, Hinduism has absorbed scores of village cults. Sacred groves formed an integral part of folk religion. Although groves steadily lost importance as worship places with the spread of the Great Tradition, there was nothing in the latter which opposed nature cults. Interestingly, the Dharmasastras, which prescribe a code of righteous life, attach great importance to planting trees. However, the

M. D. Subash Chandran and J. Donald Hughes

planted trees could never match the primeval sacred groves in species diversity and ecological functions.[33]

Incorporation of deities of the Little into the Great Tradition involves sanskritisation of their names and building of temples. The groves continued to coexist for a time with the temples, but diminished in importance. The trees were cut, initially for the temple and its repairs and subsequently for commercial and subsistence needs. Significant among the local cults to merge with Hinduism were fertility cults, especially those of the Mother goddess and Shiva.

In areas of Brahminic influence the Mother goddesses often got identified with Parvati, consort of Shiva, or her incarnations. This involves housing the Mothers of the groves in temples or small shrines. Karikanamma, the 'Mother of the Dark Forest', who hails from a beautiful hilltop grove of dipterocarps, now has a temple nearby, and is sanskritised as Parameshwari, a consort of Shiva. The uncarved rock that embodies her is covered by a metallic mask. The local lore, authored by a Brahmin, narrates how the Mother was 'saved' from the clutches of barbarians, evidently referring to the indigenous people of the woods, and housed in the temple by the Brahmins. Similar processes are seen all over India.

Archaeologists trace Shiva to the Indus Culture. An abstraction from nature, he was the lord of mountains, watershed, vegetation and animals. The male gods of the groves of Uttara Kannada, represented by vacant spots, stones or termite mounds, may well be Shiva's prototypes. He resembles Pan of late classical Greco-Roman religion. Pan was recognised as universal god of nature, primal god of herds, streams, thickets, and rocky peaks. An Orphic Hymn calls Pan 'green power in all that grows, procreator of all'.[34]

Just as local Greek gods were identified with Olympian deities, many village gods were amalgamated into the Hindu pantheon. 'The enlistment of Hanuman in the service of Rama signifies the meeting point of early nature worship and later theism.'[35] In this process, an older religion that valued nature was spiritualised, and worship that brought people directly before many forms of nature was replaced by a more sophisticated iconography. The wooded, hilly hinterlands are in the throes of a process that swept through the Indian plains in early historic times. Although Hindu tradition exhorts its followers to protect several plants as sacred and to raise groves of sacred trees in temple premises, housing of the sylvan deities in temples leads to neglect of the groves. Traces of groves are seldom seen around temples of Uttara Kannada where termite mounds stand for the original deities. An architectural environment similar to the groves is represented in temples by the dendritic form of columns, just as in Greece.

Sacred Groves in the Market Economy

Apart from changes in religion, the demands of an economy beyond local ecosystems was deleterious to preservation of the groves. Greek and Roman entrepreneurs exerted pressure for use of the resources of the groves. In spite

FIGURE 6. A small temple in Karikan, a sacred grove near Kumta,
North Canara. The grove is dedicated to Karikanamma, 'The Mother
of the Dark Forest'. Inside the temple are a spring of water and two
lingas (stone phallic symbols). Temples are often built in or beside
sacred groves, but the groves existed before the temples.

of rules against grazing, art and literature abound in pictures of domestic animals foraging in them. Xenophon mentioned that people lease 'enclosures and
sanctuaries' for removal and sale of wood and other products.[36] Strabo noted
that some sanctuaries had lost their trees.[37]

Appropriation of resources by colonial exploiters, and by proponents of
economic growth since independence, has damaged or destroyed many groves

M. D. Subash Chandran and J. Donald Hughes

in India. As early as 1633, the Portuguese in Goa made a treaty with rulers of Uttara Kannada to cut timber free of charge. British conquest in 1799 opened a period of unbridled exploitation. Uttara Kannada was well stocked with forest resources at the time of their occupation. The Indian Forest Act of 1865 asserted British ownership of forest resources. 'Reserved' forests were closed to all uses of the public. By the turn of the century, working plans were initiated for their utilisation, though lip service was given to sustained yield.[38] Local people were restricted to gather fuel and leaves and graze cattle only in the degraded 'minor forests', hastening their ruin. The Government added 769 hectares of kans in Sirsi to minor forests. In the revenue yielding spice garden villages, dominated by Havik Brahmins, the state allotted nine acres of forest per acre of garden as soppinabetta, or leaf manure patches. Some of these were kan forest. Other villagers, left to gather biomass or graze cattle in open access minor forests, lost power to protect their hitherto community forests, including many kans, from outsiders, and failed to regulate harvests by their own members. Thus former communal properties became open access resources liable to exhaustive usage, a classic case of 'the tragedy of the commons'.

Brandis noticed the widespread occurrence of the groves, calling them 'the traditional form of forest preservation':

> Sacred groves in India ... are, or rather were very numerous. I have found them in nearly all provinces ... These sacred forests, as a rule, are never touched by the axe, except when wood is wanted for the repair of religious buildings ...[39]

The state did not treat the kans as sacred, and some colonialists considered them a 'contrivance' to prevent the government from claiming its 'rightful' property. Since kans were fine patches of evergreen forest, they were often included in reserved forests. The state introduced a destructive contract system for extraction of nonwood resources. As R.T. Wingate observed,

> I am still of the opinion that the system of annually selling by auction the produce of the kans is a pernicious one. The contractor sends forth his subordinates..., who hack about the kans just as they please, the pepper vines are cut down from the root, dragged from the trees and the fruits then gathered, while the cinnamon trees are all but destroyed... I was greatly struck with the general destruction among the Kumta evergreens; they were in a far finer state of preservation fifteen years ago.[40]

Wingate noted that a proper demarcation of the kans was not conducted by the state, implying that they were merging with ordinary forests and losing their identity. The restrictions on biomass removal from reserved forests faced by the common people resulted in heavy pressure on the village kans. Felling of trees opened the rainforest canopy and permitted the spread of introduced, light-loving weeds like lantana. Encroachment on the kans by land-hungry farmers reduced their area. In Sorab in the central Western Ghats the town is expanding into the nearby kan. With increased light and heat reaching the forest floor, wild pepper has almost vanished from the Sorab kans.

Since the kans contained mostly softwoods, unmarketable at the time, they were not much exploited for timber almost to the end of the British period. Resource emergencies of the Second World War, however, prompted 'war fellings' in evergreen forests including the groves. Use for *Dipterocarpus indicus*, which survived only in a few kans, was found in the railroads and the new plywood industry.

With Indian independence in 1947, the Forest Department continued the methods of professional forestry it had inherited, so centrally directed state management continued. It was a disastrous model. In a major drive to industrialise, reserved forests were leased out to companies producing plywood, paper, matches and packing cases for heavily subsidised extraction of timber and bamboo.[41] Large tracts were submerged in hydroelectric projects or leased out to mining companies. These leases and projects adversely affected kans. A forest working plan for Sirsi and Siddapur in 1966 organised 4000 hectares of kans for extraction of industrial timbers. Another plan for Sirsi included for timber extraction 672 hectares of kans belonging to ten villages. The working plans often prescribed 'improvement felling' in kans of large trees regarded by foresters as 'overmature' specimens. A kan was clear cut and converted into a Eucalyptus plantation in Menasi village of Siddapur taluk. In 1976, with industrial demand at its peak, despite protests from the villagers, the kan of Kallabbe, which had been in an excellent state of preservation by the people, was leased to a plywood company which extracted hundreds of logs with attendant damage. Due to a series of such invasions, the area covered by kans in Siddapur taluk declined by 94.9 per cent from an estimated 6 per cent of the total area to 0.3 per cent. The sacred ponds associated with the groves are mostly silted up and covered with rank growth of plants. In some native fishes are poisoned and carp restocked.

Exclusion of local residents from the reserved forests, bans on shifting cultivation and other restrictions were seen as attacks on the community-based system of use, and provoked the first case of forest resistance in the district in 1886. The agitation for ancestral rights of forest use continued in the 1920s, and was coopted into the Gandhian Satyagraha of the 1930s. In a few cases, groves became rallying points in movements for local rights.[42] A demonstration against deforestation and commercialisation of the forests called Appiko began near Salkani, 'goddess forest'.[43] Inspired by Chipko (the 'tree hugging' movement in the Himalayas), in August 1983 the villagers of Sirsi

> requested the Forest Department to halt tree felling. When their requests failed, they marched into the forest and physically prevented the felling from continuing. They also extracted an oath from the loggers (on the local forest deity)... that they would not destroy trees in the forest.[44]

M. D. Subash Chandran and J. Donald Hughes

FIGURE 7. Church and cemetery, Messenia, Peloponnesos. The appearance of a sacred grove is suggested by the small Orthodox church, which would have been a temple in ancient times, and the grove of planted cypresses surrounded by a wall.

FIGURE 8. Two young men of Mattigar constructing a fence to demarcate and protect their grove of Devaravattikan. The barrier will prevent the entry of grazing animals; gates allow access to worshippers.

CONCLUSION

The practice of honouring sacred groves occurred for similar purposes in wide-spread parts of the world. This practice was part of a pattern that made possible a sustainable human way of life within forest ecosystems. This positive function has not disappeared; it is more important today. It is mostly too late to save sacred groves in the Mediterranean, although a few survive in the precincts of monasteries and churches. Certain surviving refugia are protected by national laws or international agreements as biosphere reserves, making them modern secular equivalents of ancient sacred groves. Places where ecosystems are relatively intact are 'holy' in the sense of possessing integrity, or natural wholeness, and are more valuable because so many groves are gone.

Sacred groves, wherever they still exist, as in India, should be preserved and restored for many reasons, including their value as historical evidence for the relationship of human beings to natural ecosystems. The groves of Uttara Kannada are living ecosystems, even if fragmentary, that relate to people of the local communities that once protected them, and can once again if they are respected as partners in the conservation effort.[45] To quote Madhav Gadgil,

> For local people, degradation of natural resources is a genuine hardship, and of all the people and groups who compose the Indian society they are the most likely to be motivated to take good care of the landscape and ecosystems on which they depend. The many traditions of nature conservation that are still practiced could form a basis for a viable strategy of biodiversity conservation.[46]

NOTES

[1] Ovid, *Fasti* 3. 295–6.

[2] M.D. Subash Chandran and Madhav Gadgil, 'Sacred Groves and Sacred Trees of Uttara Kannada (A Pilot Study)', Report Submitted to the Indira Gandhi National Centre for the Arts, New Delhi, 1993, p. 16.

[3] Pliny, *Natural History* 12.5 (9).

[4] M. Catratini, M. Fontugne, J.P. Pascal, C. Tiscot and I. Bentaleb, 'A major change at ca. 3500 years BP in the vegetation of the Western Ghats in North Kanara, Karnataka', *Current Science* vol. 61, nos 9–10 (1991), pp. 669–72.

[5] M.D. Subash Chandran, 'Shifting Cultivation in the Western Ghats and the Conflicts in Colonial Forest Policy', in Richard Grove, Vinita Damodaran and Satpal Sangwan (eds), *Nature and the Orient: Essays on the Environmental History of South and Southeast Asia*, New Delhi, Oxford University Press, 1999.

[6] Francis D. Buchanan, *A Journey from Madras through the Countries of Mysore, Canara and Malabar*, Madras, Higginbothams and Company, 1970, vol. 2.

[7] Richard Farnell, *The Cults of the Greek States*, Oxford, Clarendon Press, 1896, vol. 2, pp. 520–521.

[8] Authors such as Darice Elizabeth Birge, 'Sacred Groves in the Ancient Greek World', University of California, Berkeley, Department of Ancient History and Philosophy, 1982, Ph.D. Dissertation, p. 27, have noted the great number of groves dedicated to Apollo, but have missed Apollo's connection to hunting.

[9] François de Polignac, *Cults, Territory, and the Origins of the Greek City-State*, Chicago, University of Chicago Press, 1995, p. 26.

[10] W. Dittenberger, *Sylloge Inscriptionum Graecarum* 3, 1924, 685, A.III.81–82.

[11] D. Brandis and Grant, Joint report No. 33, 11th May, 1868, on the kans in the Sorab taluka. Forest Department, Shimoga, India.

[12] James George Frazer, *The Golden Bough: The Magic Art and the Evolution of Kings*, 3rd edn (12 vols), New York, Macmillan, 1935, Vol. 2, pp. 121–2. The work was first published in 1890.

[13] Madhav Gadgil and M.D. Subash Chandran, 'Sacred Groves', *India International Quarterly*, vol. 19, nos 1–2 (spring-summer 1992), pp. 183–7.

[14] Pausanias, *Description of Greece* 8.38.5.

[15] J. Donald Hughes, 'An Ecological Paradigm of the Ancient City', in Richard J. Borden (ed.), *Human Ecology: A Gathering of Perspectives*, College Park, MD, University of Maryland and The Society for Human Ecology, 1986, pp. 214–20. See Catherine Delano Smith, *Western Mediterranean Europe*, London, Academic Press, 1979, pp. 166–76.

[16] Aristotle, *Politics* 2.5.

[17] M.D. Subash Chandran and Madhav Gadgil, *Sacred Groves and Sacred Trees of Uttara Kannada*, pp. 17–18.

[18] R.T.T. Forman and M. Godron, *Landscape Ecology*, New York, John Wiley and Sons, 1986.

[19] N.V. Joshi and Madhav Gadgil, 'On the Role of Refugia in Promoting Prudent Use of Biological Resources', *Theoretical Population Biology*, vol. 40 (1991), pp. 211–29.

[20] Larry D. Harris, *The Fragmented Forest: Island Biogeography Theory and the Preservation of Biotic Diversity*, Chicago, University of Chicago Press, 1984.

[21] Aelian, *De Natura Animalium* 8.4; Plutarch, *Moralia* 976A; Pausanias, *Description of Greece* 7.22.4.

[22] J. Donald Hughes, *Pan's Travail: Environmental Problems of the Ancient Greeks and Romans*, Baltimore, Johns Hopkins University Press, 1994, pp. 105–8.

[23] Subash Chandran and Gadgil, *Sacred Groves and Sacred Trees of Uttara Kannada*, p. 1.

[24] Ibid., p. 38.

[25] *Myristica magnifica*.

[26] *Pinanga dicksonii*.

[27] Madhav Gadgil and M.D. Subash Chandran, 'Sacred Groves'.

[28] R.J.R. Daniels, 'A Conservation Strategy for the Birds of Uttara Kannada District', 1989, Ph.D.Thesis, Centre for Ecological Sciences, Indian Institute of Science, Bangalore.

[29] Plato, *Critias* 111D.

[30] Government of Bombay, Revenue Department Resolution, No. 7211, May, 1923.

[31] J.J. Roy Burman, 'The Institution of Sacred Grove', *Journal of the Indian Anthropological Society*, vol. 27 (1992), pp. 219–38, quotation on p. 228.

[32] Marcus Niebuhr Tod, *A Selection of Greek Historical Inscriptions*, 2 vols., Oxford, Clarendon Press, 1933, vol. 2, p. 110.

[33] Subash Chandran and Gadgil, 'Sacred Groves and Sacred Trees of Uttara Kannada', pp. 30, 36.

[34] Orphic Hymn to Pan, translated by J. Donald Hughes in 'Pan: Environmental Ethics in Classical Polytheism', in Eugene C. Hargrove (ed.) *Religion in Environmental Crisis*, Athens, GA, The University of Georgia Press, 1986, pp. 7–24, quotation on p. 11.

[35] M.D. Subash Chandran, 'Peasant Perception of Bhutas: Uttara Kannada', in Kapila Vatsyayan and Baidyanath Saraswati (eds), *Prakrti: The Integral Vision, Vol. 1, Primal Elements: The Oral Tradition*, New Delhi, Indira Gandhi National Centre for the Arts, 1995, pp. 151–66, quotation on p. 164.

[36] Xenophon, *Economicus* 4.19.

[37] Strabo, *Geography* 9.2.33.

[38] Marlene Buchy, 'Quest for a Sustainable Forest Management: A Study of the Working Plans of North Canara District (1890–1945)', in Ajay S. Rawat (ed.), *Indian Forestry: A Perspective*, New Delhi, Indus Publishing Company, 1993, pp. 141–62.

[39] Dietrich Brandis, *Indian Forestry*, Woking, Oriental Institute, 1897.

[40] R.T. Wingate, Settlement Proposals of Sixteen Villages of Kumta Taluk, No. 210, Karwar, Forest Settlement Office, December, 1888.

[41] Madhav Gadgil and M.D. Subash Chandran, 'On the History of Uttara Kannada Forests', in John Dargavel, Kay Dixon, and Noel Semple (eds), *Changing Tropical Forests: Historical Perspectives on Today's Challenges in Asia, Australasia and Oceania*, Canberra, Centre for Resource and Environmental Studies, 1988, pp. 47–58, quotation on p. 51.

[42] J.J. Roy Burman, 'The Institution of Sacred Grove', esp. p. 232.

[43] Jeremy Seabrook, 'Uttara Kannada', in *Notes from Another India*, London, Pluto Press, 1995, pp. 65–83, quotation on pp. 65–6.

[44] Madhav Gadgil and Ramachandra Guha, *This Fissured Land: An Ecological History of India*, Berkeley, University of California Press, 1992, p. 224.

[45] R.J. Ranjit Daniels, M.D. Subash Chandran and Madhav Gadgil, 'A Strategy for Conserving the Biodiversity of the Uttara Kannada District in South India', *Environmental Conservation*, vol. 20 (summer 1993), no. 2, pp. 131–8.

[46] Madhav Gadgil, 'Conserving Biodiversity as if People Matter: A Case Study from India', *Ambio*, vol. 21, no. 3 (May 1992), pp. 266–70, quotation on p. 268.

The Role of Customary Institutions in the Conservation of Biodiversity: Sacred Forests in Mozambique

Pekka Virtanen

INTRODUCTION

In many African countries the conservation of biological diversity through a centralised system of parks and reserves is threatened by problems in enforcement, as the existing conservation areas commonly suffer from various types of encroachment ranging from poaching to outright invasion. So far, remedies for the problem have been sought from economic incentives to support the enforcement of protection laws, funded by international donors and implemented through projects. But projects alone are not sufficient for comprehensive protection, as they can only cover a fraction of the vital ecological assets (Müller 2000). Concurrently, other alternatives, such as areas protected by local people according to endogenous rule-systems like sacred forests, have gained international recognition in major declarations and conventions (Nummelin and Virtanen 2000; Schaaf 1999). But despite the rekindled interest, such traditionally protected forest areas remain little studied from the conservation point of view.

According to Margules and Pressey (2000: 243) effective systems for the conservation of biological diversity must be representative and persistent. Ideally, a network of conservation areas should represent the full variety of biodiversity in the region covered. And once established, the institutional support system should be able to exclude threats and promote long-term survival of the species by maintaining ecological processes and viable populations. In Africa traditionally protected forests are often believed to represent an original, formerly much more extensive forest type, which has later been degraded into a less rich type such as grassland by human activity, typically the opening of new land for shifting cultivation or pasturage. But recently researchers have disputed the predominance of dense climax forest over most of Africa. According to them, vegetation is in continual transition, and thus it is not possible to determine 'original' or 'climax' type. Furthermore, they insist that human activity should not be considered as external to nature. For example some traditionally protected forests are the results of long-term human manipulation of local micro-environmental factors such as soil and plant characteristics, and thus at least to some extent 'non-original' (Fairhead and Leach 1995).

Environmental Values **11** (2002): 227–241.

Parts of nature are protected under traditional norms according to two different premises: the controls can be either space or species-based (Mandondo 1997: 355). Many species-based controls are linked to beliefs about spirits and their dwelling places. But in addition to religious controls, some species are protected for their utilitarian value, such as medicinal plants, fruit trees, and those species, which are believed to provide environmental services. However, if conservation is pursued at the level of ecosystems the main interest lies with space-based controls over larger areas under tree cover, such as sacred forests. There are two geographical factors, size and land type, which are important in terms of ecological representativity and viability.

The size of sacred forests varies considerably. Survey data from East Africa shows a variation in average size from over six hectares to less than one. While in some areas studied in Tanzania only one quarter of the sacred forests were under two hectares, in other areas over 70 per cent of the forests were smaller than that. In Kenya the sizes of all the sacred groves studied in a densely populated area were less than 1.2 hectares. The acreage seems to have some relationship with population density, as small groves are typical in areas with high demographic pressure. In terms of representativity for biodiversity conservation purposes, the land type where the sacred sites are situated is another important factor. Typically the protected forests are on hills and slopes, but there are also some sites on flat land and along springs and streams (Castro 1995; Gerdén and Mtallo 1990; Mwihomeke et al. 1998).

Due to their embeddedness in specific cultural systems sacred sites have also a spiritual dimension, and their persistence cannot be assessed separately from the cultural institutions which make them meaningful and valuable for the local populations. Traditionally protected forests can be divided into the following socio-cultural categories: i) burial grounds; ii) places where deities or spirits are believed to reside; iii) places for ritual; iv) sites linked to special historical events or populations; and v) forests that surround natural sacred features like rocks, caves or ponds.[1] Even though there are exceptions, all these categories are typically based on religious beliefs, and thus ultimately on supernatural sanctions for their protection. But to a varying degree these are backed up by human sanctions administered by selected members of the respective community. When there are established procedures which define who is competent to decide what is the rule and what are the human sanctions in case of its violation, the rule becomes a law (Perelman 1984: 45).

In order to grasp the socio-cultural meanings connected to traditionally protected forests we must understand the holistic relationship between man, the spirit world, nature, and society. In the context of many African cultures the concept of community includes both the living and the dead, and the spiritual is as much part of reality as the material. In this sense a sacred grove is not just a cultural relic, but an alternative way to transcend the divide between the local (or visible) and the universal (or invisible) (cf. Posey 1999: 4; Swan 1989: 383–5).

Pekka Virtanen

In patrilineal societies like the Shona a specific role in mediating the relationship between the material world and the spirit world is attributed to the elders of the apical lineages, who command privileged access to ancestral spirits of the chiefly family. The latter are believed to be the spirit guardians of the whole chiefdom, who also monitor the behaviour of the present generation. Thus, the basis of both religious and political power is transcendental (Bourdillon 1987: 253–5; Mandondo 1997). But such a system is viable only in a relatively homogeneous community which has a common history based on joint cultural and religious values. Failure to maintain consensus over such essential values within the political community is likely to result in lack of authority to enforce the rules in other way than by force (Perelman 1984: 53–4).

Even though the basis of traditionally protected forests is in local customary law, it does not exclude state law, and in some cases the two systems coincide. For example some of the tree species protected by state law (like the African black ebony, *Dalbergia melanoxylon*) are also present in sacred groves, and riverine forests are officially protected for example in Tanzania and Zimbabwe. The situation is somewhat different when local by-laws are explicitly made to reinforce existing customary rules, like in some villages in Tanzania (Gerdén and Mtallo 1990). In Kirinyaga in Kenya the local Native Council had ordered the registration and subsequent protection of all places of worship, including sacred groves, during the colonial era (Castro 1995). The difference between the two approaches is that unitary state law treats every citizen on the basis of equality irrespective of their membership in a local community, while recognition of the co-existence of various local laws with different cultural and religious bases presupposes legal pluralism. The question is whether the special rights enjoyed by members of the local community are compatible with the principle of equity between citizens (Perelman 1984: 46–7).

STUDY AREA AND METHODS

In this article the conservation value of the sacred forest institution is discussed on the basis of a case study from Mozambique. The article is based on fieldwork carried out during 1998–9 in the Chôa highlands in Barué District, which lies along Mozambique's western border with Zimbabwe in Manica Province. The Chôa highlands form a series of north-south ridges with a central plateau area. Climatically the area falls within the 1400 mm isohyet, with marked alteration of annual wet and dry seasons. Vegetation consists of extensive wooded or open grasslands with some deciduous woodlands and a few patches of evergreen forest near perennial streams (Mussanhane et al. 2000: 90–5). The Afro-montane habitats in the escarpment region are characterised by high species endemism (Hatton and Munguambe 1997: 14). With only about ten inhabitants per square kilometre, the population density of this remote area is low. Almost all the people are Shona, and speak the ChiManica dialect. The main livelihood is subsistence agriculture, which is supplemented by cattle and some cash crops (Mussanhane et al. 2000: 95–6).

In addition to the ecological representativity of the two sacred forests studied, I will examine the socio-cultural basis of their continued functioning at the juncture of changing state law and customary law. The methods used include participant observation, semi-structured interviews with key informants, and visits to the sacred forests.[2] Aside from my own fieldwork the results are based on a forest inventory carried out by J. Mussanhane and L. Nhamuco from the Eduardo Mondlane University[3]. The study is part of a multi-disciplinary research project on the local management of natural resources in Mozambique, Zimbabwe and Tanzania.

CUSTOMARY AUTHORITY AND TRADITIONAL RELIGION IN MO-ZAMBIQUE

Mozambique gained independence in 1975 after a protracted liberation war, and subsequently a new radical government was established by Frelimo, the front organisation which had directed the armed struggle. In the colonial period customary authorities had played an important role in local administration, especially with respect to labour recruitment, law enforcement and collection of taxes. In line with their Marxist doctrine, the new rulers viewed customary authority as 'feudal', geared to serve its colonial masters and its own selfish class interests. Along with others who had collaborated with colonial authorities, representatives of customary authority were systematically excluded from positions of responsibility in the new local structures (Alexander 1997: 2). The government policy even warned against recuperation of such 'negative historical values' as veneration of the graves of the chiefs, which could be used for 'reactionary and obscure purposes' (Resolução sobre questões sociais e culturais 1980: 9). But despite the hostile government rhetoric and practice, socio-cultural institutions grounded in traditional religion never really ceased to exist (West and Kloeck-Jenson 1999: 457–9).

One factor in this process was a new civil war, which started only two years after independence, and lasted up to 1992. During the war both sides were forced to seek the support of customary authorities. Even though such efforts by the Frelimo government were rather informal and localised, the new guerrilla movement Renamo sought actively to incorporate traditional authorities in their rudimentary civil administration of the zones they occupied. But while chiefs and headmen were increasingly involved in the everyday management of community affairs in many areas, the war tended to disrupt traditional religious life, such as the annual rain ceremonies. This was due to various reasons like translocation of population which fled the war zones to neighbouring countries and urban areas, problems with transport and material resources needed for organising the ceremonies, and lack of security. The last aspect was a serious problem for the chiefs and their spirit mediums, who had become a target group (to be either recruited or eliminated) for both belligerent parties (Alexander 1997: 5–15).

Pekka Virtanen

Ending of the war moved the dispute to the political arena, and by the late 1990s recognition of such local institutions as sacred forests was already incorporated into new legislation (see e.g. Lei no. 10/99: Art. 1, 13).

The legitimacy of customary socio-political institutions is based on traditional religion, in which spirits of the ancestors have a crucial role. Among the Shona the spirits of the dead are preserved in a burial ceremony, and the burial place is considered to be sacred. The spirits of dead chiefs and other members of the ruling elite have a special status, for they are considered as the founding spirits who are the custodians of the land and its fertility. The most sacred burial sites are those of the first settled chiefly lineages of the area, and they are usually also the sites for the most important rituals like rain ceremonies. There are also specific places and trees connected to *mhondoro* spirits, or chiefly ancestors, which can take the form of a lion acting as messenger (Mandondo 1997).

In Chôa the principal sacred forest is Chinda, which is the main ritual site of the area, as well as the burial place of an important early chief. There are other sacred forests with local value like Mungwa, where the rain ceremonies of the Nhacapanga ward are held, and where one of its apical ancestors is buried (Mussanhane et al. 2000: 100–1). Besides such localised burial sites used for ceremonies, and graveyards for commoners, there are other types of traditionally protected sites like forests around springs and streams, and some caves where the spirits of some past warriors are believed to reside. Especially in the lowland area at the northern part of Chôa a number of individual trees (typically *Ficus* spp., *Afzelia quanzensis* or *Khaya nyasica*), which are traditionally protected ceremonial sites were also identified. Most of the sacred forests indicated were situated on ridges or hills (Barauro, 18 March 1999; Tewetewe, 29–30 April 1999).

The rules governing access to and use of resources from sacred sites vary considerably, but usually those that are linked to founding spirits and which are of regional significance are protected more strictly than individual or village level sites. In general the rules that apply in Chôa are relatively tolerant: cutting of live trees or branches, setting fires and opening fields are forbidden. On the other hand collecting dead wood, fruits and mushrooms, and even grazing cattle are allowed except in the most sacred core area where the ceremonies are held. There are also various behaviour related controls regarding indecent language, sexual behaviour, and colours that can be worn. The sanctions are mainly of a supernatural type, as the ancestral spirits are believed to maintain constant vigil over these sites (Chinda, 14 November 1999).

With respect to Chinda, local elders related various stories about natural calamities that had befallen guerrilla groups which had broken the customary norms upon entering the forest during the recent civil war. The misfortunes included extensive rain and fog, which made the perpetuators lose their sense of direction, as well as encounters with spirit-beasts in the form of *mhondoro* lions and huge pythons. According to customary law those caught are also subject to different kinds of human punishment, ranging from reparatory ceremonies to ejection of the culprit from the community. However, supervision is rather lax as the sites are not visited regularly, and nobody has been punished for such offences during the post-war period. The force of this kind of idealised rules grounded on religious beliefs is difficult to evaluate, as their enforcement relies mostly on internalisation and voluntary observation (Chôa-Sede, 3 November 1998; Chinda, 14 November 1999; cf. Perelman 1984: 45).

Control over ritual sites is not a politically neutral issue. In African societies religion and politics are not separate domains, as various case histories testify. Schoffeleers (1979) has recorded the intense struggles for the control over religious shrines between autochtonous populations and the conquerors in pre-colonial and colonial Malawi, while Mukamuri (1995: 72–97) has documented the political manipulation of sacredness by the ruling lineages to gain and maintain control over immigrant groups in post-colonial Zimbabwe. Similar struggles have also been reported from Mozambique, where the prolonged political transition has revived dormant disputes over customary power (Alexander 1997: 9–10; West and Kloeck-Jenson 1999: 476–9).

In the study area people of the *shato* (python) clan,[4] to which the present chief Macufa also belongs, is considered as the apical lineage of longest known residence in the area. Consequently the lineage is also in control of the main ritual forest at Chinda, and it holds overall religious power. But in pre-colonial times the lineage did not command political power, which belonged to the Hatziro lineage of *gwai* (sheep) clan. According to the Hatziro family the two roles were united only in the 1940s when the ruling chief Sahatziro came into conflict with the colonial authorities, and the head of the Macufa family managed to get himself nominated as the new chief by the Portuguese. But even though their ritual pre-eminence in the highland area clearly helped the Macufa family to grab political power too, their spiritual power was subsequently contested in part of their new political territory. Presently Mr. Sabadza, a local rain maker of the *shava* (eland) clan claims that the Macufas' spiritual territory ends at the escarpment on the Mozambican side of the Caeredzi valley, while the valley and its ritual sites belong to his area (Nhaterere, 12 May 1999; Barauro, 21 May 1999; Inyazonia, 18 November 1999; Nyamaropa, 12 December 1999; Hadabi, 13 December 1999).

Along with such disputes over legitimacy, there are signs that respect for customary norms is weakening with regard to sacred forests, which are threatened by uncontrolled bush-fires and illegal cutting of trees (Chôa Sede, 18 November 1998; Tewetewe, 19 April 1999; Barauro, 21 May 1999). One factor in the apparent decline of customary institutions is the disruptive effect of the civil war, which dispersed the local communities and subjected them to various new influences. However, in some ways the period of indiscriminate violence also reinforced the status of traditional religion, and ceremonial sites like the Chinda forest in Chôa have been used as venues for officially sanctioned healing rituals to appease the spirits of those wronged during the war (Chôa-Sede, 3 November 1998).

Pekka Virtanen

SACRED FORESTS AND BIOLOGICAL DIVERSITY

The National strategy and action plan for the conservation of biological diversity in Mozambique, which was prepared in 1997, seeks to 'establish and manage a representative system of areas for the protection of habitats and maintenance of viable populations species in natural surroundings' (Hatton and Munguambe 1997: 40). This is a rather demanding task in a large (784,755 km²) and under-developed country like Mozambique. The situation is further aggravated by the wide variety of ecosystems, including patches of Afro-montane habitats, different types of woodlands/forests, edaphic grasslands, and a variety of wetlands, coastland, and marine habitats. While about seven per cent of the territory is formally under conservation areas (national parks and wildlife reserves), in reality most of them lack effective protection (Hatton and Munguambe 1997: 2–4).

The problems are acknowledged in the national biodiversity strategy, which emphasises the need to involve local communities and other stakeholders in the management of protected areas, and to develop new measures for the protection of natural habitats outside the state parks and reserves. Special emphasis should be placed on identification of sensitive ecosystems. Recommended actions include the promotion of community-based management schemes, and integration of traditional knowledge and management practices into scientific research programmes and conservation initiatives (Hatton and Munguambe 1997: 40–3).

This project carried out a forest inventory in the Chôa highlands. It covered the sacred forests of Chinda and Mungwa, and two non-protected forests/woodlands for comparative purposes.[5] The objective of the inventory was to assess the conservation value of traditionally protected forests in the area. Both of the protected forests are relatively large: the size of Chinda is about eight hectares, and Mungwa is approximately nine hectares. The two non-protected forests, Dani and Njere, are about 20 and 25 hectares respectively. Protected Chinda, which is the only evergreen forest in the sample, is located at an altitude of 1,600 metres, while the three deciduous woodlands are situated lower, at 1,300 to 1,400 metres. As the few remaining evergreen forests in the area are all traditionally protected, it was not possible to find a comparable non-protected site for Chinda (Mussanhane et al. 2000: 101–3).

The results show that the two non-protected woodlands have higher frequencies in the lower diameter classes, a higher regeneration rate, and a lower basal area than the protected areas. The forest structure is also simpler in the former, where it consists of only one stratum with an average height of five to nine metres. In protected Mungwa the average height varies between 8 and 18 metres, and consists of two strata in places. In protected Chinda the whole forest is in two strata, and trees in the dominant stratum attain a height of 18 metres. The species composition is also different: in unprotected Njere 85 per cent of the trees consist of one species (*Uapaca kirkiana*), and in unprotected Dani two species (*U. kirkiana* and *Brachystegia spiciformis*) make up 95 per cent of the tree population.

While *U. kirkiana* (74%) is also the dominant tree in protected Mungwa, there are no other clearly dominating species. The proportion of mature individuals of *U. kirkiana* is also much higher in Mungwa (21%) than in Njere (3%) or Dani (6%). In Chinda there is no clearly dominant species, even though *Garcinia* spp. is common in the regeneration stratum (Mussanhane et al. 2000: 104–7).

The relatively higher tree diversity of the sacred forests is also indicated by the two biodiversity indices, Berger & Parker and Shannon & Wiener,[6] which were used in the study. In Chinda and Mungwa the Shannon & Wiener index indicated higher biodiversity, than in Njere and Dani. The values of the Berger & Parker index confirm these results, even though the value for Mungwa was closer to those of Njere and Dani than those of Chinda. The results are similar to those obtained by the Zimbabwe team for the sacred forest of Dzete Mountain and the surrounding woodlands, which they studied (Table 1).

Table 1. Tree species diversity in selected traditionally protected and non-protected forests and woodlands in Mozambique and Zimbabwe (according to Mussanhane et al. 2000; Tyynelä and Mudavanhu 2000; Tyynelä 2001).

Locality	no. of plots	Berger & Parker index	Shannon & Wiener index
Chinda (protected)	4	0.21	2.22
Mungwa (protected)	4	0.74	1.45
Njere (non-protected)	8	0.84	0.90
Dani (non-protected)	8	0.86	0.65
Dzete (protected)	8	0.20–0.36	1.63–2.40
Mukarakate (non-prot.)	34	0.50	1.36

An important aspect with regard to conservation of biological diversity is the species composition. While the number of tree species is not considerably higher in the protected forests, there are various species which are rare or absent from the surrounding non-protected areas. For example, of the 23 tree species identified in protected Mungwa 11 were also found in the non-protected woodlands, while almost all of the species encountered in Chinda were absent from the other areas (Mussanhane et al. 2000: 104). Results from two case studies in Tanzania and Zimbabwe also indicated that certain tree species such as African black ebony, *Dalbergia melanoxylon*, are found only in the protected areas (Lukumbuzya 2000:166–7; Tyynelä and Mudavanhu 2000: 61–2).

The findings of a Tanzanian study about plant diversity supports the above finding that a large proportion of the species found in traditionally protected forests are not found in the surrounding non-protected areas (Mwihomeke et al. 2000: 181–9). This is linked to the fact that in many areas practically all the

Pekka Virtanen

remaining patches of evergreen forest are traditionally protected. The species that are typical to these forests require shade, permanent water and protection from fire. In favourable conditions trees develop a closed canopy, which maintains a humid habitat inside the forest, and makes it less susceptible to fire. The number of species is also elevated, but there is no obviously dominant species. On the other hand the surrounding woodlands are typically dominated by one or two fire-tolerant species like *U. kirkiana*. Thus, even though the protected forests might not represent a previously dominant climax type, they do provide an important fire refuge, which makes it possible for a number of typical evergreen forest species to survive (Mussanhane et al. 2000; cf. Mwihomeke et al. 2000: 191).

But what is the conservation value of the forests' status as sacred according to local beliefs in Chôa? We should recall that other than socio-cultural factors also have an impact on the rate of exploitation and damage by fire. These include the low population density of the area, as well as the long distance and difficult accessibility from the nearest human settlement. In Chinda the area is protected from fire by perennial rivers and the forest's humid micro-climate. Soils and topography are also important factors, which condition both fire sensitivity and initial establishment of a vegetation type. All these factors are inter-linked, but it is unlikely that geographical factors alone would have been sufficient to preserve the sacred forests of Chinda and Mungwa in their present form. The sacredness of the forests has contributed substantially to the preservation of their relatively high biological diversity.

CONCLUSION

There are no previous studies which measure the biological diversity of tradition-ally protected forests in Mozambique, or assess their ecological representativity. Even though the scope of this study was very limited and our knowledge of the study area's ecological history is deficient, the results show that the sacred forests studied are valuable for conservation purposes. They serve as important fire refuges for plants and animals with low fire tolerance, and can create a net-work of 'green islands' serving as stepping stones for patchy metapopulations of endangered species (see Hanski and Ovaskainen 2000). Even though the sites that are traditionally protected are rather small, and their distribution tends to be biased towards certain landscape categories like hills and streamsides, they remain important especially in the Mozambican context, where conservation of government-controlled parks and reserves is poorly enforced. Without them the biological diversity of many areas would be much poorer, and various endemic species would already have disappeared.

In addition to assessing the ecological representativity of the traditionally protected forests, we must also consider their institutional basis and its persis-tence. Various researchers have emphasised empowerment of those endogenous institutions, which are presently *de facto* managing local resources. Replacing or duplicating them with new structures has seldom worked, even though the new institutions might appear more democratic by Western standards (Fairhead and Leach 1995). With respect to sacred forests the responsible endogenous structures often combine both religious and political authorities, which control the performance of ceremonies and access to sacred sites together. But their present legitimacy varies considerably from place to place. In Mozambique they still wield considerable autonomous power in sparsely populated areas like Chôa, but they have often been undermined by state and party structures in other, more populous areas (see e.g. West and Kloeck-Jenson 1999). As noted by Falconer (1999: 370), the diversity of traditionally protected forests and especially their socio-cultural contexts makes it difficult to rely on unitary state law for their use in nature conservation. In Mozambique the concept of 'cultural-historical protected zones' of the new Forest Law represents a move towards pluralist legislation, which supports the use of local customary institutions for the conservation of nature.

However, if local institutions are promoted from above merely as a tool to preserve biodiversity for its own sake, separating the institutions from their socio-cultural bases, they will soon lose their legitimacy. Sacredness is a powerful means of conservation only when it is linked to a broadly respected belief system, with adequate normative controls and means for their enforcement. The problem with norms, which rely on internalisation of local socio-cultural values and voluntary observation, is that they depend upon consensus on what is proper and reasonable (Perelman 1984: 53). When faced with an influx of outsiders with different belief systems, or penetration of new social doctrines, the situation can remain under control only if the endogenous local institutions are given appropriate legal authority to enforce the local norms. At present granting such authority to customary chiefs is problematic even in Chôa, where a long dormant dispute over the identity of legitimate office-bearers has re-emerged in the new political context. Similar situations have been reported from other areas in Mozambique (Alexander 1997; West and Kloeck-Jenson 1999).

Instead of concentrating on one privileged institution like chieftainship, we should perhaps look into the whole constellation of local management practices. Sustainable management of natural resources is essentially a process, not a static condition. Rather than placing one local institution above the others for administrative convenience, the emphasis should be on creating enabling legislation and socio-economic conditions to support efficient local management. The problem is how to combine respect for different community-based rule-systems with equal rights of individual citizens, who might not share the local cultural and religious traditions. Historical development in the West has led to a secular state based on religious and ideological pluralism, but with relatively high socio-cultural homogeneity and a uniform legal system (Perelman 1984: 54). In Mozambique the heterogeneity of the socio-cultural and ethnic composition of the population has made smooth development of a unitary state difficult, and the recent

Pekka Virtanen

transition to multi-party democracy has brought up the latent tensions, which the centralised single-party rule had attempted to suppress. During the process such local institutions as sacred forests with high symbolic value have also become entangled in national political struggles, which can impair their local legitimacy. In this context any external interventions should be considered carefully.

NOTES

[1] The classification is based on ten studies which contain some information about traditionally protected forests, namely Castro 1995 for Kenya; Gerdén and Mtallo 1990, Kajembe 1994, Mwihomeke et al. 1998, and Ylhäisi 2000 for Tanzania; Matowanyika 1991 and Mandondo 1997 for Zimbabwe; Falconer 1999 for Ghana; and Laird 1999 and Swan 1989 for more general overviews.

[2] Reference to my own fieldwork data is made by indicating the place and date when it was collected; the fieldnotes are in my personal archives. The fieldwork for this study was funded by the Academy of Finland through the project 'Popular participation in the management of local natural resources: the role of endogenous institutions in Tanzania, Zimbabwe and Mozambique'.

[3] In each forest/woodland studied sample plots were placed at a distance of 50 and 100 metres from the area's geographical centre to the north, east, south and west (eight plots in the non-protected forests and four plots in the smaller protected forests). Within these plots trees were sampled according to diameter classes. For trees with a diameter at breast height (dbh) larger than five centimetres a sample area of 1000 m^2 was used. Trees in that area were identified and counted, and their height as well as dbh were measured. For trees with a dbh less than five centimetres or height under 1.3 metres (regeneration) the number of individuals and stubs were counted within a reduced area of 100 m^2 (Mussanhane et al. 2000: 96). The species nomenclature follows Coates-Palgrave (1996).

[4] The Shona kinship system is patrilineal, and each person inherits a traditional clan name (*mutupo*) from his/her father. These exogamous clans are further divided into sub-clans (*zvidao*) which can have their own totems. The clan name is often the name of an animal, and members are not allowed to eat the flesh of that animal, or at least there is a token taboo on some part of it. For example the Hatziro family which is prominent in Chôa is of *kamba* (tortoise) *zvidao*, and of *gwai* (sheep) *mutupo*. The spirit of the common ancestor of the whole group becomes a prominent spirit guardian (*mudzimu*), and is frequently honoured with ceremonial beer. (Bourdillon 1987: 23–8.)

[5] In this study a standard Zimbabwean classification system was used to identify classes of woody cover. It consists of the following five classes: natural forest (canopy cover > 80% of the ground surface, and tree height > 15m); woodland (canopy cover 20–80%, and tree height 5–15m); bushland (canopy cover 20–80%, and height 1–5m); wooded grassland (canopy cover 2–20%, and height 1–5m); and grassland. The study sites were selected from the first two classes. (Mussanhane et al. 2000: 96.)

[6] The index of Berger and Parker (D') was calculated from: $D' = N_{max}/N$, in which N_{max} stands for the number of trees of of the most common tree species in the sample, and N for the total number of trees in the sample. The index of Shannon and Wiener (H') was defined as:

$$H' = \sum_{i=1}^{k} \ln(p_i)p_i,$$

where p_i is the proportion of species i in the sample (N_i/N), and k is the total number of tree species in the sample. Low values of D' and high values of H' indicate a high species diversity.

REFERENCES

Alexander, J. 1997. 'The local state in post-war Mozambique: political practice and ideas about authority', *Africa* **67**: 1-26.

Bourdillon, M. 1987. *The Shona peoples*. Gweru: Mambo Press.

Castro, A.P. 1995. *Facing Kirinyaga: a social history of forest commons in southern Mount Kenya*. London: Intermediate Technology Publications Ltd.

Coates-Palgrave, K. 1996. *Trees of Southern Africa*. Cape Town: Struik Publishers.

Fairhead, J. and M. Leach 1995. 'False forest history, complicit social analysis: rethinking some West African environmental narratives', *World Development* **23**: 1023-35.

Falconer, J. 1999. 'Non-timber forest products in Ghana: traditional and cultural forest values', in D.A. Posey (ed) *Cultural and spiritual values of biodiversity*, pp. 366-70. London: Intermediate Technology Publications Ltd.

Gerdén, C.Å. and S. Mtallo 1990. *Traditional forest reserves in Babati District, Tanzania. A study in human ecology*. Working paper 128. Uppsala: Swedish University of Agricultural Sciences.

Hanski, I. and O. Ovaskainen 2000. 'The metapopulation capacity of a fragmented landscape', *Nature* (Lond.), **404**: 755-8.

Hatton, J. and F. Munguambe (eds) 1997. *First national report on the conservation of biological diversity in Mozambique*. Maputo: Impacto Ltd.

Kajembe, G.C. 1994. *Indigenous management systems as a basis for community forestry in Tanzania: a case study of Dodoma urban and Lushoto Districts*. Tropical Resource Mangement Papers 6. Wageningen: Wageningen Agricultural University.

Laird, S.A. 1999. 'Forests, culture and conservation', in D.A. Posey (ed) *Cultural and spiritual values of biodiversity*, pp. 366-70. London: Intermediate Technology Publications Ltd.

Lei no. 10/99 de 7 de Julho de 1999. Lei dos recursos florestais e faunísticos. *Boletim da República*, I série, no. 27.

Lukumbuzya, K. 2000. 'The effects of different management regimes on the diversity and regeneration of woody vegetation under a joint forest management model in Tabora, Western Tanzania', in P. Virtanen and M. Nummelin (eds) *Forests, chiefs and peasants in Africa: local management of natural resources in Tanzania, Zimbabwe and Mozambique*. Silva Carelica 34, pp. 159-77. Joensuu: University of Joensuu.

Margules, C.R. and R.L. Pressey 2000. 'Systematic conservation planning', *Nature* (Lond.) **405**: 243-253.

Matowanyika, J.Z. 1991. *Indigenous resource management and sustainability in rural Zimbabwe: an exploration of practices and concepts in commonlands*. A Ph.D. thesis in Geography. Waterloo: University of Waterloo (mimeo).

Mandondo, A. 1997. 'Trees and spaces as emotion and norm laden components of local ecosystems in Nyamaropa communal land, Nyanga District, Zimbabwe', *Agriculture and Human Values* **14**: 353-372.

100

Pekka Virtanen

Mukamuri, B. 1995. *Making sense of social forestry: a political and contextual study of forest practices in South Central Zimbabwe*. Acta Universitatis Tamperensis 438. Tampere: University of Tampere.

Müller, F. 2000. 'Does the Convention of Biodiversity safeguard biological diversity?', *Environmental Values* 9: 55-80.

Mussanhane, J., L. Nhamuco and P. Virtanen 2000. 'A traditionally protected forest as a conservation area: a case study from Mozambique', in P. Virtanen and M. Nummelin (eds) *Forests, chiefs and peasants in Africa: local management of natural resources in Tanzania, Zimbabwe and Mozambique*. Silva Carelica 34, pp. 89-115. Joensuu: University of Joensuu.

Mwihomeke, S.T., C. Mabula and M. Nummelin 2000. 'Plant species richness in the traditionally protected forests of the Zigua, Handeni District', in P. Virtanen and M. Nummelin (eds) *Forests, chiefs and peasants in Africa: local management of natural resources in Tanzania, Zimbabwe and Mozambique*. Silva Carelica 34, pp. 178-193. Joensuu: University of Joensuu.

Mwihomeke, S.T., T.H. Msangi, C.K. Mabula, J. Ylhäisi and K.C.H. Mndeme 1998. 'Traditionally protected forests and nature conservation in the North Pare mountains and Handeni District, Tanzania', *Journal of East African Natural History* 87: 279-90.

Nummelin, M. and P. Virtanen 2000. 'Local forest management by traditional and introduced means in Southern Africa - a synthesis and recommendations', in P. Virtanen and M. Nummelin (eds) *Forests, chiefs and peasants in Africa: local management of natural resources in Tanzania, Zimbabwe and Mozambique*. Silva Carelica 34, pp. 220-9. Joensuu: University of Joensuu.

Perelman, C. 1984. *Le raisonnable e le déraisonnable en droit: au-delà du positivisme juridique*. Paris: Librairie Générale de Droit et de Jurisprudence.

Posey, D.A. 1999. 'Introduction: culture and nature - the inextricable link', in D.A. Posey (ed) *Cultural and spiritual values of biodiversity*, pp. 3-16. London: Intermediate Technology Publications Ltd.

Resolução sobre questões sociais e culturais. In: Documentos da Primeira reunião nacional das aldeias comunais, no dia 27 de Março de 1980, na aldeia comunal "3 de Fevereiro" (mimeo).

Schaaf, T. 1999. 'Environmental conservation based on sacred sites', in D.A. Posey (ed) *Cultural and spiritual values of biodiversity*, pp. 341-2. London: Intermediate Technology Publications Ltd.

Schoffeleers, J.M. 1979. 'The Chisumphi and Mbona cults in Malawi: a comparative history', in J.M. Schoffeleers (ed) *Guardians of the land. Essays on Central African territorial cults,* pp.147-86. Gwelo: Mambo Press.

Swan, J. 1989. 'Sacred sites: cultural values and management issues', in *International perspectives on cultural parks*. Proceedings of the First World Conference, Mesa Verde National Park, Colorado, 1984, pp. 383-9. U.S. National Park Service: USA.

Tyynelä, T. 2001. Species diversity in Eucalyptus camaldulensis woodlots and miombo woodland in Northeastern Zimbabwe. Unpublished manuscript.

Tyynelä, T. and H.T. Mudavanhu 2000. 'Management and species diversity of a sacred forest in a deforested area: the case of Dzete Mountain, North-East Zimbabwe', in P. Virtanen and M. Nummelin (eds) *Forests, chiefs and peasants in Africa: local manage-*

ment of natural resources in Tanzania, Zimbabwe and Mozambique. Silva Carelica 34, pp. 55-64. Joensuu: University of Joensuu.

West, H.G. and S. Kloeck-Jenson 1999. 'Betwixt and between: 'traditional authority' and democratic decentralisation in post-war Mozambique', *African Affairs* **98**: 455-484.

Ylhäisi, J. 2000.'The significance of traditional forests and rituals in Tanzania: a case study of Zigua, Gweno and Nyamwezi ethnic groups', in P. Virtanen and M. Nummelin (eds) Forests, chiefs and peasants in Africa: local management of natural resources in Tanzania, Zimbabwe and Mozambique. *Silva Carelica* 34, pp. 194-219. Joensuu: University of Joensuu.

Local Environmental Conservation Strategies:
Karanga Religion, Politics and Environmental Control

B.B. Mukamuri

INTRODUCTION

This article concerns environmental religion and the local institutions of environmental control. It is based on research in semi-arid Zvishavane, Chivi and neighbouring districts, which are located in the heartland of the people who now refer to themselves as 'Karanga'. Information was collected through interviews, by attending rituals, visiting places of religious significance and observing and participating in rural development and conservation.[2]

The notion of 'conservation' and an economically managed environment are not the main emphases of Karanga rituals. This does not mean that environmental religion does not have major ecological effects; just that the practices do not generate a 'conservation strategy' in the scientific sense. 'Conserved' resources are often used by the ruling lineage at the expense of the wider community of outsider lineages (*vatogwa*). More frequently, resource control by ruling lineages is a tool to legitimise their leadership. People's practices clearly modified their ecosystem and even enhanced the value of their land to some extent. However, the politicisation of resource management institutions meant that resources were not always being conserved optimally for the benefit of society as a whole. However, a participatory approach to research and development can transform local institutions for the better.

Rural communities do not have a single belief system and common set of perceptions and behaviour. Though individuals share an environment, they compete for natural resources: they have well defined personal interests as well as group concerns. People are in different situations, with different opportunities for enhancing their own status and access to resources. Chiefs, lineage elders, the young and old, rich and poor, men and women all try to manipulate the supposedly shared belief system to their own advantage. Individuals' arguments about what should be done based on 'traditional' values, were not the same as what they appeared to believe and practice themselves. Chiefs, for example, have shaped the whole concept of 'conservation' to their own advantage: rather than acting as the humble guardians of society's common interests, they have sought personal gain through preferential access to rural resources. This is not to suggest that rural society is completely dominated by selfish political and economic motives, but rather that indigenous conservation institutions should be regarded as centres of inherent conflict.

Environment and History **1** (1995): 297–311

CONTROL OF RAINFALL AND WETLAND ACCESS

The study area is within natural regions officially categorised 'four' and 'five'. It is a semi-arid zone, with rainfall of about 400 to 600mm per annum, and frequent droughts are experienced. During bad droughts many farmers run short of food, which in the past was purchased from good farmers (*hurudza*). For such farmers droughts were a blessing in disguise: in areas such as Bungowa and central Chivi, where there are extensive areas of productive wetlands (*makuvi*), the good farmers have benefitted much in past droughts from this trading. Within the region there are two woodland types. On sandveld, often associated with patches of granitic hills, there is the diverse miombo type. Heavy soil plains support mopane woodlands, with areas of *Combretum* spp. and *Acacia* spp.

The provision of rainfall is a central issue for environmental religion – a fact that is hardly surprising in such a dry area with very variable rainfall. Controlling rainfall greatly enhances the status of ruling lineages, who achieve control through access to land spirits who are their own ancestors, or by attempting to communicate with autochthonous spirits. As dynasties have recently expanded in southern Zimbabwe,[3] immigrant lineages have virtually suppressed autochthons over only a few centuries. This process seems to have been particularly rapid during the colonial period. What seem to have been autochthonous Rozvi-related institutions, such as *mhondoro* (lion spirits), have declined almost to insignificance.[4] Maybe this is because the lions were killed by white settlers.

The nineteenth century saw people confined to patches of defensible hills, and factionalism was limited by the need for military alliances. The colonial period subsequently encouraged and enabled a spreading out of people. Spatial expansion was coupled with an increasing tendency for chiefly lineage to divide into separate 'houses', and with the growth of 'wards' as more important political institutions. Division was caused by the combination of collateral succession, population increase, and demilitarisation. These emerging 'traditional' rulers have been very concerned to enhance their religious standing. This could be because they were losing legitimacy under colonial rule where the administration installed them and gave them new legal powers so as to use them as a tool of white authority.[5]

It is important at this stage to introduce the concept of Zame. Zame is the Karanga rain-god, or the so-called *mwari we kumabwa* (god of the rocks/hills). However, field studies showed that Zame is not the only god who provides the Karanga with rain. Rainfall comes from four directions, and each points to the providing god. The others are Fupajena of the Duma, Musikavanhu, and Muchembere of the eastern region (who provides the drizzle [*guti*] in February-March). Furthermore these gods can fight, resulting in droughts.

My interest in Zame was to find out whether he offered any form of control over natural resources. But in fact it seems he rarely even mentions them at his oracle in the Matopos Hills. He is more involved in settling political disputes,[6]

B.B. Mukamuri

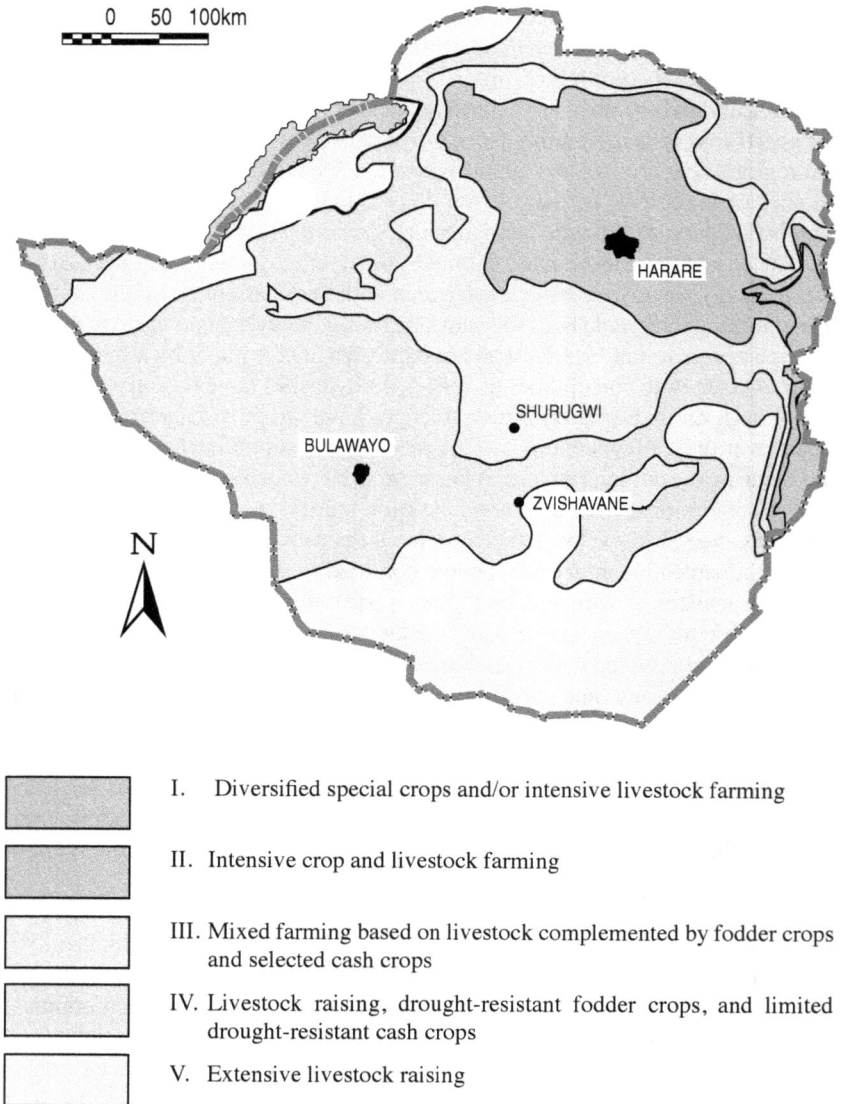

I.   Diversified special crops and/or intensive livestock farming

II. Intensive crop and livestock farming

III. Mixed farming based on livestock complemented by fodder crops and selected cash crops

IV. Livestock raising, drought-resistant fodder crops, and limited drought-resistant cash crops

V. Extensive livestock raising

FIGURE 1. 'Natural farming regions' of Zimbabwe (after Nelson 1983) and particularly in solving 'domestic' problems. He does not actively protect trees in particular, except fruit trees in a general way. It can be difficult to know whether a statement is from Zame or the government. For example, during the 1982-4 drought Zame was reported to have urged people to plant drought-resistant crops, whilst government extension agents were campaigning for the planting of the same crops and government advice was also being broadcast over the radio. Probably one could say Zame is also moving with the times. But grain varieties may not reflect Zame's concerns at all. The attribution of this advice to Zame may reflect peasant consciousness and peasants' long history of resistance to foreign interventions. Their acceptance of this new crop variety required some legitimacy from Zame.

To most Karanga, Zame is just like any of the great land spirits. He is the senior *jukwa* spirit, the *jukwa guru*. He works within a specified territory with boundaries, which cannot be an attribute of a high god. At these boundaries of his area he competes with other land spirits. Some people hold the extreme notion that Zame is just an ancestral spirit which is used by a clan for economic benefits, of which tribute money for rainfall (*rusengwe*) is the most important. In this context an ex-chief maintained:

> Zame and company are now broke, what you can see are just old scrap cars ... and I only went there when I was chief because I feared people would kill me if I refused to go and ask for rain ...

Legitimacy for many ritual, political and resource management decisions is attributed to Zame. For example, apical lineage ancestors have their own rainmaking festivals known as *mitoro* or *mikwerere*. These rituals are held so that people can ask for rainfall from Zame through their lineage ancestors. People mostly claim that *mitoro* rituals are done under instruction from Zame. *Mitoro* tend to be held at the beginning of the rainy season, though if rains come early they can be forgotten altogether, and during droughts they may be done more than once. The ritual is held under big trees such as *muchakata* (*Parinari curatellifolia*). During the festival the spirits that possess people are said to be descending from these *muchakata* trees. To cut such a tree would be regarded as a crime, or even as sabotage, by the owning clan. Even normal ecological 'die-back' of some of these trees was attributed to witchcraft by malicious clan-rivals. In one case in which a tree used in rain-making ceremonies died, Zame had to decree that there was nothing wrong with the site.

The *mutoro* beer is brewed by women of post-menopausal age, and in theory, the grain to be used is soaked in natural water-filled depressions in granite outcrops (*makawa*). Zame is supposed to send rain to fill these depressions and soak the grain within a few days of it having been put there. Thereafter, the brewed beer is supposed to be carried to the ritual by old women and young girls. A goat or sheep is killed, usually at the top of a hill near the ancestor's grave, and what is not consumed is thrown into the fire. The grave is then swept, and a pot of beer left at the grave-side so that the *mbada* (lion/leopard) can come and consume it. The truth is that the beer is drunk by a member of an outsider-lineage (*mutogwa*), who carries the pot secretly to the ruling elder. The latter then claims to have visited the grave early in the morning and generally reports seeing the footprints of the lion or leopard.

The role of the members of subject-lineages is only to contribute the *rusengwe* tribute to Zame, and beer and grain for the festival. During the ritual they have to shout the praise-names (*zvidawo*) of the ruling lineage and of Zame. Where *mitoro* are organised by vassal lineages themselves, however, paramount lineages are not so subordinate. One *mutoro* holder in Chivi grumbled that 'they only come so as to drink the large amount of beer at the ritual'. Where the ceremony is actually being held at the grave of someone important who is not in fact an ancestor of the ruling lineage this information is not freely disclosed.

Only after detailed research did it become clear that most of the *mitoro* being held in this region were actually established during the colonial period, as created rather than maintained traditions. They are mostly dedicated to people who died in the last century. The spirits chosen may be recent ancestors of the elders who are trying to promote the *mutoro*, or they may be the spirits of *myusa* (*mwari*-cult messengers), which appeal to a different legitimacy. Immigrant peoples, both within areas of ancient settlement and also in newly opened areas (claimed by more distant chiefs), have started their own *mitoro* to enhance their autonomy. Chiefly lineages accuse them of insubordination. The immigrant elders who establish them aim to strengthen their standing and prestige. The *mitoro* in any given area can often be seen as competing: one *mutoro* has often captured the ancestral spirit of its more ancient neighbour.

In 1946 one population group was evicted from part of the neighbouring district (Chivi-Central), including two family-heads. Before eviction, they had observed one *mutoro* dedicated to their father Musvuvugwa. When the District Commissioner changed in 1959-60 they were allowed to return: the former mutoro-owner 'A' returned to find that 'B' had already taken over the *mutoro* and moved the site. The previous messenger (actually a female mbonga) and the *svikiro* (medium) for Musvuvugwa had both died. 'B' captured the *mutoro* by taking his own *nyusa* (messenger) and *svikiro* from the same house. Moving the site enabled him to get it under the name of his own faction, but it is only one of the *mitoro* presently being directed to Musvuvugwa.

Commissioned *mitoro* are commonly held in areas where immigrants have become well-established. They are organised by fairly autonomous immigrants, but are held at the grave of, and directed to, a member of the ruling lineage. Sometimes supplication is done by a member of the ruling lineage. There is likely to be serious dispute in cases where immigrants have set up their own *mitoro* to their own ancestral settler (usually quite recently deceased), without the permission of the ruling lineage.

Similarly, some emergent headmen have established an independent link for their own *nyusa* (messenger) with the *mwari* cult centre. Important *nyusa* lineages have been the subject of competition; headmen who cannot get appropriate people have been forced to appoint elders from within the chiefly lineage faction on an ad hoc basis.

Mitoro are idealised as strengthening social homogeneity and communal values, but in fact they tend to be focal points for disputes. Attendance at *mitoro* is effectively a barometer of the strength of headmen in different areas. Other members of the ruling lineage and immigrant (*vatogwa*) people react differently to this call for unity in attendance. Ruling lineage members tend to fall into three categories according to the strategy they adopt. Those in distant 'houses' can only get the *mutoro* if they can capture the political leadership of the ward as a whole. They tend to therefore distance themselves, weakening the impressiveness of the *mutoro*. But they cannot withdraw completely as they need the *mutoro* to remain a legitimate institution. A second strategy is used to subordinate members of the faction in power. They strengthen the *mutoro*, perceiving long term benefits. Thirdly some leading people are close enough to power that they try actively to capture the *mutoro* during the performance. Outsider lineages can resist attendance and the paying of *rusengwe* (tribute). During mitoro they can break regulations to demonstrate their insubordination.

People in different positions in regard to any particular *mutoro* would make different belief-claims about the functioning of the rituals. Organising elders sometimes admitted privately that they did not believe in the rituals causing rain. Numerous *nyusa* messengers have been convicted of embezzling funds donated to *mwari*. Taboos presented at length to the public, such as on sexual abstinence, may be privately flouted. The nature of the taboos has evolved recently: for example, the soap called *jerimani* is now being treated as an ancient and hence ritually pure substance.

CONSERVATION AS ECONOMIC PRIVILEGE FOR THE RULING LINEAGES

Environmental religion provides ruling lineages with economic privileges, as chiefs struggle to legitimise their control of land. *Badza* payments for opening fields were often paid by immigrants. Though this was with the support of the colonial regime, based on some inventing of history, the chiefs remained anxious to make it seem purposeful for the preservation of the ecological order.

Ruling lineages maintained special rights to certain game, such as the pangolin (*hhambakubvu*), and to portions of all big game eaten (e.g. a forelimb, *bandavuko*). Whilst this did secure and bolster their political status, it also had clear economic benefits when hunting was frequent. The fields of the ruling lineages used to be cultivated by tributary labour (*zunde*). The chiefs could justify this as essential for maintaining the productivity of the land, for example, to ensure rainfall.

Since independence the rise of the party and the village development committee (Vidco) structure means that chiefly lineages have suffered. Chiefly lineage leaders have argued that the people belong to the government, but the land belonged to the ancestors, and hence should be managed by themselves. The bad droughts 1982-4 and 1986-7 have certainly bolstered the chiefs' claims, and the government has now decided to reinstate some of their authority. But land allocation remains outside their jurisdiction. During the late 1987-8 rainy season the dispute over whether pangolins should be given to President Mugabe or local chiefs was on the television and in the national press almost daily.

Examples can also be found where Zame and chiefs did not manage environmental rules for their own benefit. An example is the *mavhenekera nyika* fine which was imposed on anyone who caused (unwanted) veld fires. The chief was supposed to be the one who set alight the first hunting season fire. Chiefs also resisted development that increased rural differentiation on religious lines. Burnt bricks and the use of cement were not allowed for building homes in Mazvihwa until quite recently. They were described as dead, and *kuvuraya nyika* (killing the country). Abandoned homes, (it was said) would be devoid of grass: the community as a whole would lose a piece of territory.

SACRED WET PLACES: PERCEPTIONS AND CONSERVATION

Perceptions of conservation and sacredness of wetland are a fascinating subject.[7] Sacredness (*kuyera/kuzira*) can be translated as abstinence, and the sacred wetland can only be approached carefully and observing avoidance taboos. It is clear that at the base of any rural society the environment is not inert. There are forces embedded within the environment of the living and the dead. This may sound illogical but it is not possible to use an iron tin to fetch water from a sacred pool. A gourd is required, and this creates a special relationship.

Research on this topic has been carried out in the districts already referred to. However, Mwenezi, which was settled by people from many areas after the promulgation of the Land Apportionment Act also adhered to the same beliefs and rituals. Referring to southern Zimbabwe, water is immediately brought to mind. Water keeps Zame alive, and there is a lot of evidence that in a drought Zame gains particular prominence. Places with water are often revered and protected. Sacredness becomes more common and strongly appreciated when the site is natural.

For the Karanga, wetlands are guarded by live 'natural' animals or rather beings whose survival in a place means the continued availability of water. It is these that *kuchengeta* (keep/look after) a place, and make it sacred. These pools represent the world of the *vari pasi* (those below), and this world reflects our world. Homes are established and voices can be heard calling cattle for milking while dogs bark, indicating that there is a whole world active beneath the pool. Sounds are often followed by the filling of the pool. To Karanga these are not just geophysical adjustments but also the work of *njuzu* (water sprites), which are one of the types of pool guardians. *Njuzu* have been described as half-human, and often pull wrong-doers into the pool. Those dragged in are most commonly lighter brown in complexion and those who wash with soap or use metal utensils to draw water.

The people taken by *njuzu* often find themselves below the deep pools being fed on mud and worms. Those who refuse are often killed, and those released become *n'anga* (herbalists) of great repute using the powerful *jukwa* spirits. That this happens is subject to doubt as there seem to be no recent occurrences. The frequent deaths at these pools (usually of children) are said by some rural sceptics to be due to python-attacks, as indicated by blood coming out of their noses.

In some cases snakes act as guardians: some are pythons, but there may also be a strange variety called *mvuvamacheche*. Catfish (*mhatye*) can also be guardians of sacred wetplaces. As long as the guardians remain present then the continued existence of the pool is guaranteed.

Similar sacred pools can be found on top of granite hills and these may be called *ninga*, a term which also refers to the *njuzu* (water sprite).[8] Again it is the presence of the *njuzu* that controls the amount of water. These places can be called *hozhobwe*. One I visited in 1987 was dry, and the headman explained that the water in it had been called by the water in the sky and would return when it wanted. The use of soap at the site may have explained the lack of water, but the headman observed that *jerimani* soap (the earliest soap brand introduced) is preferred by the *njuzu*. Water from this pool is used for the soaking of grain to brew the beer for rituals such as the *mitoro*. The water is said to be clean in its own domain: it has not been tainted by contact with the human domain.

These wetlands cannot be taken out of context because they are found at the centre of societies in dry regions. Water from these places is used by *n'anga* to initiate possession and the water is said to have healing powers. In some cases avenging spirits (*ngozi*) have been cooled using the water. It is not surprising that enterprising people have created such ponds, like one made by an immigrant healer I interviewed. Catfish were put in to be the guardians, but were fished out by a chiefly lineage member, a teacher, due to his *kudherera* (lack of respect). The pool dried up and the dispute reached the point where the teacher had to be transferred.

Two other examples can be given of the importance of wetland areas. In Chivi central a borehole that had been constructed near a sacred site had to be removed as it was suspected of making the pool dry up. This place slowly developed into something of a cult centre during the recent droughts. Even the independent churches (zionists and apostolics) came to collect 'holy water' and carried out baptism at the pond. This ended up with one of the zionists being

B.B. Mukamuri

'swallowed' by the pond only to emerge naked through the *mhino* ('nose': two small pools further down). Both the church groups and traditional lineage elders want to take control. Using cement to up-grade wells in this area is another matter that has led to disputes.

The processes by which a pool becomes sacred are interesting. Some are even made by people and belong to them for which they get economic benefits (see above). Most are communally owned and provide water to the rural folk near to them. Sacredness of wetlands and ponds can be proclaimed by individuals for historical spirits, for example a traditional healer with a healing spirit can say that a pond belongs to his spirit. Control of an important sacred site enhances that particular lineage faction's authority. This is clear in Chivi where there is a VaNgowa enclave in the Mhari chiefdom, which controls a famous *chidziva* (pond). There are also agricultural benefits attached to these wet places, because farming near them makes for good yields, especially in drought years. This increases social standing and brings power. This is why access to wetlands is so unequal and why they are mainly held by leaders of ruling lineages.

SACRED MOUNTAINS

Sacred mountains and hills are also places revered in Karanga religion. These are mostly connected with burial sites of the Rozvi people and also of clan founders. Most of them also have wet places which are also sacred. In Mazvihwa each hill is associated with one ruling section of the ruling lineage. It is in these hills that clan rituals are held, for example the mitoro. These hills serve as symbols for the clans they are associated with.

Mount Bupwa (Buchwa) is one of the most famous sacred mountains in south central Karangaland. There has been a lot of debate about this mountain.[9] It is rich in iron ore, and mining of the site could have started as early as the thirteenth century. The people who stayed around the mountain are the so-called VaMizha (craftsmen) who made iron hoes, axes and spears for the Mwenemutapa or Changamire. They were subdued by the Mgowa tribe. The mountain is believed to have graves of some important Rozvi. But of late the Ngowa have wanted them to be called Ngowa grave-sites.

Bupwa mountain is presently being mined by ZISCO (Zimbabwe Iron and Steel Company), and this has been a thorn in the flesh of the local ruling lineage. Many stories have been told about how the mountain refused to be mined for a long time. In these accounts a lot of people disappeared and are said to have been taken away by the mountain spirits. These sprits are generally described as *mapa* (given by Zame).

Another feature of Mount Bupwa is that the people a distance away think that the *kuyera* or sacred things on the mountain are Rozvi, because the Rozvi were really the owners of the mountain. The VaNgowa who stay nearby claim that the spirits there belong to Ngowa of the praise-name 'musaigwa'. Generally Bupwa is referred to by them as *Gomo raMataruse* (Ngowa chief Mataruse's mountain).

Truly people believe that Bupwa is *zame's* mountain. The mining of the mountain is seen as killing the country by causing droughts. Spirits withhold rain due to their anger at being disturbed. Miners are said to experience a lot of problems with ore extraction. Low quality ore can be found after massive and expensive blasts, and this is said to cause most problems. In order to get the good ore, it is said they have to call an Ngowa person to come and propitiate and supplicate the Ngowa spirits. This has been denied by the mine management (but this could be for professional reasons), who say these are simply Ngowa claims. The truth is that the freedom fighters killed Mataruse Dzingai, the Ngowa chief in the 1970s, on the accusation that he had 'sold the mountain to the capitalists', by supplicating his Ngowa ancestors to make the way for the mining. He is also said to have been given a lot of money after the ceremony. According to oral interviews this kind of supplication, called *kufupira* (to make the place quiet), has not stopped. It is said that the mine management often secretly consult ruling lineage elders whenever 'their things are not moving properly'.

Although the mining has been going on for a long time at Bupwa, and some parts of the mountain have been completely destroyed, the people around the mountain believe that the place is still sacred. The parts of the mountain described as the most sacred have changed from time to time as the mine advances. Originally the whole mountain was sacred but these days only the southern end is said to be the sacred part. I was shown a mound near the mine pit, which the informant claimed was left because the dynamite failed to blow it up. But the mine management denies that its sacredness protected it. They claim that the mound is of low quality iron ore. Accidents on the mine are explained by locals and workers as the works of ancestral spirits who are against the mining of the mountain.

Karanga do not think that they are controlling the mining process, but through being consulted by the mine management the chiefs feel that their position is recognised and hence their power enhanced. The money they get for the supplications should be seen as marginal to the value of sacredness in maintaining their political hegemony.

The mountain should burn at the start of a good rainy season. Zame is supposed to set it alight, but rumour has it that some of the chiefs secretly start fires so that they can legitimise their power and hence maintain recognition. If the mountain does not burn, the chief's rule is questioned: Zame is saying that he should not be chief and therefore the rains will be poor. Of late the mountain has burnt but the rains have been poor, and this reinforces the questioning of local cynics as to whether it is being illicitly burnt. Continued droughts maintain mountain burning as an issue.

B.B. Mukamuri

It can be concluded that the 'conservation' of sacred mountains is in fact similar to that of wetland conservation, and the sacred woodlands described below. The issue is actually one of 'resource control' rather than 'management'. Sacred places act as supporting pillars to the political hegemony of ruling lineages, in the sense that they function as symbols of identification and legitimisation.

TRADITIONAL PERCEPTIONS AND MANAGEMENT OF INDIGENOUS WOODLAND

Karanga societies have set management strategies for indigenous trees. The bulk of these strategies were never recorded but were carefully passed on from generation to generation. Detailed research had to be undertaken to identify what people were doing and what they knew. Though local management of woodland management gives an encouraging picture of what can be achieved by indigenous resource management, some aspects of tree conservation are also tools of political authority.

Sacred woodlands (called *rambotemwa*: refuse to cut) contribute with other sacred places to the spiritual and hence political hegemony of the ruling elite. These were left for Zame and linked to autochthonous Rozvi spirits. Each one was controlled by a chief, who is to ensure that it is not cut by outsiders or locals. There was some hunting allowed, and women could collect dead wood for firewood.

The *rambotemwa* also served as a burial ground for some of the important people of the clan. Oral sources pointed to the Rozvi in particular. In Mapanzure, for instance, the *rambotemwa* is found at the base of a hill with graves, acting as a buffer zone to this religious site. Lion spirits are also a characteristic feature of *rambotemwa*, and these are said to chase people flouting the taboos. Those with evil intentions can also encounter leopards (*mbada*), of the variety symbolic of the ancestors. People could get lost after breaking a taboo, and would only be found after propitiation of the angry spirits. Growling could indicate impending rains or dissatisfaction with rulers.

In Mazvihwa there appears to have been only one *rambotemwa*, representing all the lineage sections descending from Mazvihwa. A small rural business centre has been opened within it (called Rambotemwa Township) and the son of the late chief has opened a field in one section. Live trees are also being cut within it. These changes followed a visit by the late chief to Matonjeni to request that Zame hear the plight of Mazvihwa's children and allow them to encroach into the holy grove. This suggests that political and religious factors are more behind these woodlands than economic motivations.

It seems that *rambotemwa* legitimise authority. Each branch of the ruling lineage can have one which becomes important when they capture the chieftainship. A leader without one is prone to ridicule if he has to propitiate at a competing section's grove. There may have been cases where *rambotemwa* were started quite recently by such a local leader anxious for authority. There are competing claims about the one in Mototi, Mazvihwa. Like *mitoro* ritual sites described above, *rambotemwa* have risen and fallen with the clans associated with them, and ruling groups have tried to create or enhance the status of the sites they can control.

Many *rambotemwa* were destroyed by the government's imposed land-use plans in the period 1920-1960. Chiefs lost control of the land. Some were also destroyed in areas turned over to settler or mission farms. During the 1970s when the Smith regime lost control of the rural areas, some were destroyed by young people opening new land in the period known as 'madiro' (freedom, or doing what one will). The recent wave of droughts have made people less happy about the destruction of the *rambotemwa*. Traditional leaders are reviving the punishment for those who cut.

Big trees are associated with rainfall. This link is interpreted both religiously and ecologically. Trees are said to catch clouds and initiate rainfall. But such trees are also said to harbour *hwaya*, cuckoos that migrate into the area, whose calls are said by western informants to bring on rainfall. Deforestation and drought have heightened concern to preserve these trees. The current change in wind direction has been associated with less trees and is said by many to be the cause of droughts. It is also said that there has been a general increase in wind, which carries away the clouds and the rain. As this is not a tradition it can be felt by everyone and not just chiefs.

People are not allowed to cut certain valuable fruit trees, for example *mishuku*, *misumba*, *mitobwe*, *mitamba*, *michakata*, *miwonde*.[10] According to tradition, Zame, as well as local ruling lineages promote these cutting taboos. Cutting such a tree could involve a fine of a goat. The justification is the protection of trees to provide people with fruit, etc. This is especially important to small children and in drought years. Stories are told of travellers surviving on miwonde (fig) fruits. These stories are told to children so that they can know the value of trees. Everybody benefits from such rules, and they can also be seen to improve the area in the long term. Many people in south-central Zimbabwe have left trees in fields despite the call by agricultural demonstrators to fell them. Today this seems non-religious, because the farmers explain the benefits of the trees in economic terms: providing shade, fruit and improving soil fertility. But resistance could also be articulated in the 'traditional' idiom of environmental religion, in which the chiefs and the farmers could feel relatively united.

Attempts to conserve trees around homes and in grazing areas were strategies that were not seen as chiefly impositions. They benefitted people as a whole. This is linked to the many perceived benefits from trees: that they increase rains, improve the soils, provide wood and browse for livestock. Because conserving trees was contrary to government policy people were able to mobilise environmental religion to better husband their natural resources.

B.B. Mukamuri

Generally people perceive that though Zame and specific ancestors do not own individual fruit trees (though they do own forests), they do in some way provide them for people. For example drought years tend to be times of heavy fruiting, which helps survival.

DEVELOPING ENVIRONMENTAL RELIGION FOR CONSERVATION

The above discussion illustrates the different interests and institutions involved in Karanga environmental conservation strategies. There are limitations in the factionalistic ways these currently operate, as with chiefly institutions, which have monopolistic and exploitative conservation strategies. Control over resources is not all inequitable to the same degree, however. For example, trees have not been monopolised by one group of people at the expense of others, at least in the Zimbabwean case. Tree conservation does not have a history of elitist imposition and this means people are more open to engage in it.

The rural people under discussion are not bourgeois, but their idiom of conservation is. The patch of woodland which becomes a *rambotemwa*, the ordinary grave which becomes a *mutoro* site and the small hole in the rock which becomes an *njuzu* pool can all be regarded as commodities – and everybody wants one. The owner/controller can accumulate a little prestige by association, and (perhaps) a little cash or produce, which may enable him to rise slightly above the near subsistence level his cash cropping allows.

Local Karanga environmental religious institutions do not have the economic motive of benefitting the environment for the community as a whole. Yet both individuals and institutions are engaged in resource management and have a great deal of knowledge about their environment. Since groups and individuals have divergent interests, development projects should not aim at homogeneity, but at benefitting as wide a range of people as possible. By facilitating meetings in which conflicting views are brought to the fore and challenged, development organisations can enable local communities to draw up management plans, and local development committees can make decisions that benefit the community as a whole. The most important role for development agencies in this context is to change attitudes and motives to be purely conservationist in a scientific sense. Institutions have to be redirected away from monopolistic and exploitative conservation strategies, but not alienated. The question remains, can these indigenous institutions provide a useful basis for natural resource development projects?

Local Environmental Conservation Strategies

NOTES

[1] Schoffeleers 1979, pp.2-3.

[2] This part of Zimbabwe is my home area. I started research in 1985, assisting Ken Wilson. From 1986-7, I did the research for my by B.A. Thesis (Mukamuri 1987). I later worked for ENDA-Zimbabwe, implementing a community based woodland management project which grew out of research in Mazvihwa undertaken by a team including myself, K. Wilson, M. Chakavanda, O. Chikamba, B. Higgs, Z. Phiri, I. Scoones and others. Although hundreds of people assisted in this research, particularly helpful interviews were given by: Old Dewa, VaTangwena, acting chief Mazvihwa, Old Dzviti, England Dzviti, VaKunjani, Va Chibidi, VaMagaya, OldBwoni, Old Bunga (late), Old Mohobele, ex-chief Msizibi, Old Masinire, the late Maruvure, the late Madyzkuseni, VaMabomba, Mr and Mrs Saul Jim, VaSpikita, VaSangatowa, VaJokonya, VaChibagwe, VaHoto, Councillor Bwoni, chief Madyangove, VaPhiri, VaTsaurayi, VaChikombeka, Mrs. Choshamba, VaShoko, VaMechanika (VaMudhomori), VaShilongoma. I would like to thank: UZ for loaning a tape recorder; K. Wilson and I. Scoones for loaning their motor bikes; J. Madyakuseni for recording some interviews; my BA supervisors D. Moyo and Dr. Mackay; D. Gumbo and K. Wilson for helping prepare this article; government officials and local authorities for allowing field research.

[3] Beach 1980.

[4] These *mhondoro* should not be interpreted as being identical to *mhondoro* amongst other 'Shona' groups, as reviewed by Bourdillon 1982.

[5] Lan 1985; Ranger 1985.

[6] Daneel (1970) describes a political transaction. The cult is also analysed by Werbner (1977) and Ranger (1986).

[7] More detailed descriptions of wetland sacredness are given by Wilson (1986, 1988).

[8] *Ninga* can also refer to a pothole in a rock, or to the actual point of a spring.

[9] A fuller account of the sacredness of Bupwa Mountains is given in Mukamuri 1987.

[10] Wilson (1989) discusses the ecological and religious issues connected with leaving trees in fields iin detail.

REFERENCES

Beach, D.N. 1980. *The Shona and Zimbabwe, 900-1850.* Gwero, Zimbabwe.

Bourdillon, M.F.C. 1982. *The Shona Peoples,* 2nd ed.

Daneel, M.L. 1970. *The God of the Matopos: An Essay on the Mwari Cult in Rhodesia.* The Hague: Mounton.

Lan, D. 1985. *Guns and Rain: Guerrillas and Spirit Mediums in Zimbabwe.* London: James Currey.

Mukamuri, B.B. 1987. Karanga religion and environmental protection. BA Thesis, University of Zimbabwe.

Nelson, Harold D. (ed.) (1983) *Zimbabwe: A Country Study.* Washington, D.C.: U.S. Government Printing Office.

Ranger, T.O. 1985. *Peasant Consciousness and Guerrilla War in Zimbabwe.*

B.B. Mukamuri

Ranger, T.O. 1986. Religious studies and political economy: the *mwari* cult and the peasant experience in Southern Rhodesia, in W.M.J. van Binsbergen and J.M. Schoffeleers (eds) *Theoretical Explorations in African Religion,* pp.287-321. London: Kegan Paul.

Schoffeleers, J.M. 1979. *Guardians of the Land: Essays on Central African Territorial Cults.* Gwelo.

Werbner, R.P. 1977. Continuity and policy in southern Africa's High God cult, in R.P. Werbner *Regional Cults.* A.S.A. Monograph.

Wilson, K.B. 1986. Aspects of the history of vlei cultivation in southern Zimbabwe. Paper presented at the University of Zimbabwe workshop 'The Use of Dambos in Zimbabwe's Communal Lands', August.

Wilson, K.B. 1988. Indigenous perceptions of wetland sacredness and conservation. Unpublished manuscript.

Wilson, K.B. 1989. Trees in southern Zimbabwe. *Journal of Southern African Studies* **15**(2).

From Myths to Rules:
The Evolution of Local Management in the Amazonian Floodplain

Fabio de Castro

INTRODUCTION

Local management systems[1] (LMS) are locally crafted institutions, based on prescriptions that define how a given resource or ecosystem should be used. Such systems rely upon information obtained from local and repetitive experiences and are passed down to younger generations through learning and imitation processes.[2] Usually, LMS are described as community-based, enduring, and conservation-prone institutions.[3] Yet, broader social and ecological processes directly influence how users shape their local management by formulating and adjusting rules-in-use.[4] Similarly, LMS are not necessarily long-lived institutions. New local management initiatives may arise whenever incentives for collective organisation are at hand.[5] Finally, outcomes of LMS may vary according to ecological features of the resource, social features of the users, the set of rules-in-use[6] and the motivations driving the collective action.[7]

Given the complex environment in which LMS may operate, changes in their structure and organisation over time are to be expected. Prescriptions are replaced and redefined according to how incentives for and goals of resource management modify through time. As a result, LMS are better described as mosaics of prescriptions created throughout the history of the users according to a range of incentives and goals. The mixed and dynamic nature of these institutions provides a theoretical scenario for testing hypotheses concerning human responses to socioenvironmental changes. The emergence of institutions is a costly process, and thus more likely to occur under specific circumstances related to the resource features and social attributes of the users.[8] In general, unless radical changes in the social structure transpire, institutional change tends to manifest itself as a modification of the old structure.[9] Thus, the history of LMS can unveil the factors leading to institutional adjustments and, eventually to institutional emergence.

By sorting out the major management prescriptions during the history of a given region, one can relate the socioenvironmental context to agency with regard to the pattern of resource use. Management prescriptions vary according to the source of influence. Stocks describes three main categories of prescription.[10] Ecological prescriptions are individual or group foraging behaviours

Environment and History **8** (2002): 197–216.

Fabio de Castro

characterised by short-term economic efficiency, directly affected by attributes of the resources such as spatial distribution and abundance, predictability, and mobility. Cultural prescriptions are common-sense practices defined from social interactions that may indirectly enhance or maintain resource productivity. Such practices include things like rituals and taboos. Political prescriptions are developed from a 'conscious' problem-solving process carried out by a user group.

The co-existence of three types of prescriptions – ecological, cultural, and political – shaped during different historical periods raises theoretical questions regarding levels of compatibility and conflict between the conservation goal of LMS and the other factors that shape these local institutions. Ecological prescriptions result from constant interaction between users and the resource, and they change according to the biophysical features of the system. Conversely, cultural and political prescriptions are influenced by the social interactions among users, and they change according to the social attributes of the users. While the conservation outcomes of the ecological and cultural prescriptions can represent an epiphenomenon of other processes, the primary conservationist goal of political prescriptions is usually quite obvious.

The Amazonian floodplain is an ideal place to explore these questions. The Amazon has been site of major social transformations due to factors such as cultural change,[11] urbanisation,[12] technological innovation,[13] governmental policies,[14] economic pressure,[15] and local organisation.[16] In the floodplain, where resources were intensively used during the Pre-Conquest period,[17] a new set of local prescriptions has recently emerged to regulate fishing activity – the fishing accords.[18] The fishing accords represent a conscious effort of floodplain residents to limit lake access to themselves and, thus, maintain resource productivity and ensure local control over the fishing system. Although the political prescriptions stand out in the current LMS, they have been combined with other ecological and cultural prescriptions developed in the past.[19] In this paper, I analyse the history of the LMS in the Amazonian floodplain, highlighting the connections between the fishing accords and other prescriptions developed in the past, in order to explore the institutional adjustments through time in addition to the motives behind the emergence of this new local institution.

A general historical description of ancient resource use is followed by a more detailed description of resource management in the contemporary period. The discussion is focused on the local and regional processes influencing prescription change, and how the structure of the LMS is related to conservation or other goals.

STUDY SITE AND METHODOLOGY

The Amazonian floodplain is a complex landscape with high spatial and temporal heterogeneities. Four major landscape forms – lake system, grasslands, natural levees and river channel – are influenced by annual variations in river

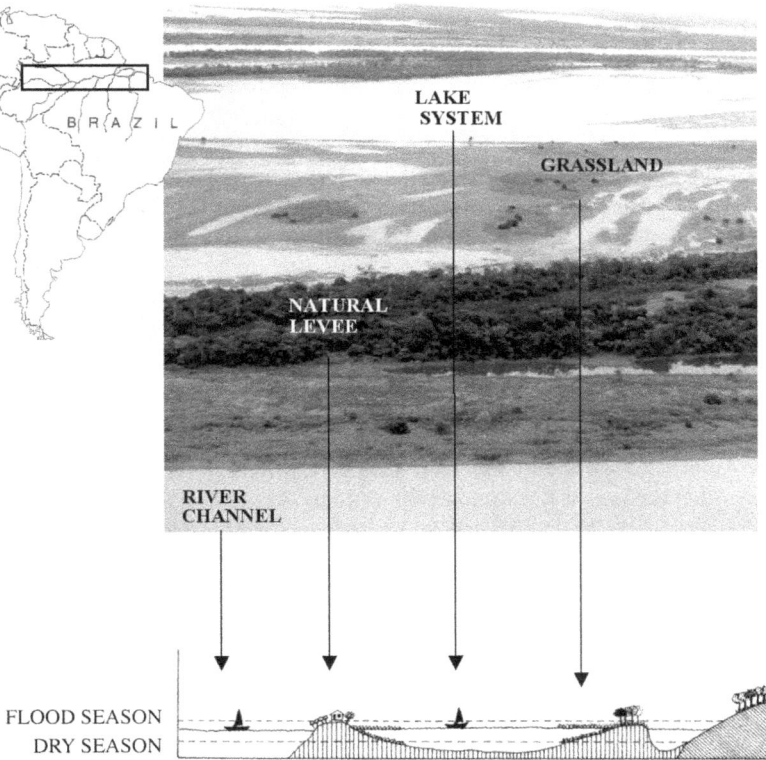

FIGURE 1. Aerial and cross-section views of the Amazonian floodplain subsystems.

level that can range from a few metres in the Lower Amazon to twenty metres in the Upper Solimões (Figure 1).[20] Despite the risks inherent in residing along the river, human populations have historically been attracted by its nutrient-rich soil, which sustains a rich biota and provides natural transportation pathways.

For the purpose of this article, human occupation on the floodplain is divided into three major historical periods: Amerindian, Migrant, and *Caboclo*. Each period is marked by a different pattern of technological endowment, resource values and social interactions. Consequently, prescriptions regulating resource use vary considerably across each period with respect to the target resources, user groups and major rulers. The description of the Amerindian Period was based on a review of the ecological anthropology literature. Information on the Migrant Period was drafted from the local and scientific literature as well as from a survey of fourteen newspapers published in the Lower Amazon region from 1911 through 1960. The *Caboclo* Period was based on a survey of six newspapers published in the Lower Amazon between 1960 and 1980, and on seventy-seven documents of fishing accords established between 1981 and 1996, found in the archives of the Fishers' Union and the governmental office IBAMA[21]. In-depth interviews with floodplain residents were carried out between 1991 and 1999 to contextualise the local perception of the fishing accords.

LOCAL MANAGEMENT SYSTEMS IN THE AMAZONIAN FLOOD-PLAIN

Amerindian Period

The Amerindian Period spans the Pre-Colombian period, when Indian populations controlled the use of the Amazonian system. According to Roosevelt, human occupation in the Amazon dates back to at least 12,000 years ago, and is divided into four stages according to the social organisation and pattern of resource use: 1) nomadic hunter-gatherers before 10,000 B.C.; 2) early transitional sedentary groups starting in 8,000 B.C.; 3) widespread ceramic sedentary society of early horticultural villagers subsisting on root crops, fish, and game, starting in 3,000 B.C.; and 4) highly dense agricultural chiefdoms supported by intensive seed cropping and supplemented by intensive fishing and hunting until 1,000 A.D. [22]

Sociocultural development on the floodplain was influenced by ecological opportunities, such as natural pathways, fertile soils and relatively abundant protein. Technological innovations, such as domestication of plant species and a shift from game to crops as major sources of protein, enabled local populations to overcome limitations to population growth and set the stage for sedentary behaviour by 3,000 B.C.[23] The seed-cropping agricultural societies that developed following this stage supported up to six million inhabitants.[24] As a result, socially complex chiefdoms, such as the Omagua in the Upper Solimões, the Tapajó in the Middle-Lower Amazon, and the Marajó in the estuary, existed in several parts of the Amazonian floodplain prior to the arrival of Europeans.[25]

Contrary to popular romantic visions of this time period as being marked by peaceful, homogeneous, conservationist societies, archaeological studies reveal conflicts within and between highly diverse groups who fought for territories and, sometimes, were able to exhaust local resources.[26] Similarly, the ecological environment was not stable throughout the Amerindian period of human occupation. Human groups experienced strong changes in climate, which affected their social organisation, including patterns of settlement and of resource use.[27] Despite the high population density, social complexity within and between groups, and the climate changes, ecological and social interactions were mostly at the local level and for subsistence and exchange purposes. In addition, the relatively slow pace of environmental change seems to have enabled the adaptive process of resource management with social development, demonstrated by the humanised landscape evidenced by historical ecologists.[28]

Prescriptions of LMS during the Amerindian period are hard to assess due to the limited nature of archaeological data. Roosevelt discusses some evidence of political prescriptions to regulate resource use as a response to declines in fish productivity.[29] Studies of contemporary Amerindians, however, suggest some ecological and cultural prescriptions that may have been inherited from their ancestors.[30] For example, the influence of ecological attributes of the resource/system, such as distribution and abundance of the fish resource, landscape diversity and level of predictability, is observed in fishing strategies among the Pumé,[31] Bari[32] and Cocamilla.[33] Cultural prescriptions are described in terms of social mechanisms related to two major strategies: 'altered ecosystems' and 'resource avoidance'. Altered ecosystems imply active human manipulation of the environment, such as the alteration by the Kayapo of the composition of plant species in fallows (*apetes*) on the upland to attract game.[34] In the floodplain, Stocks argues that the Cocamilla habit of dumping organic waste into lakes may help to increase fish productivity.[35] Resource avoidance implies prohibiting individuals from using a specific resource (food taboo) or an ecosystem (sanctuary). Food taboos of aquatic animals relate mainly to carnivores. Dolphins, for example, are considered an enchanted species[36] while catfish are considered *remoso*, an Indigenous term referring to certain food categories that have the effect of creating or aggravating health problems.[37] Dolphins and catfish are top predators, and their protection has ecological implications in regulating fish communities.[38] Sanctuary implies resource avoidance through beliefs or social norms shared among a user group. Chernela, for example, discusses the customary agriculture system by Uanano Tukano in a nutrient-poor floodplain system (*igapó*), which consists of protecting the vegetation on the river's edge.[39] The author argues that such a strategy enhances the productivity of this poor aquatic system, since fish productivity strongly depends upon terrestrial nutrient sources. The use of nutrient-rich floodplain systems (*várzea*) may also be restrained through fear of mythical creatures such as the *Tapiré-iauara,* a tapir nymph which patrols the flooded forest to keep fishers away.[40] Similarly, the giant water snake that lives in marshy areas scares fishers away throughout the Amazon floodplain.[41] These myths are usually related to highly productive floodplain lakes surrounded by dense flooded forest and macrophytes where fries and juvenile fish grow, much like the 'dying lakes' in Peruvian Amazon.[42] Although we cannot assume conservation functionality of the ecological and cultural prescriptions, we can suppose that under low external influences, low technological endowment and slow pace of change, these practices were likely to have enabled local maintenance of floodplain productivity as an epiphenomenon.

Fabio de Castro

Migrant Period

Ironically, the ecological opportunity of natural pathways which facilitated the floodplain occupation played against the Amerindians who were assaulted by European colonisers in the early seventeenth century. In a century, Europeans nearly annihilated the floodplain Amerindians through war, the spread of disease, and slavery.[43] Catholic missionaries protected Amerindian populations until the mid-eighteenth century, after which they were thrown out of the country. As a way to incorporate the native labour force into their production system, the Crown created incentives for intermarriage between Amerindians and Europeans. This policy resulted in the formation of a new ethnic group, the *Caboclos*,[44] who later re-occupied the floodplain.

The *Caboclos* differ from Amerindians in two major social aspects: their nuclear household organisation[45] and engagement in the regional market.[46] From their Amerindian ancestors, the *Caboclo* populations retained some ecological prescriptions based on traditional ecological knowledge and some cultural pre-scriptions based on their cosmology.[47] From the Europeans, they assimilated skills to deal with the broader socioeconomic realm. The floodplain reoccupation by the *Caboclo* populations during the nineteenth century was disturbed by three major external factors. In the early 1800s, a basin-wide political movement led by the *Caboclos* called *Cabanagem* resulted in the loss of thousands of lives on the floodplain.[48] In the mid-1800s, the land tenure policy of the Portuguese Crown granted migrants land titles to floodplain farms, and overrode the *Caboclo* ownership system.[49] In the late 1800s, the rubber boom drew a large number of people away from the floodplain to work on the upland rubber-tree groves.

Land conflict between migrant landholders and *Caboclos* was commonplace throughout the Migrant period, as observed in several articles in local news-papers published in the Lower Amazon. However, due to the small economic importance of the floodplain, landholders maintained their property rights but did not close access to *Caboclos* residents. Landholders were involved with cocoa plantations and extensive cattle ranching, while the *Caboclos* practised a mixture of subsistence activities including agriculture, fishing, extractivism, hunting and small livestock rearing (chickens and ducks). *Caboclo* populations interacted with others to reach the market; they sold wood for steamboats, and traded floodplain products for basic urban products such as salt, sugar, oil and soap with itinerant boat traders (*regatões*).[50] They also teamed up with landhold-ers to carry out commercial fishing of a few aquatic species, such as *pirarucu* (*Arapaima gigas*), turtle, caiman, and manatee.[51] In this partnership system (*feitoria*), landholders provided access to their private lakes and to infrastructure for catching and processing fish, while *Caboclos* provided the labour force.[52]

Old residents of communities in the Lower Amazon described ecological and cultural prescriptions of LMS during the Migrant period. They recalled fishing strategies carried out by their fathers and grandfathers according to productive fishing spots, appropriate fishing technologies for each fish group, and fishing seasons. Beliefs concerning 'dangerous' places and visions of 'giant snakes' are also present in their memories.[53] After 1930, however, the introduction of jute into the floodplain by Japanese migrants increased the importance of the floodplain soils, and consequently, the interest of landholders in their properties. Jute was rapidly assimilated in the floodplain economy due to the available labour force, market demand, low technological requirement, and ecological adaptation to the flooding cycle.[54]

The pattern of land tenure and lack of financial aid forced *Caboclo* populations to seek support from the landholders and *regatões* to engage in the jute market[55]. As a result, the *aviamento* system, which first emerged in the upland region during the Rubber Boom,[56] was established in the jute production system.[57] The *aviamento* system is a patron-client relationship in which the patron provides financial support for production and basic needs, and the client is committed to sell their harvest exclusively to their patrons for a low price. The *aviamento* system developed under other social ties between the two parties. Co-parenthood is an example of a social relation that transcended economic dependency. This religious-based system is grounded in a social commitment whereby patrons are blessed as co-parents of the clients' children.[58] In some regions, the patrons facilitated the engagement of *Caboclos* in cattle activity through an informal system of 'cattle partnership' like that in the Lower Amazon.[59] This system still exists today and consists of an agreement in which *Caboclos* take care of a herd for a landholder and receive half of the calves in return. In other words, rather than being dependent and hierarchical, the relationship between landholders and *Caboclos* has oscillated between conflict and co-operation. On one hand, the higher access to land and to political support by landholders created disputes between patrons and clients. On the other hand, land use strategies bound the two actors in an economic collaboration.

In contrast to the Amerindian Period, the Migrant Period was shorter and was strongly influenced by two external forces – the establishment of a private tenure system of floodplain management (which informally persisted after its statisation), and by an economic change (the jute boom). The maintenance of ecological and cultural prescriptions was subordinated to a higher social structure. The unbalanced power relationship and relatively stable local politics between the two emergent actors – *Caboclos* and landholders – led the LMS to be shaped mostly by political prescriptions based on the patronage system. The Migrant Period was cut short by a new set of external factors that developed in the region and enabled *Caboclos* to claim political control of the floodplain.

Caboclo Period

While soil was extensively used for jute cultivation and cattle ranching, lakes did not become commercially appealing until the 1960s, when another wave of social transformations took place in the floodplain. Fishing efficiency increased owing to technological innovations in fishing equipment (manufactured gillnets), transportation (oil-motor boats), and fish storage (ice factories and Styrofoam). In addition, a new wave of migration to the Amazon, propelled by road-building projects launched by the government in the upland region, led to rapid urbanisation along the rivers.[60] As a result, the combination of better fishing production combined with an increasing fish market filled the economic gap in the floodplain created by the decline of jute production.[61]

In contrast with other floodplain production systems, commercial fishing by *Caboclos* during this period was not based on any relationships of dependence with other actors such as motorboat owners, landholders or itinerant boat traders. Fishing systems have remained small-scale artisanal in most of the Basin today, based on wooden paddle canoes with one or two fishers.[62] Ecological and cultural prescriptions developed in the Amerindian period persisted during the *Caboclos* period. In many parts of the Basin, fishing systems were related to ecological prescriptions[63] and cultural prescriptions.[64] However, the major feature of the LMS during this period is the fishing accord, an explicit system of rules based on the consciously made decision to conserve. A list of fishing rules with the community signatures are prepared in community meetings and converted into a document. The analysis of the fishing rules listed in seventy-seven separate accord documents observed reveals that local ecological knowledge is used to define 'how to fish' (fishing spot, season, fishing technology) while social factors influenced the formulation of rules to define 'who is allowed to fish'.[65]

Fishing accords are sent to regional offices (Fisher's Union or IBAMA) and broadly publicised on the radio. The monitoring system is carried out by a local armed patrol which applies the sanctions, including sometimes physical confrontation, retaliation, and destruction of the offenders' fishing devices. The fishing accords have gradually evolved into a more complex structure according to the emergence of distinct actors, the level of empowerment of the *Caboclo* population, and the level of formalisation of the local decisions. This process can be divided into three stages: the local organisation stage, the integrated organisation stage, and the participatory organisation stage.

The *local organisation stage* comprises the period between the mid-1960s and the mid-1980s and emerged from conflicts between floodplain residents and motorboat fishers. These two fisher groups differed in terms of their level of attachment to the lake system and to their fishing-related technologies. While the lake system was part of the community fishers' livelihood, outside fishers rotate their activity across different lakes. Owing to their access to bank credits, motorboat fishers demonstrated higher transportation efficiency (motor boats), storage capacity (ice boxes), and catching capability (amount of gillnets). The technological externality generated by the motorboat fishers led to deadly conflicts between floodplain residents and outsiders in the Upper Solimões [66] and Lower Amazon.[67]

The role of the Catholic Church in fomenting local political organisation was fundamental in setting the stage for the response of community fishers to the intensification of fishing. During the Catholic outreach programmes, household units scattered throughout the basin were organised into community-based settlements in order to facilitate their work. Usually, a floodplain community encompasses a group of kin-related households settled in a contiguous area along a stream, with collective facilities such as a church, a common shelter, and a school.[68] In the 1970's, the catholic organisations FASE (Organisation for Social and Educational Assistance) and CPT (Pastoral Land Commission), promoted educational programmes, political training and information networks (radio station, bulletins and regional meetings), which favoured the development of local leaders. The church also introduced the concept of Lake Zonation by assigning particular functional categories for each lake, such as preservation or subsistence.[69] *Caboclo* residents also relied upon political support from landholders in the Lower Amazon, who intensified the cattle ranching after the end of the jute boom and did not compete over the lake resource. The landholders' interest in the fishing accord was primarily an attempt to keep outsiders away in order to prevent property poaching and cattle piracy. In sum, the local organisation stage was marked by conflicts between local residents and motorboat fishers, the emergence of a *Caboclo* leadership, and, in some areas, the support of landholders. Fishing accords emerged in different regions of the Basin, but remained isolated initiatives with rough formal structure and no regional organisation. Documents analysed from this stage rarely spelled out all the prescriptions shared by the group, and usually were defined among residents of a single community.

The *integrated organisation stage* started in the mid-1980s, when small-scale fishers wrested control of the Fisher's Unions from non-fishers as part of the nationwide democratisation process.[70] During this stage, a more heterogeneous group of fishers emerged in the commercial fishing arena, escalating the fishing conflict between community fishers and other fisher categories (e.g., urban fishers and fishers from neighbour communities). The Unions played a major role in bridging the gap between community fishers and the governmental offices in order to find solutions to these conflicts. Documents from this stage reveal structural and organisational improvements, such as the foundation of community associations upon which the accords were established, multiple-community accords, and the participation of representatives of the Fishers' Union in community decisions. In 1984, several Fisher's Unions in the region met in a workshop to discuss their problems, and they generated the *Carta de Óbidos* (Óbidos Statement) – an influential document in the government's decision to launch a participatory management strategy.[71] In sum, fishing accords developed into a regional issue, with improved formalisation as the documents became increasingly better structured. Yet, the aggravated conflicts resulting from the illegal status of this local institution called for a formal recognition of their management system.

In the early 1990s, the formal engagement of grassroots organisations, governmental offices, and NGOs in the fishing accords issue opened the third stage of fishing accords, the *participatory organisation stage*. The ultimate goal of this stage has been legal recognition of the fishing accord as part of a co-management enterprise. An example of this enterprise is the joint programme launched in 1992 by the government (state and federal), international donors and a local NGO to develop a participatory management plan for the Mamiraua Reserve. With the goal of combining conservation of a hot spot with local development of the rural population, three ecozones with distinct resource use restrictions – settlement zone, sustainable use zone, and preservation zone – were defined between researchers and residents according to ecological and socioeconomic criteria.[72] In 1994, the Brazilian and German governments funded the IARA Project in the Middle-Lower Amazon '…aiming at sustainable use [of fish resource], compatible with the interests and needs of the local population and the society as a whole, as well as the regional and national economy'.[73] Later, NGOs engaged in the process with research programmes supported by international development agencies, to support LMS through fomenting social organisation and developing local economic alternatives.[74] In the late 1990s, the government created a fishing committee represented by different stakeholders to create a collaborative management system of the Maicá Lake (Lower Amazon). The Provarzea Project from the Pilot Program G-7 is the most recent government-based initiative to support promising local initiatives, as well as to provide scientific and technical subsidies to implement co-management enterprises.[75] This stage has achieved fundamental structural improvements of fishing accords, which link them to a broader co-management system. On the one hand, the *Caboclo* populations lost their exclusive control over the floodplain system. On the other hand, they became visible in the decision-making process. The fishing accords are now evolving into formalised structures with legal grounds. Despite some problems regarding stakeholders' representation and the extension of power in decisions, the participatory approach for the management of the floodplain system represents a turning point from a top-down, Amazonia-wide, government-based management system to a bottom-up, local-based, participatory management system.

DISCUSSION

During 12,000 years of human occupation, the development of the LMS in the Amazonian floodplain has gone through major transformations, due to modifications in the structure of opportunities and constraints at local and external levels. Although local populations enjoyed access to a wide range of resources provided by this heterogeneous environment, the combination of prescriptions influenced by ecological, cultural and political factors have operated differently in each historical period according to the social features of the users and the ecological features of the target resources.

During the Amerindian Period, the LMS was mostly centred on subsistence goals. Increase in short-term energy reward and long-term food security were major motivations influencing local prescriptions. Ecological features of the resources directly influenced foraging behaviour, and individual failure to achieve optimal energy returns played as a sanction. Cultural prescriptions were based on socially constructed behaviour (taboos, myths, and social pressure), while sanctions were based on individual emotions such as fear, self-respect, and shame. Thus, ecological prescriptions defined mostly 'how to use the resource' while cultural prescriptions guided 'what should not be used'.[76] During the Amerindian Period, ecological and cultural prescriptions as well as sanctions to violators were internalised and self-controlled. While prescriptions were shaped from unrelated conservation-driven motivations, strong food reliance on natural resources, low technological endowment, local-based social and ecological interactions, and the relatively slow pace of environmental change were major factors influencing the conservation outcome of the ecological and cultural practices during the Amerindian Period.

In contrast to the Amerindian Period, the prescriptions that developed during the Migrant Period were heavily influenced by external institutions, such as changes in land property rights and the broader market system, which allowed landholders to dictate strategies of resource use based on market demands. Floodplain *Caboclo* populations enjoyed access to resources during this period, but had little control over their management system. As a result, the ecological and cultural prescriptions that developed during the Amerindian period were overridden by new political prescriptions shaped from the patronage system. Increased economic and political returns became the two major goals of the LMS in this period, while the low reliance on floodplain resources for subsistence by the landholders created little incentive to adjust resource use patterns to sustainable levels. Landholders sought to increase economic efficiency through changing the production system according to the market demand for products like cocoa, jute and, recently, cattle. The fast pace of socioenvironmental change, the fragile ecological conditions of the floodplain, and the broader social complexity between resource use and local users during the Migrant period inhibited the development of responses addressing ecological concerns. As a result, the floodplain landscape was heavily modified from land-use activities in just a few decades.[77]

Just after external and internal changes related to the social structure and economic patterns of resource use occurred, a new set of prescriptions emerged. Intensification of commercial fishing, development of community leadership, and external support for local decisions created new conditions for the floodplain *Caboclo* populations to claim their right to rule the LMS. Fishing, then, was

Fabio de Castro

carried out mostly by floodplain *Caboclo* populations, at sustainable levels due to the combination of low technological assets[78] and low market demand. The introduction of more efficient devices by motorboat fishers combined with the increased fishing market spurred other fisher groups to engage in the commercial fisheries. Threats to their stable fish food availability and to their exclusive access to the lake system motivated floodplain *Caboclos* to respond promptly with violent attacks on motorboat fishers, and, later, with the formulation of political prescriptions. The increased interest of individuals in joining the fishing accord, on one hand, and the improved ability of the group to organise on the other, enabled *Caboclo* populations to reclaim their control over management decisions regarding the floodplain lakes. While fishing accords have succeeded in protecting the lakes from motorboat fishers, in a few cases improved fishing productivity has been reported.[79] Thus, more than a conservationist-driven institution, fishing accords represent a new venue for expanding power to control the use of local resources.

The emergence of fishing accords is part of a broader social movement among Amazonian natives such as the Indigenous peoples,[80] rubber tappers[81] and black communities,[82] who have articulated their power to negotiate with the broader society for the legitimacy to conserve natural resources.[83] The increasing power to claim rights to nature has been a major factor motivating local populations to resist external pressures based on the conservationist discourse. The recognition of this political process driving the LMS during the *Caboclo* period is fundamental to assessing the potential of fishing accords for conservation purposes. Rather than being a conservation-oriented institution, the fishing accords are part of a historical process of change in the LMS to 'increase return', whether ecological (e.g., energy), cultural (e.g., respect), economic (e.g., money), or political (e.g., control). Whether or not strategies to increase efficiency will be consonant with resource conservation is a matter of how the LMS structure is related to the ecological system.

CONCLUSION

Local management systems are dynamic and complex institutions whose prescriptions are reshaped and created according to socioenvironmental influences at different scales. The outcome of this process is a multi-layered set of prescriptions, defined under distinct circumstances existing during different historical periods. The common emphasis on current structure of the LMS and on local social relations has limited the analysis of these local institutions. Paying more attention to the historical dimension of the LMS and the external factors influencing local decisions has unveiled sources of motivations embedded in the local institutional crafting process. The Amazonian floodplain case reveals that recent LMS were strongly influenced by past social experiences and by broader socioenvironmental factors. Five main issues have been raised in this case study as contributions to improving the analysis of LMS.

First, the LMS may evolve from many incentives. Sometimes, the combination of prescriptions are consonant with resource conservation and prove to support resource sustainability, as during the Amerindian Period; at other times, the prescriptions may lead to unsustainable ecological and social outcomes, as during the Migrant Period. Thus, rather than assuming positive or negative outcomes, the LMS must be analysed in terms of how the prescriptions are compatible with changing social and ecological systems.

Second, LMS retain both social capital and social costs accumulated throughout their history. Social capital, such as traditional knowledge, group ethics, social organisation, and administrative skills reveals the potential of LMS to support broader management strategies. Social costs, such as power relationships (e.g., patronage), local conflicts, and non-conservationist habits, may create barriers in the development of socially effective local institutions. Therefore, rather than emphasising social capital or social costs of LMS, a focus on the incentives to support the former and to abate the latter is fundamental in building co-management schemes.

Third, although LMS may encompass many different prescriptions, those actors enjoying local power dominated the decisions on what prescriptions should prevail (e.g., migrants during the Migrant Period and *Caboclos* during the *Caboclo* period). Therefore, instead of assuming good-for-all outcomes by returning power to the local populations, it is important to understand the local politics in the decision-making process.

Fourth, LMS become more coercive as the social system becomes more complex. The faster pace of change and higher diversity of user groups diminishes the level of commitment, trust, and mutual interest, and provides incentives for free-riding and rent-seeking behaviour. From more individual-oriented sanctions based on cultural and ecological prescriptions during the Amerindian Period, LMS evolved into more explicitly oriented sanctions based on economic and political prescriptions during the following periods. Thus, LMS can survive even in a very complex social environment if the local populations succeed in keeping control of the ecosystem.

Finally, recently established LMS, such as the fishing accords, are by no means isolated systems. They are closely related to former LMS, as the prescriptions capture ecological and social influences from the past. After all, fishing accords represent a social innovation to overcome an old problem – ensuring resource access. Thus, instead of assuming the conservationist discourse, any analysis of LMS should account for other hidden agendas.

In sum, LMS carry along motivations and goals developed throughout the history of the user groups. Often, prescriptions are not based on conservation ethics. The historical dimension of these local institutions is fundamental to unveiling the social context in which prescriptions emerge, resist, modify or vanish.

Fabio de Castro

NOTES

I would like to thank the Brazilian Agencies FNO, Conselho Nacional de Desenvolvimento Científico e Tecnológico (CNPq), and FAPESP (Fundação de Amparo à Pesquisa do Estado de São Paulo) for research grant and doctoral and post-doctoral scholarships, respectively. I also thank the Várzea Project (IPAM) for logistic support, the Boanerges Senna Library in Santarém for making accessible its historical collection of newspapers of the region, and the two anonymous referees. I am deeply thankful also to the residents of several communities I visited during this research. I am solely responsible for any errors of fact or interpretation that might remain in the final product.

[1] Local management takes several names, such as folk management, community-based management, communal management, traditional management. I chose the term 'local' because it includes the concept of management of a specific area by a local population, regardless of its cultural origin (folk), political boundary (community-based), or historical presence (traditional).

[2] Boyd and Richerson 1985, Berkes 1999.

[3] Berkes 1989, Dyer and McGoodwin 1994.

[4] Ostrom et al. 1994.

[5] Castro 2000.

[6] Ostrom 1990.

[7] Edwards and Steins 1999.

[8] Ostrom 1999.

[9] North 1990.

[10] Stocks 1987.

[11] Parker 1985.

[12] Browder and Godfrey 1997.

[13] Roosevelt 1980.

[14] Moran 1981, Schmink and Wood 1984.

[15] Weinstein 1983.

[16] Allegretti 1990.

[17] Roosevelt 1989.

[18] McGrath et al. 1993.

[19] Smith 1981, Begossi 1998, Castro 2000.

[20] Junk 1997.

[21] Brazilian Institute for Renewable Natural Resources and the Environment.

[22] Roosevelt 1989.

[23] Ibid.

[24] Denevan 1992. In areas of nutrient-poor soils, social reorganisation, including a complex regional exchange system were adopted in order to overcome environmental constraints (Moran 1991).

[25] Meggers 1971; Roosevelt 1980, 1989.

[26] Meggers 1971.

[27] Meggers 1995.

28 Balee 1994; Denevan 2001.
29 Roosevelt 1989.
30 Beckerman 1994.
31 Gragson 1993.
32 Beckerman 1983.
33 Stocks 1987.
34 Posey 1985.
35 Stocks 1987.
36 Goulding et al. 1996.
37 Smith 1981.
38 Jackson et al. 2001.
39 Chernela 1989.
40 Smith 1996.
41 Smith 1981.
42 Stocks 1987.
43 Denevan 1992.
44 The term *Caboclo* is used in this article to refer to non-Indian native populations, with no further social connotation.
45 Parker 1985.
46 Lima 1992.
47 Wagley 1953, Moran 1974.
48 Di Paolo 1990.
49 Benatti 1996.
50 Bates 1892, Ross 1978.
51 Verissimo 1895.
52 Furtado 1984.
53 Field notes.
54 Zimmerman 1987.
55 The floodplain became state property with the Constitution of 1934, but land tenure did not effectively change, owing to the lack of monitoring by government agencies.
56 Weinstein 1983.
57 Gentil 1988.
58 Wagley 1953.
59 Castro 2000.
60 Browder and Godfrey 1997.
61 Smith 1985, Chapman 1989, McGrath et al. 1993.
62 Petrere 1978, Smith 1981, Almeida et al. 2001.
63 Smith 1981, Goulding 1980, McGrath et al. 1998.
64 Smith 1981, Begossi 1998.
65 Castro 2000.
66 Goulding 1983.
67 Hartmann 1989.
68 Castro 2000.

Fabio de Castro

[69] CPT 1992.
[70] Leroy 1988, Breton et al. 1996.
[71] Furtado 1993.
[72] http://www.siamaz.ufpa.br/cgi-bin/folioisa.dll/Mami.nfo?
[73] IBAMA 1995.
[74] McGrath et al. 1999.
[75] http://www.worldbank.org/html/extdr/offrep/lac/ppg7/docs/participation_full.pdf
[76] As noted earlier, cultural prescriptions are not assumed to have emerged from functional process of adaptation. Yet, as an epiphenomenon, some prescriptions may have affected the ecological sustainability.
[77] Goulding et al. 1996.
[78] Low technological asset does not necessarily imply low impact (e.g., poison fishing). Yet, in this case it describes low-impact fishing devices (e.g., gig, harpoon, castnet).
[79] Castro 2000.
[80] Cocklin and Graham 1995.
[81] Allegreti 1990.
[82] Verán *apud* O'Dwyer 2002.
[83] Schmink and Wood 1992.

REFERENCES

Allegretti, M.H. 1990. Extractive reserves: An alternative for reconciling development and environmental conservation in Amazonia. In *Alternatives to Deforestation: Steps Toward Sustainable Use of the Amazon Rain Forest*, ed. A. B. Anderson, 252–64. New York: Columbia University Press.

Almeida, O.T., D.G. McGrath, and Ruffino, M.L. 2001. The commercial fisheries of the Lower Amazon: An economic analysis. *Fisheries Management and Ecology* 8(3): 253–70

Balee, W. 1994. *Footprints of the Forest: Ka'apor Ethnobotany – The Historical Ecology of Plant Utilization by an Amazonian People*. New York: Columbia University Press.

Bates, H.W. 1892. *The Naturalist on the River Amazon*. London: Murray.

Beckerman, S. 1983. Carpe diem: An optimal foraging approach to Bari fishing and hunting. In *Adaptive Responses of Native Amazonians*, ed. R. Hames and W. Vickers, 269–98. New York: New York Academic Press.

Beckerman, S. 1994. Hunting and fishing in Amazonia: Hold the answers, what are the questions? In: *Amazonian Indians: From Prehistory to the Present: Anthropological Perspectives*, ed. Anna Roosevelt, 177–200. Tucson: The University of Arizona Press.

Begossi, A. 1998. Resilience and neo-traditional populations: The *caiçaras* (Atlantic Forest) and *caboclos* (Amazon, Brazil). In *Linking Social and Ecological Systems: Management Practices and Social Mechanisms for Building Resilience*, ed. F. Berkes and C. Folke, 129–57. Cambridge, U.K.: Cambridge University Press.

Benatti, J.H. 1996. *Posse Agro-Ecológica: Um Estudo das Concepções Jurídicas sobre os Apossamentos de Camponeses Agro-Extrativistas na Amazônia*. Master's thesis, Federal University of Pará, Belém, Brazil.

Berkes, F. 1999. *Sacred Ecology: Traditional Ecological Knowledge and Resource Management*. Philadelphia: Taylor and Francis.

Berkes, F. (ed.) 1989. *Common Property Resources: Ecology and Community-Based Sustainable Development*. London: Belhaven Press.

Boyd, R. and Richerson, P. 1985. *Culture and the Evolutionary Process*. Chicago: University of Chicago Press.

Breton, Y., Benazera, C., Plante, S. and Cavanagh, J. 1996. Fisheries' management and the Colonias in Brazil: A case study of a top-down producers' organization. *Society and Natural Resource* 9: 307–15.

Browder, J.O., and Godfrey, B.J. 1997. *Rainforest Cities: Urbanization, Development, and Globalization of the Brazilian Amazon*. New York: Columbia University Press.

Castro, F. 2000. *Fishing Accords: The Political Ecology of Fishing Intensification in the Amazon*. PhD Dissertation. Indiana University, Bloomington, IN.

Chapman, M.D. 1989. The political ecology of fisheries depletion in Amazonia. *Environmental Conservation* 16: 331–7.

Chernela, J.M. 1989. Managing rivers of hunger: The Tukano of Brazil. *Advances in Economic Botany* 7: 238–48.

Cocklin, B.A. and Graham, L.R. 1995. The shifting middle ground: Amazonian Indians and eco-politics. *American Anthropologist* 97(4): 695–710

Comissão Pastoral da Terra (C.P.T.) 1992. *Os Ribeirinhos: Preservação dos Lagos, Defesa do Meio Ambiente e a Pesca Comercial*. Manaus, Brazil: CPT, Regional Amazonas e Roraima.

Denevan, W.M. 1992. The aboriginal population of Amazonia. In *The Native Population of the Americas in 1492*, 2nd edn, ed. W. Denevan, 205-234. Madison, Wisc.: University of Wisconsin Press.

Denevan, W.M. 2001. *Cultivated Landscape of Native Amazonia and the Andes*. Oxford: Oxford University Press.

DiPaolo, P. 1990. *Cabanagem: A Revolução Popular da Amazônia*. CEJUP, Belém.

Dyer, C.L., and McGoodwin, J.R. (eds) 1994. *Folk Management in the World's Fisheries: Lessons for Modern Fisheries Management*. Niwot, Colo.: University Press of Colorado.

Edwards, V.M. and Steins, N.A. 1999. A framework for analyzing contextual factors in common-pool resource research. *Journal of Environmental Policy and Planning* 1: 205–22.

Furtado, L.G. 1984. Pesca artesanal: Um delineamento de sua história no Pará. *Boletim do Museu Paraense Emilio Goeldi, Antropologia* 79: 1–50.

Furtado, L.G. 1993. 'Reservas pesqueiras', uma alternativa de subsistência e de preservação ambiental: Reflexões a partir de uma proposta de pescadores do Médio Amazonas. In *Povos das Águas: Realidade e Perspectiva na Amazônia*, ed. L. F. Gonçalves, W. Leitão and A. F. Mello, 243–76. Coleção Eduardo Galvão. Belém, Brazil: MCT/CNPq, MPEG.

Gentil, J.M.L. 1988. A juta na agricultural de várzea na área de Santarém - Médio Amazonas. *Boletim do Museu Paraense Emílio Goeldi, Série Antropologia* 4: 118–99.

Goulding, M. 1980. *The Fishes and the Forest*. Berkeley, Calif.: University of California Press.

Fabio de Castro

Goulding, M. 1983. Amazonian fisheries. In *The Dilemma of Amazonian Development*, ed. E. F. Moran, 189–210. Special Studies on Latin American and the Caribbean. Boulder, Colo.: Westview Press.

Goulding, M., Smith, N.J.H. and Mahar, D.J. 1996. *Floods of Fortune: Ecology and Economy along the Amazon*. New York: Columbia University Press.

Gragson, T.L. 1993. Human foraging in lowland South America: Pattern and process of resource procurement. *Research in Economic Anthropology* 14: 107–38.

Hartmann, W. 1989. Conflitos de pesca em águas interiores da Amazônia e tentativas para sua solução. In *III Encontro de Ciências Sociais e o Mar no Brasil*, ed. A. C. Diegues, 103–18. São Paulo, Brazil: Instituto Oceanográfico, Ford Foundation, Universidade de São Paulo.

IBAMA (Instituto Brasileiro do Meio Ambiente e dos Recursos Naturais Renováveis). 1995. Projeto IARA - Administração dos Recursos Pesqueiros do Médio Amazonas: Estados do Pará e Amazonas. *Coleção Meio Ambiente*. Série Estudos de Pesca 15. Brasília: Brazil: IBAMA.

Jackson, J.B.C. et al. (19 authors) 2001. Historical overfishing and the recent collapse of coastal ecosystems. *Science* 293: 629–38.

Junk, W.J. 1997. General aspects of floodplain ecology with special reference to Amazonian floodplains. In *The Central Amazon Floodplain: Ecology of a Pulsing System*, ed. W. J. Junk, 3-20. Ecological Studies 126. Berlin, Germany: Springer.

Leroy, J.P. 1988. Pescadores e entidades de apoio: Experiências e lutas. *Proposta: Experiências em Educação Popular* 38: 37–51.

Lima, D. 1992. *The Social Category Caboclo: History, Social Organization, Identity and Outsider's Social Classification of the Rural Population of an Amazonian Region (Middle Solimões)*. Ph.D. dissertation, King's College, Cambridge, UK.

Meggers, B. 1971. *Amazonia: Man and Culture in a Counterfeit Paradise*. Chicago: Aldine.

Meggers, B. 1995. Judging the future by the past: The impact of environmental instability on prehistoric Amazonian populations. In: *Indigenous Peoples and the Future of Amazonia: An Ecological Anthropology of an Endangered World*, ed. L. E. Sponsel, 15–43. Tucson: The University of Arizona Press.

McGrath, D., Castro, F., Câmara, E. and Futemma, C. 1999. Community management of floodplain lakes and the sustainable development of Amazonian fisheries. In *Várzea: Diversity, Development, and Conservation of Amazonia's Whitewater Floodplain*, ed. C. Padoch, J. M. Ayres, M. Pinedo-Vasquez and A. Henderson, 59–82. New York: The New York Botanical Garden Press.

McGrath, D., Silva, U.L. and Crossa, N.M.M. 1998. A traditional floodplain fishery of the Lower Amazon river, Brazil. *The ICLARM Quarterly*: 4–11.

McGrath, D., Castro, F., Futemma, C., Amaral, B.D. and Calabria, J. 1993. Fisheries and the evolution of resource management on the Lower Amazon floodplain. *Human Ecology* 21: 167–95.

Moran, E.F. 1974. The adaptive system of the Amazonian *caboclo*. In *Man in the Amazon*, ed. C. Wagley, 136–59. Gainesville, Fla.: University of Florida Press.

Moran, E.F. 1981. *Developing the Amazon*. Bloomington: Indiana University Press.

Moran, E.F. 1991. Human adaptive strategies in Amazonian blackwater ecosystems. *American Anthropologist* 93(2): 361–82.

North, D.C. 1990. *Institutions, Institutional Change and Economic Performance*. Cambridge, UK: Cambridge University Press.

Ostrom, E. 1990. *Governing the Commons: The Evolution of Institutions for Collective Action: The Political Economy of Institutions and Decisions*. Cambridge, UK: Cambridge University Press.

Ostrom, E. 1999. Coping with tragedies of the commons. *The Annual Review of Political Science* 2: 493–535.

Ostrom, E., Gardner, R. and Walker, J. 1994. *Rules, Games, and Common-Pool Resources*. Ann Arbor, Mich.: The University of Michigan Press.

Parker, E. 1985. Caboclization: The transformation of the Amerindians in Amazonia 1615-1800. In *The Amazon Caboclo: Historical and Contemporary Perspective*, ed. E. P. Parker, 1–50. Williamsburg, V.I.: Studies in Third World Societies no.32.

Petrere Jr., M. 1978. Pesca e esforço de pesca no Estado do Amazonas II – Locais, aparelhos de captura e estatística de desembarque. *Acta Amazonica* 8: 1–54.

Posey, D. 1985. Indigeneous management of tropical forest ecosystems: The case of the Kayapo Indians of the Brazilian Amazon. *Agroforestry Systems* 3: 139–58.

Roosevelt, A. 1980. *Parmana: Prehistoric Maize and Manioc Subsistence along the Amazon and Orinoco*. New York: Academic Press.

Roosevelt, A. 1989. Natural resource management in Amazonia before the conquest: Beyond ethnographic projection. In *Resource Management in Amazonia: Indigenous and Folk Strategies*, ed. D. A. Posey and W. Balee, 30–62. Advances in Economic Botany 7.

Ross, E. 1978. The evolution of the Amazon peasantry. *Latin American Studies* 10: 193–218.

Schmink, M., and Wood, C.H. 1984. *Frontier Expansion in Amazonia*. Gainesville, Fla.: University of Florida Press.

Schmink, M., and Wood, C.H. 1992. *Contested Frontiers in Amazonia*. New York: Columbia University Press.

Smith, N. 1981. *Man, Fishes, and the Amazon*. New York: Columbia University Press.

Smith, N. 1985. The impact of cultural and ecological change on Amazonian fisheries. *Biological Conservation* 32: 355–73.

Smith, N. 1996. *The Enchanted Amazon Rain Forest: Stories from a Vanishing World*. Gainesville, Fla.: University Press of Florida.

Stocks, A. 1987. Resource management in an Amazon *várzea* lake ecosystem: The Cocamilla case. In *The Question of the Commons: The Culture and Ecology of Communal Resources*, ed. B. J. McCay and J. M. Acheson, 108–20. Tucson, Ariz.: The University of Arizona Press.

Véran, J. 2002. *Quilombos* and land rights in contemporary Brazil. *Cultural Survival Quarterly* (Winter): 20–25.

Verissimo, J. 1895. *A Pesca na Amazônia*. Rio de Janeiro, Brazil: Livraria Clássica.

Zimmerman, J. 1987. Manaus importa alimentos e nas várzeas se produz fibras. Como explica a contradição? *Tübinger Geographische Studien* 95: 207–19.

Wagley, C. 1953. *Amazon Town: A Study of Man in the Tropics*. New York: Macmillan.

Weinstein, B. 1983. *The Amazon Rubber Boom: 1850–1920*. Stanford: Stanford University Press.

Reflexive Water Management in Arid Regions: The Case of Iran

Mohammad Reza Balali, Jozef Keulartz and Michiel Korthals

1. INTRODUCTION

Today, there is a broad consensus that we are facing a growing global water crisis. Not surprisingly, there is less consensus with respect to the question of the causes and consequences of this crisis. Most people seem to be convinced that the main cause of the water crisis is water shortage or water stress, resulting from population pressures coupled with industrialisation and urbanisation, and, more recently, with global climate change and the disastrous combination of lower precipitation and higher evaporation. While the world's population tripled in the twentieth century, water use has grown six-fold. This massive rise in the consumption of water, which went hand in hand with an increase in the contamination of this finite resource, was made possible by relatively recent technological advances in dam-building, well-drilling and pump technology. Consequently, people who attribute the global water crisis to water scarcity primarily look for technical solutions, and promote the design and development of more adequate or appropriate technologies like desalination, drip irrigation, rain water capture and storage, and water-free toilets.

There is, however, a growing number of people, who do not attribute the global water crisis merely to the growing scarcity of finite water resources, but mainly to 'a crisis in governance', as was declared at the Second World Water Forum of 2000 in The Hague. The very same year, the World Water Council made the following statement: 'There is a water crisis today. But the crisis is not about having too little water to satisfy our needs. It is a crisis of managing water so badly that billions of people – and the environment – suffer badly' (Cosgrove and Rijsberman, 2000: xix). In his keynote address at the Fourth World Water Forum of 2006 in Mexico, HRH Prince of Orange Willem-Alexander of The Netherlands also highlighted the fact that the water crisis is in fact a management crisis (WWF, 2000: 16). The second edition of the UN's World Water Development Report from 2006 likewise claimed that the water crisis is one of water governance, essentially caused by the ways in which we mismanage water, and outlined many of the leading obstacles to sound and sustainable water management: sector fragmentation, poverty, corruption, stagnated budgets,

Environmental Values **18** (2009): 91–112.

declining levels of development assistance and investment in the water sector, inadequate institutions and limited stakeholder participation.

Yet another group of people, among whom are many environmental philosophers, wag their fingers at our unsustainable and 'water-intensive' lifestyles. Globally, consumption preferences and patterns show an increasing desire and demand for products that require large amounts of water. Water consumption is also bound to increase as long as people are not facing water scarcity directly and physically, and believe that access to water is an obvious and natural thing.

While the first group of people stresses the – partially technologically induced – scarcity and shortages of our limited water resources, and the second group focuses on unsound governance and mismanagement, the third group draws our attention to public perceptions and preferences. There is, however, growing awareness among environmental social scientists that every single one of these perspectives is important and relevant for sustainable solutions to the global water crisis. But these different perspectives should not be treated separately, these scientists claim, because technological developments, governance regimes and personal belief systems and lifestyles are strongly interconnected.

A recent collection of papers from environmental social scientists who examine the ways that technology, governance and people shape each other is Joseph Murphy's edited book, entitled *Governing Technology for Sustainability* (2007). In this book, the challenge of sustainable development is explored by 'rethinking the relationship between people, technology and governance. In fact, understanding and recasting the people-technology-governance nexus might be two of the most important challenges associated with sustainable development' (Murphy, 2007: 207).

> This nexus is a web of relationships, with each element constantly reproducing or reshaping the other two. Governance for example, leads to strategic decisions about technology, based in part on assumptions about people. At the same time, however people can resist those assumptions and the way they are used to justify some technologies and not others. (ibid.: 217)

We will use this framework – the people-technology-governance nexus – to explore and examine the problems and possibilities of a transition to sustainable water management in Iran. The main challenge confronting Iran is how to continue the expansion of food production to meet future demand without imposing negative effects on the environment.[1] Since the country has a long history of agriculture, its inhabitants have already occupied almost all the fertile land. In recent times, however, there has been a slight increase in the total area under cultivation. This was achieved by bringing under cultivation the barren lands that have only a marginal agricultural potential. A comparison of the 1973 and the 1998 agricultural censuses shows that in a quarter of a century only 483,000 ha of new land was brought under cultivation, 2.8 per cent of the total. On the other hand, the negative water balance[2] implies that no more new

Mohammad Reza Balali, Jozef Keulartz and Michiel Korthals

land can be brought under cultivation, and that the country is already facing a critical situation regarding the management of water resources and sustainable food production in existing cultivated lands.

The case of Iran is also relevant for other countries of the Middle East and North Africa (MENA) region, which not only have similar (arid and semi-arid) environments, but also, to a large extent, share the same religion and history. The transition to sustainable water management is especially urgent for this region, because data from a major report published on 11 March 2007 by the World Bank shows that all countries in the region are facing a severe water crisis. Nearly 80 per cent of all precipitation in the region is used, compared with only two per cent in such regions as Latin America, the Caribbean and Sub-Saharan Africa. The water crisis is expected to get worse in light of high population growth and climate change. In fact, it is estimated that per capita water availability in the region will fall by half by 2050.

To address the challenge of sustainable development, we will focus on the transition from industrial modernity to what sociologists like Ulrich Beck, Anthony Giddens and Scott Lash (1994) have called 'reflexive' modernity. Reflexive modernity does not indicate a break with modernity, but stands for a radicalisation within modernity – a 'modernisation of modernity'. An important aspect of this 'second order' modernity is the reevaluation and rehabilitation of tradition. That is why we will start with a description of the pre-modern technology-governance-people water nexus. Instead of water nexus we prefer to use the term 'water paradigm', from Tony Allan (2006).

The traditional or pre-modern water paradigm can by characterised by its key technical system (the *Qanat* system of underground irrigation channels), its main governance institution (the *Buneh* cooperative organisation of agricultural production) and its ethico-religious belief system (Zoroastrianism and Islam) (section 2). The current paradigm of industrial modernity can be identified by the partial replacement of *Qanats* by deep wells and large dams, the substitution of the *Buneh* by a system of smallholding, and the emergence of a mechanistic worldview with important ethical ramifications (section 3). In the North, since the 1960s and the 1970s, industrial modernity has gradually given way to what has come to be known as reflexive or second modernity. The paradigm of reflexive modernity can be characterised (1) by the integration of traditional (indigenous, small scale) and modern (scientific, large scale) technological infrastructure, (2) by participatory water resources management in the form of multi-stakeholder platforms or water user associations, and (3) by its recognition of the diversity of cultural values of water (section 4). We will conclude our paper with a sketch of what we consider to be the main contours of reflexive water resources management in Iran and other countries of the MENA region (section 5).

2. TRADITIONAL WATER MANAGEMENT

The Qanat irrigation system – a brief history

More than 3000 years ago, the inhabitants of the dry, mountainous regions of Iran perfected a system for conducting snowmelt through underground channels, the so-called *Qanat*, which began in the mountains and carried water downwards to the plains by gravity, to farms, country gardens and towns (Foltz, 2002). The conduits – which are usually 50 to 80 centimetres wide and 90 centimetres to 1.5 metres high – vary between several hundred metres to more than 100 kilometres in length. In Iran alone, there are some 22,000 of them, comprising more than 273,500 kilometres of underground channels.

The *Qanat* irrigation system rests on indigenous knowledge and experimental hydrology. It was widely used for several reasons. First, unlike other traditional irrigation devices, such as the counterpoised sweep, *Qanats* require no power source other than gravity to maintain a flow of water. Second, water can be moved over substantial distances through these subterranean channels with minimal evaporation losses and little danger of pollution. Finally, the flow of water in a *Qanat* is proportionate to the available supply in the aquifer and, if properly maintained, these irrigation canals could provide a reliable supply of water for centuries (Haeri, 2006).

Qanats are built by specialists called *muqanni* (*Qanat* diggers), who transmit their knowledge from father to son. A windlass is set up at the surface and the

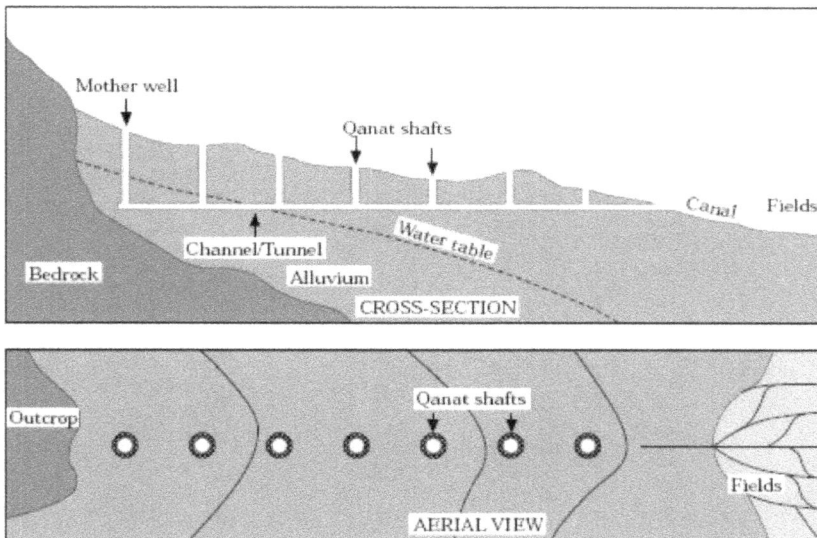

FIGURE 1. Qanat irrigation system (Lightfoot, 1996)

140

Mohammad Reza Balali, Jozef Keulartz and Michiel Korthals

excavated soil is then hauled up in leather buckets. A vertical shaft of about three feet in diameter is dug out, one man working with a mattock and the other with a short-handled spade. A gently sloping tunnel is thus constructed which conducts water from an infiltration section beneath the water table to the ground surface by gravity flow.

The *Qanat* works were built on a scale that rivalled the great aqueducts of the Roman Empire, but, whereas the Roman aqueducts now are only of historical interest, the *Qanat* system is still in use after 3000 years. The advantage of the *Qanats* over the Roman open air aqueducts is that less water is lost by evaporation on the way from hill to plain.

There is little doubt that ancient Iran (Persia) was the birthplace of the *Qanat*. Greek historian Polybius credits the Achaemenids (550 to 331 B.C.) for bringing water to remote areas throughout the Persian Empire through the use of *Qanats*. The Achaemenid rulers provided a major incentive for *Qanat* builders and their heirs by allowing them to retain the profits from newly-constructed *Qanats* for five generations. As a result of this water supply, thousands of new settlements were established and others expanded.[3]

Three centuries later, when the Parthians invaded Iran, *Qanats* were in widespread use on the Iranian plateau. To the west, *Qanats* were constructed from Mesopotamia to the shores of the Mediterranean, as well as southward into parts of Egypt and Arabia. To the east of Iran, *Qanats* came into use in Afghanistan, the Silk Road oases settlements of Central Asia, and the Chinese province of Sinkiang (now Xinjiang) (English, 1997).

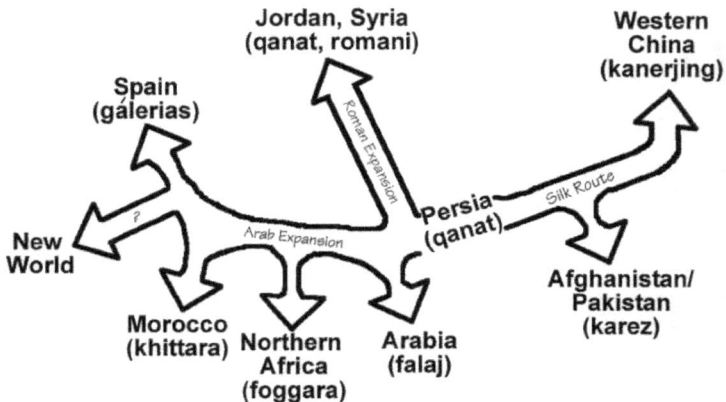

FIGURE 2. Qanat technology diffusion (Qanat, waterhistory.org).

During the Roman-Byzantine era (64 B.C. to A.D 660), many *Qanats* were constructed in Syria and Jordan. From here, the technology appears to have diffused north and west into Europe. There is evidence of Roman *Qanats* as far away as the Luxembourg area.

The expansion of Islam initiated another major diffusion of *Qanat* technology. The early Arab invasions spread *Qanats* across North Africa into Spain, Cyprus and the Canary Islands. Finally, evidence of New World *Qanats* can be found in western Mexico, in the Atacama regions of Peru, and Chile at Nazca and Pica. The *Qanat* systems of Mexico came into use after the Spanish conquest.[4]

Buneh – the Qanat system as a socio-technical system

Technological systems cannot be separated from the human activities and social institutions that make them work. In other words, technology is part of a nexus that also includes governance. The *Qanat* system is a socio-technical system. It is not only an engineering wonder, but also a remarkable social phenomenon. *Qanats* reflect collective and cooperative work. Because individual peasants possessed neither the capital nor the manpower that was needed for construction and maintenance of the *Qanat* system, independent production was at a disadvantage compared to other systems of production such as the multi-family collective or the *Buneh* in Iran. The major function of the *Buneh* was the efficient exploitation of productive land and the careful use of scarce water resources. Although *Buneh* had some disadvantages (e.g., an internal unequal division of labour and crop), it strengthened the socio-economic position of the peasants (Lahsaeizadeh, 1993).

Basically, each *Buneh* has six main members. It was under the charge of one peasant known as the *sarBuneh* (*Buneh* head) or *abyar* (Irrigator). He was chosen by the landowner or his bailiff. Experience and expertise in agricultural affairs were necessary qualifications for the *sarBuneh or abyar*. Each *sarBuneh* had two assistants, known as *varBuneh*, chosen by the *sarBuneh* from among his friends and relatives. Finally, sharecroppers formed the foundation of a *Buneh* structure.

At the beginning of each agricultural year, all the *sarBunehs* of the village gathered to decide how the fields should be distributed among *Bunehs*. Once these basic decisions had been made, the important tasks of each *sarBuneh* included marking off the boundaries of his *Buneh*'s field and plots, determining the type of crop for each plot, assigning tasks for each member, coordinating irrigation, sowing seed, contracting seasonal workers, supervising threshing the grain, controlling the division of crops, and, finally mediating between the *Buneh*'s members and the landlord.

The *Buneh* also included some groups others than peasants. The first group consisted of those craftsmen who worked directly for the *Buneh*. Members of this group included *muqanni* (well and *Qanat* diggers), *ahangar* (blacksmith)

Mohammad Reza Balali, Jozef Keulartz and Michiel Korthals

and *najjar* (carpenter). They were paid in kind at harvest time and carried out repairs for the *Buneh* throughout the year. The second group included barbers and bath keepers. Members of the *Buneh* were allowed to go to the public bath regularly without payment during one agricultural year. Also, the village barber went to the *Buneh* field weekly and cut *Buneh* members' hair and shaved their beards free of charge. In return, both bath keepers and barbers received a share of the crop at harvest time. Finally, each *Buneh* needed some extra hands during harvest time. For this purpose, daily wage labourers were hired. They were temporarily employed by *Bunehs* and paid either in cash, in kind, or a combination (Safinejad, 1989).

Ethico-religious frameworks: Zoroastrianism and Islam

To complete our sketch of the pre-modern water paradigm, we should draw attention to the belief systems that have supported the traditional socio-technical irrigation system morally as well as legally, Zoroastrianism and Islam.

Zoroastrianism, the dominant religion in the pre-Islamic era, rests on three pillars: *Humata* (Good Thoughts), *Hûkhta* (Good Words) and *Hvarshta* (Good Deeds). By 'Good Thoughts', a Zoroastrian is able to concentrate his mind in divine contemplation of the Creator, and live in peace and harmony with his fellow man. By 'Good Words', he is obliged to observe honesty and integrity in all commercial transactions, to prevent hurting the feelings of others, and to engender feelings of love and charity. By 'Good Deeds', he is directed to relieve the poor, to irrigate and cultivate the soil, to provide food and fresh water in places where needed, and to devote the surplus of his wealth in charity to the well-being and prosperity of his fellow man.

Nature is central to the practice of Zoroastrianism and many important Zoroastrian annual festivals are in celebration of nature; new year on the first day of spring, the water festival in summer, the autumn festival at the end of the season, and the mid-winter fire festival (Jafarey, 2005). In the *Avesta*, the holy book of Zoroastrianism, there is strong emphasis on the protection of water and soil.

Like Zoroastrianism, from its very origins fourteen centuries ago, Islam offers a basis for ecological understanding and stewardship. According to the *Qur'an*, the universe and everything in it has been created by God and is considered a sign (*āyāt*) of God. Human beings, although at the top of creation, are only members of the community of nature. Humankind is just considered as a trustee for the planet: humans are entitled to live on the Earth and benefit from it but they are not entitled to pollute or destroy the environment. Any behaviour that can jeopardise the future of the natural resources is seen as an act against God and His creation (Abdel Haleem, 1989).

Nature has been created in order and balance, and with extraordinary aesthetic beauty, and all these aspects of nature, while enhancing humankind's life should be honoured, developed and protected accordingly. All patterns of human production and consumption should be based on this overall order and balance of nature. The rights of humankind are not absolute and unlimited: we should not simply consume and pollute nature as we wish, carelessly (Özdemir, 2003).

Water is a pivotal issue in Islam, not surprisingly since it is a religion that originated in a desert area and spread mainly to other arid or semi-arid territories. It is evident from numerous verses in the *Qur'an* that water is a major theme in Islamic cosmogony and iconography as well as a recurrent topic in liturgy and daily life (Gilli, 2004). One of the most famous verses pertaining to water is taken from the '*Sura* of the Prophets' and it states, 'We made from water every living thing'. This is not the only verse where the word *Ma'* (water) appears, since it occurs more than sixty times in the *Qur'an*.

In Islam, all water is sacred and sent as a gift from Allah. It is one of the three things that every Muslim is entitled to; grass (pasture for cattle), water and fire. Water should be freely available to all and any Muslim who withholds unneeded water sins against Allah. Mohammad attached great importance to the moderate use of water and forbade its excessive use even when performing ablutions, saying that to do so was 'detestable' (*makrūh*). He even prevented people from using too much water for ablutions when preparing to enter the Divine Presence for prayer.

There is a fundamental difference in the valuing of water between Islam and Christianity. Whereas Islamic doctrine ascribes holiness to all water, in Christianity only water that has been blessed in the name of Christ is sacred. Francesca de Châtel gives some striking examples of the undervaluation of water in Christianity. Saints of early Christianity boasted that water had never touched their feet except when they had to wade across a stream. St. Jerome denounced bathing as a pagan practice and affirmed that 'He who has bathed in Christ [i.e. has been baptised] does not need a second bath' (Châtel, 2005a: 56).

Islamic law, the *Shari`ah*, goes into great detail on the subject of water to ensure its fair and equitable distribution within the community. The word 'Shari`ah' itself is closely related to water. Originally it meant 'the place from which one descends to water'. Before the advent of Islam in Arabia, the *Shari`ah* was, in fact, a series of rules about water use. The term later evolved to include the body of laws and rules given by Allah. There are two fundamental precepts that guide the rights to water in the *Shari`ah*: *shafa*, the right of thirst, establishes the universal right for humans to satisfy their thirst and that of their animals; *shirb*, the right of irrigation, gives all users the right to water their crops.

It should be obvious by now that the technical, social and ethical aspects of the traditional system of land and water management were highly interconnected. The *Qanat* underground irrigation system was dependent on the social institution of the *Buneh* to operate properly, while Zoroastrianism and Islam can be considered as an adequate ethico-religious framework for this socio-technological arrangement. But around the middle of the twentieth century the 'Age of *Qanats*' came to an end.

Mohammad Reza Balali, Jozef Keulartz and Michiel Korthals

3. INDUSTRIAL MODERNITY AND THE END OF THE AGE OF QANATS

The hydraulic mission – the replacement of Qanats by deep wells and large dams

From the late nineteenth century until the 1970s, Northern industrialised economies were dominated by the vision and politics of what has been termed the 'hydraulic mission' (Allan, 2002). This mission, involving hydraulic mega-projects like gigantic dams and large-scale irrigation systems, was inspired by the belief that nature, including water, can be controlled and should be subjected to the mastery of science and industry. This mission was implemented in liberal western economies, first and foremost in the United States (Worster, 1992; Reisner, 1986), but also in the centrally planned economies of the Soviet Union.

In the second half of the twentieth century, the hydraulic mission was introduced to the developing countries of the South, especially in India but also in Egypt and other countries of the MENA region. In Iran too, it was assumed that arid regions could be industrialised by making the necessary water resources available through building dams, pumping up groundwater and bringing in water from remote sources in order to 'make the desert bloom'. To pave the way for industrial modernity, the Iranian authorities tried to belittle all traditional irrigation and production systems. Most Iranian scholars and politicians exaggerated the technical deficiencies of the *Qanats* to justify their own programmes and to convince farmers to use pump extraction instead of *Qanats* (Khaneiki, 2007).

At first, modern devices such as pumps and drilling machines received no warm welcome, but after some pumped wells were drilled, farmers started to express their admiration for these new technologies. After all, while the construction of a *Qanat* would sometimes take tens of years, drilling a well took less than one month. If the farmers wanted to increase the discharge of a *Qanat* even a little bit, they had to extend the tunnel, which would take two or three years, whereas it was easy to increase the discharge of a pumped well by two times just through changing the diameter of the pump or adding some units or parts (Yazdi and Khaneiki, 2007).

Electric and diesel-pumped wells offer advantages over *Qanat* irrigation by allowing water to be brought to the surface on command, but over-pumping has caused water tables to fall, aquifers to be depleted and *Qanats* to be abandoned at an accelerating pace. The role of *Qanats* in securing all the functions of water in Iran has decreased from 70 per cent prior to 1950, to 50 per cent around 1950 and to 10 per cent in the year 2000 (Haeri, 2006).[5]

The substitution of the Buneh by a system of smallholding

The use of mechanically-pumped wells was heavily encouraged as a result of the Land Reform Act of 1962, which broke up the large estates and re-distributed land to the peasants. The general pattern of land ownership in Iran prior to the land reform was a combination of large-scale feudal landownership with small-scale absentee and peasant proprietorship (Lahsaeizadeh, 1993). Because of the importance of artificial irrigation to Iranian agriculture, sharecropping (*muzara-eh*) was dominant among the different types of relation between the peasant and landowner. This traditional system of land ownership and tenure, and the socio-economic organisation of villages (*Buneh*), were well adapted to the optimal use of the *Qanat* system. The land holdings given to the peasants following the Land Reform were too small to maintain the *Qanats*, while many landowners and farmers now prefer pumped wells and allow their *Qanats* to languish. In effect, the traditional sense of water resources management for the benefit of the community seems to be giving way to an 'every man for himself' mentality. In addition to the mostly privately owned and constructed wells, the public sector is engaged in the construction of many large-scale dams.

The emergence of a mechanistic worldview

The new water resource management regime of deep wells and large dams is more in tune with a mechanistic worldview than with the ethico-religious frameworks of the past. Critics of the mechanistic worldview fear that if man sets himself up as the measure and master of all things, nature will appear solely as 'material' that he can control and command as he pleases. Nature, including water, ceases to be an independent source of value and turns into a mere resource to be disposed of at will instead. To quote Donald Worster's 1992 book *Rivers of Empire* on the advent of the hydraulic society in the American West:

> The most fundamental characteristic of the latest irrigation mode is its behaviour towards nature and the underlying attitudes on which it is based. Water in the capitalist state has no intrinsic value, no integrity that must be respected ... It has now become a commodity that is bought and sold and used to make other commodities ... It is in other words, purely and abstractly a commercial instrument. All mystery disappears from its depths, all gods depart, all contemplation of its flows ceases ... Where nature seemingly puts limits on human wealth, engineering presumes to bring unlimited plenty. Even in the desert, where men and women confront scarcity in its oldest form (...) every form of growth is considered possible. (Worster, 1992: 52)

Modern water technologies have deeply affected the way people perceive, value and use water. In her paper on the conversion of rainwater into tap water, Nicole Stuart argues that industrial technologies dissociate people from the natural environment upon which they depend. 'Urban water infrastructure allows people to "take water for granted" ... The urban water infrastructure provides an "illusion of abundance" – enabling twenty-four hour access to clean and potable water, seven days a week' (Stuart, 2007: 419).

Based on four years of field research in 11 countries of the MENA region, Francesca de Châtel (2005b) came to a similar conclusion with respect to public awareness of water scarcity. The sheer size of dam reservoirs and the huge amount of water that is transported through pipelines leads the general public to believe that water supplies are endless and conceal the reality of water scarcity. Moreover, through the development of modern water distribution systems, the link that used to exist between the individual user and his water is severed. As soon as water starts flowing from a tap, it is taken for granted. People forget that a fluctuating river or an erratic weather system lies at its origins. 'By making its source invisible, water's existence is divorced from the elements and the seasons, and it becomes paradoxically omnipresent. The user can comfortably assume that it flows from an endless supply' (Châtel, 2005a).

4. REFLEXIVE MODERNITY

The ideas underpinning industrial modernity were challenged during the 1960s and 1970s, when some of its disastrous effects – 'the hydraulic society's worsening headaches' (Worster, 1992: 324) – such as salinity, sedimentation, pesticide contamination, diminishing hopes of replenishment and the dangers of aging, collapsing dams, begun to appear, not only in the U.S. and other Northern countries but also in Southern countries like Iran, where, over the past four decades, farmers and others close to the land have watched water tables drop as one well after another dried up, and formerly fertile lands were inevitably taken out of production (Foltz, 2002).

As a response to these challenges a new paradigm has emerged, the paradigm of 'reflexive modernity'. As already mentioned in the introduction, reflexive modernity does not imply a break with modernity, but refers to a radicalisation within modernity – a 'modernisation of modernity'. Radicalised or reflexive modernisation is a process whereby modernisation has become directed at itself, at the destructive and continually expanding side-effects and risks that are systematically produced by industrial society. While nature in 'first' modern societies is conceived of as a neutral resource, which can and must be made available without limitation, nature in 'second' modern societies 'is no longer solely perceived as an outside that can be adapted to one's purposes, but increasingly as part and parcel of society' (Beck et al., 2003: 7). Beck argues for 'ecological enlightenment', which requires a reorientation from a focus on economic growth to one of sustainable development (Beck, 1995).

According to Tony Allan (2006), reflexive modernity in the area of water management can be shown to have three phases. In the first phase, from the 1960s until the 1980s, changes in water policy were inspired by the growing awareness of the *environmental* costs of the hydraulic mission. In the second phase, from the early 1990s onward, the idea that water is an *economic* resource gained currency, paving the way for the concept of the water market. In the third phase, which emerged at the turn of the century, the notion that water management is a *political* process seized the North. This notion is central to the concept of Integrated Water Resource Management (IWRM). IWRM is an intensely political process which includes stakeholder consultation and participation to enable the mediation of conflicting interests of water users and water management agencies.

Allan believes that, by and large, the semi-arid North can be shown to have passed through all three stages of water management and water policy. In the South, by contrast, the professional community generally, and all water users and politicians, have resisted the adoption of the paradigm of reflexive modernity. Especially in the MENA region, the hydraulic mission of industrial modernity is still alive and flourishing. 'The big players, Egypt, Turkey and Iran, are all engaged in major hydraulic projects' (Allan, 2002: 145).

Allan's sketch of the course of reflexive modernity in water policy in the North is not only meant as a purely empirical description, but also as a normative prescription for water policy reform in the South. This is, however, very problematic on two accounts. In the first place, it seems to place too much faith in a unilinear model of institutional evolution. According to Frances Cleaver (2002), such a model fails to recognise that decision making and cooperative action are deeply embedded in the web of local livelihood networks and practices. To understand the complex and dynamic nature of institutional change we should see it as a process of 'bricolage', i.e., a process operating by trial and error and using a diverse range of social and cultural resources.

In the second place, Allan fails to recognise that the course that second modernity has taken within a European constellation will differ considerably from its course within non-European constellations, where the dynamic of reflexive modernisation displays its effects not on first modern societies but rather on the distorted constellations of post-colonialism. '*Different non-European routes to and through second modernity* still have to be described, discovered, compared and analysed' (Beck et al., 2003: 7). The next section will give a rough sketch of a possible trajectory to reflexive water management in Iran and other countries of the MENA region.

5. TRANSITION TO REFLEXIVE MODERNITY IN IRAN

Restoration and integration of Qanats

At the Fourth World Water Forum of 2006 in Mexico there was general agreement that nations should consider both small-scale decentralised solutions and large-scale approaches involving dams and reservoirs to meet their needs at the lowest possible social and environmental costs. Furthermore, the forum remarked that, regrettably, local knowledge and adaptive technology develop-

Mohammad Reza Balali, Jozef Keulartz and Michiel Korthals

ment have been neglected historically, and recognised that knowledge coming from several sources could be complementary and might reinforce each other in solving water issues locally. In the context of Iran's transition to reflexive water management the forum's recommendation to try for 'a proper mix of science, technology and local knowledge' would imply a rehabilitation of the traditional *Qanat* underground irrigation system and its integration with modern water supply systems.

The rehabilitation of the *Qanat* system is important because this system represents one of the most ecologically balanced water recovery methods available for arid and semi-arid regions. *Qanats* tap the groundwater potential only up to, and never beyond the limits of natural replenishment and, as a consequence, do not upset the hydrological and ecological equilibrium of the region. As the *Qanats* are often dug into hard subsoil, there is little seepage, no rising of the water table, no waterlogging, no evaporation during transit – and hence no salination in the area surrounding the conduits. Moreover, *Qanats* rely entirely on passive tapping of the water table by gravity only, whereas the extractive pumps consume an enormous amount of fuel per year.[6]

However, the rehabilitation of the *Qanat* irrigation system can only succeed with the help of modern technology. Modern mining technologies can be used to enhance the water efficiency of the *Qanat* system, whereas water productivity can be improved by combining *Qanats* and modern irrigations systems.[7] Such a revitalisation of the *Qanat* system by modern technological means can result in a substantial decline of our dependency on deep wells.

What is required in addition to the restoration of the *Qanat* system is its integration in a modern environment. The rapidly increasing demand for water due to population growth and agricultural expansion in Iran cannot be accommodated by *Qanats* only. Therefore, what is called for is a complementary system of all three methods of water provision. Among other things, this implies that existing *Qanat* systems should no longer be ignored during the building of large dams and the excavation of deep wells. Islamic water law ensures that new irrigation systems or wells are not constructed too close to an existing one. However, with the emergence of the pumped tube well, the traditional *harim*-area (usually between 100 and 300 metres) does not suffice any longer and should be enlarged considerably.

Fortunately, there is a revival of interest in several countries where ancient underground irrigation systems have been declared as national heritage. The Government of Iran is lately giving much attention to the *Qanat* system. The first international Qanat Research Conference was held in Yazd in 2000. As a result of the recommendations of this conference, the Government of Iran has established the International Qanat Research Centre in Yazd in collaboration with Afghanistan and Pakistan, and with support from UNESCO.[8] Another example is China, where the Song Yudong group, in close collaboration with Xinjiang local authorities, has come up with several practical suggestions to revitalise existing *Karez* systems, including an overall plan for protection and improvement of the existing system in Turpan Prefecture in China. Recently international organisations such as UNESCO, and the United Nations University (UNU) have also shown interest in promoting studies on the *Qanat* system through the International Hydrological Program and the Traditional Technology in Drylands Program that supports young researchers from countries with a long tradition and heritage of *Qanat* systems such as Syria, Oman, Tunisia and Yemen to undertake systematic studies (Kobori, 2005).

New participatory arrangements

Because the *Qanat* system as a socio-technical system can only operate within a suitable social context, its restoration is impossible without renewal of the traditional social infrastructure. The traditional organisation of villages (*Buneh*) was well adapted to the optimal use of the *Qanat* system. A major disadvantage of the *Buneh*, however, was its hierarchical structure and the unequal division of labour and crops. The land reform of 1962 brought an end to this feudal situation, but at the same time it sounded the death-knell of the *Qanat* system.

What is needed in order to restore the *Qanat* system under present-day circumstances is some form of water resources management that encourages collective action with a participatory rather than a hierarchical character. Here, the concept of Multi-Stakeholder Platforms, that has become popular as an institutional framework for resolving complex resource management problems, could be helpful (Warner, 2007). The idea is that multiple stakeholders, who have different interests and needs with respect to water, should organise and arrange water use and conservation issues amongst them through some form of cooperation, including the building of capacity for collective learning and decision making. Today, such water user associations are emerging in many countries of the Muslim world.

Towards an Islamic land ethic?

To round off our sketch of a possible pathway to a more reflexive water management in Iran and other countries of the MENA region, we need to focus on the belief systems that could facilitate or hamper such a transition. As we have previously argued, with Francesca de Châtel (2005a; 2005b) and Nicole Stuart (2007), industrial water technologies tend to dissociate people from their natural environment. Restoration of the *Qanat* irrigation system could help to reconnect people with nature and to encourage greater ecological awareness and activism. To achieve this, however, more is needed than the purely technical restoration of the *Qanat* system and the creation of water-user associations. Presently, the general public in Iran tends to perceive this sustainable water supply system as outdated and backward.

Mohammad Reza Balali, Jozef Keulartz and Michiel Korthals

Since religion still exerts a very big influence on Iranian society and because water plays a pivotal role within Islam, awareness campaigns based on religious principles could be very useful to counterbalance the mechanistic worldview underlying industrial modernity and the undervaluation of water due to the influence of modern water supply technologies. As Holmes Rolston has recently remarked, 'Christianity, together with other faiths that influence human conduct, needs again to become "a land ethic"' (Rolston, 2006: 312).

According to Muslim teaching, water is a gift from God that should be freely available to all. At present, this creed leads to gross underpricing of water, which in turn results in widespread wastage. What seems to be forgotten is that the *Qur'ran* also incites believers to use water sparingly. Mankind is not entitled to ruin, corrupt, pollute or destroy the environment. Any behaviour that can jeopardise the future of the natural resources, water included, is seen as an act against God and its creation. Preventing the corruption of natural resources or the pollution of water is not simply an ethical and civilised behaviour but it is also an act of worship. In fact, saving water is a religious duty.

In the last decade these Islamic principles have been widely implemented in the Muslim world, including Iran, through awareness campaigns. Mosques were used as platforms for these campaigns, and imams have been properly trained in drawing the attention of the believers on water scarcity during the Khutbah, the Friday sermon. Posters, leaflets, booklets and stickers using religious terminology and imagery have also been used to promote awareness of water issues (Gilli, 2004).

CONCLUSION

To highlight the problems and perspectives of water resources management in Iran and comparable arid and semi-arid countries in the MENA region, three subsequent paradigms have been identified, the traditional, the industrial and the reflexive paradigm Each of these paradigms consists in a specific technology-governance-people nexus: within every paradigm, technologies, social institutions and environmental mentalities are linked and are constantly shaping each other.

We have focused on the *Qanat* irrigation system as one of the oldest feats of human engineering, which was established about 3000 years ago and has been diffused throughout arid and semi-arid regions globally ever since. After a brief sketch of the nature and history of this ancient irrigation system, we have outlined its institutional context (*Buneh*) and the underlying ethico-religious belief system (Zoroastrianism and Islam). However, due to the land reform of 1962, with the introduction of the pumped tube well and the construction of large-scale dams, the 'Age of Qanats' came to an end: the *Buneh* system of sharecropping gave way to a system of smallholding, and a mechanistic worldview emerged in which society and nature became more and more disconnected.

Recently, with the growing danger of water scarcity and the manifestation of some negative effects of modern hydraulic technology, we witness the transition from the 'hydraulic mission' of industrial modernity to a more reflexive approach to water management. An important aspect of such a transition in Iran and other MENA countries is the rehabilitation of the traditional *Qanat* underground irrigation system and its integration with modern water supply systems. By enhancing the water efficiency and water productivity of *Qanats*, e.g. with the help of modern mining technologies and modern drip irrigation systems, such rehabilitation can lead to a significant decline of the dependency on deep wells. It can also help people to reconnect with nature and to encourage greater ecological awareness and activism.

At this moment, however, rehabilitation of the *Qanat* system is hampered by the general public's perception of this sustainable water supply system as outdated and backward. Since religion is still of great importance in Iran and because water plays a pivotal role within Islam, awareness campaigns based on religious principles could be very useful in overcoming this unfavourable image.

NOTES

The current study is part of a PhD research project aiming at the development of a reflexive ethical framework of land and water management in Iran. The work on this paper was supported by a grant from the Iranian Ministry of Science, Research and Technology (MSRT) and the Agricultural Research and Education Organization (AREO).

[1] Like other MENA countries, Iran can never be fully self-sufficient, but will remain dependent on importation of water-intensive staple food commodities. While in 1999 Iran was the ninth highest wheat importer and the third highest rice importer in the world, in the year 2001, due to a prolonged drought, Iran rose to the position of fifth-highest wheat importer in the world. (Alizadeh and Keshavarz, 2005: 97). To increase efficient water use domestic food production should be balanced with food importation.

[2] With an annual rainfall recharge reaching the aquifers of 46.6 billion cubic metres and an annual discharge from the groundwater of 49.7 billion cubic metres Iran has a negative groundwater balance of over 3 billion cubic metres (Moameni, 2000).

[3] The largest known *Qanat* is in the Iranian city of Gonabad; after nearly 2,700 years it still provides drinking and agricultural water to nearly 40,000 people. Its mother well is 360 metres deep and the channel is 45 km long (Kobori, 2007).

[4] The system has been variously named – *Qanat* in Iran; *Qanat Romani* in Syria and Jordan; *Karez* in Afghanistan, Pakistan and Turkmenistan; *Kahn* in Baluchistan; *Kanerjing* in China; *Falaz* in Oman and other parts of the Arabian Peninsula; *Foggera* in Algeria and other North African countries; *Khattara* in Morocco; and *Galleria* in Spain (Kobori, 2007).

[5] However, it is important to know that today the traditional *Qanat* systems continue to provide water for as much as one third of irrigated land (Foltz, 2002).

[6] In the Yazd area there are 4,340 wells with extractive pumps, which totally consume 205,854,880 litres of gas oil a year in order to obtain 926,350,000 cubic metres water. But in the same area there are 2,948 *Qanats*, which withdraw 329,870,000 cubic metres water a year without any fuel (Khaneiki, 2007: 81).

[7] In Syria Wessels and Hoogeveen (2006) have seen that combining ancient *Qanats* and modern drip irrigation systems for fruit trees might prolong the life of some *Qanats* and encourage the younger generation to commit to their upkeep. Another option they mention is to encourage eco-tourism based around *Qanats* to provide alternative income for the farmers.

[8] http://www.qanat.info/en/sitemap.php (accessed March 10, 2008).

REFERENCES

Abdel Haleem, M. 1989. 'Water in the Qur'an'. *Islamic Quarterly* **33**(1): 34–50.

Allan, T. 2002. *The Middle East Water Question: Hydropolitics and the Global Economy.* London/New York: I.B. Tauris Publishers.

Allan, T. 2006. 'Millennial water management paradigms: making IWRM work', http://www.mafhoum.com/press/53aE1.htm (accessed 10 March 2008).

Alizadeh, A. and A. Keshavarz. 2005. 'Status of agricultural water use in Iran'. Proceedings of the Iranian-American workshop: Water conservation, reuse, and recycling. National Academy of Sciences, http://www.nap.edu/catalog/11241.html (accessed 10 March 2008).

Beck, U., A. Giddens and S. Lash. 1994. *Reflexive Modernization: Politics, Tradition and Aesthetics in the Modern Social Order.* Cambridge: Polity Press.

Beck, U. 1995. *Ecological Enlightenment: Essays on the Politics of the Risk Society.* Amherst, NY: Prometheus Books.

Beck, U., W. Bonss and Chr. Lau. 2003. 'The theory of reflexive modernization: problematic, hypotheses and research programme'. *Theory, Culture and Society* **20**(2): 1–33.

Châtel, F. de. 2005a. 'Perceptions of water in the Middle East: the role of religion, politics and technology in concealing the growing water scarcity'. http://www.ipcri.org/watconf/papers/chatel.pdf (accessed 10 March 2008).

Châtel, F. de. 2005b. *Het water van de profeten* [Water of the Prophets]. Amsterdam, Antwerpen: Uitgeverij Contact.

Cleaver, F. 2002. 'Reinventing institutions: bricolage and the social embeddedness of natural resource management'. *European Journal of Development Research* **14**(2): 11–30.

Cosgrove, W.J. and F. R. Rijsberman (World Water Council). 2000. *Making Water Everybody's Business.* London: Earthscan

English, P. 1997. 'Qanats and lifeworlds in Iranian plateau villages'. Proceedings of the Conference: Transformation of Middle Eastern natural environment: legacies and lessons. Yale University, http://environment.yale.edu/documents/downloads/0-9/103english.pdf (accessed 10 March 2008).

Foltz, R.C. 2002. 'Iran's water crisis: cultural, political, and ethical dimensions'. *Journal of Agricultural and Environmental Ethics* **15**: 357–380.

Gilli, F. 2004. 'Islam, water conservation and public awareness campaigns', http://www.ipcri.org/watconf/papers/francesca.pdf (accessed 10 March 2008).

Haeri, M.R. 2006. 'Kariz (Qanat); an eternal friendly system for harvesting groundwater', http://unfccc.int/files/meetings/workshops/other_meetings/application/pdf/121103_iran.pdf (accessed 10 March 2008).

Jafarey, A.A. 2005. 'Zoroastrian ethics and culture'. http://www.vohuman.org/Article/Zoroastrian%20Ethics%20and%20Culture.htm (accessed 10 March 2008).

Khaneiki, M.L. 2007. 'Traditional water management: an inspiration for sustainable irrigated agriculture in central Iran'. Proceedings of the International History Seminar on Irrigation and Drainage, Tehran, 2–5 May, 2007, pp. 73–84.

Kobori, I. 2005. 'Lessons learned from Qanat studies: A proposal for international cooperation', in R. Coopey, H. Fahlbush, N. Hatcho and L. Jansky (eds.), *A History of Water Issues: Lessons to Learn* (Tokyo: United Nations University), pp. 187–194.

Kobori, I. 2007. 'Role of traditional hydro-technology in dryland development: *Karez, Qanat* and *Foggera*'. Proceedings of the International History Seminar on Irrigation and Drainage, Tehran, 2–5 May, 2007.

Lahsaeizadeh, A. 1993. *Contemporary Rural Iran*. Aldershot: Averbury.

Lightfoot, D.R. 1996. 'Syrian Qanat Romani: history, ecology, abandonment'. *Journal of Arid Environments* **33**: 321–336.

Moameni, A. 2000. *Production Capacity of Land Resources of Iran*. Soil and Water Research Institute of Iran. Pub. no. 1110. Tehran, Iran.

Murphy, J., ed. 2007. *Governing Technology for Sustainability*. London: Earthscan.

Özdemir, İ. 2003. 'Environmental ethics from a Qur'anic perspective', in R.C. Foltz, F.M. Denny and A. Baharuddin (eds.), *Islam and Ecology, a Bestowed Trust* (Cambridge MA: Harvard University Press), pp. 3–37.

Reisner, M. 1986. *Cadillac Desert. The American West and its Disappearing Water*. New York: Viking Penguin Inc.

Rolston III, H. 2006. 'Caring for nature: What science and economics can't teach us but religion can'. *Environmental Values* **15**: 307–313

Safinejad, J. 1989. *Buneh, Traditional Farming Systems in Iran*. Tehran: Amirkabir.

Stuart, N. 2007. 'Technology and epistemology: environmental mentalities and urban water usage'. *Environmental Values* **16**: 417–431.

UNESCO. 2000. The Second UN World Water Development Report, *Water, a Shared Responsibility*. New York: Berghahn Books.

Warner, J. ed. 2007. *Multi-Stakeholder Platforms for Integrated Water Management*. Aldershot: Ashgate.

Wessels, J. and R.J.A. Hoogeveen (2006), 'Renovation of Qanats in Syria', http://www.inweh.unu.edu/inweh/drylands/Publications/Wessels.pdf (accessed 10 March 2008).

World Water Form. 2000. Final report of the 2nd WWF, *Vision for Water, Life and the Environment in the 21st century*, The Hague

World Water Forum. 2006. Final Report of the 4th WWF, *Local Actions for a Global Challenge*, Mexico.

Worster, D. 1992. *Rivers of Empire. Water, Aridity, and the Growth of the American West*. New York/Oxford: Oxford University Press.

Mohammad Reza Balali, Jozef Keulartz and Michiel Korthals

Yazdi, S.A.A. and M.L. Khaneiki. 2007. 'Exploitation of groundwater in Iran from the last century to the present'. Proceedings of the 5th IWHA conference on the past and future of water, 13–17 June 2007, Tampere, Finland.

III. Indigenous Subjectivities:
Perception, Myth, Memory

Deforestation, Erosion, and Fire: Degradation Myths in the Environmental History of Madagascar

Christian A. Kull

Tany mena tsy mba mirehitra
The red earth isn't on fire, it just looks like it

(Malagasy proverb)

INTRODUCTION

Madagascar receives enormous amounts of attention as a hot spot of biological diversity, environmental degradation, and conservation action. As a result of rapid deforestation and species extinctions, the islands hosts a frantic effort by development and environmental organisations to establish protected areas and improve resource management (Kull 1996). Accompanying this conservation effort is a specific narrative, a story of exotic nature and environmental destruction, which is used to justify conservation fundraising, policies, and actions. This story – promulgated in the media, travel guides, television documentaries, song lyrics, environmentalists' writings and agency documents – rests on several problematic assumptions and outdated facts (Jarosz 1993; Hardenbergh et al. 1995), yet it persists due to its compelling story line and its usefulness in gaining public and government support. As a result, historical and current processes of environmental transformation in Madagascar are often misunderstood. The task of this paper is to highlight six of the errors, assumptions and rhetorical ploys in the story of Madagascar as an environmental hotspot. Deforestation rates, species extinctions and soil erosion are dramatic enough that the story of environmental degradation need not be overstated.

The past decade has seen a growing academic concern with re-evaluating received wisdoms about the human role in environmental change and analysing the importance of such stories in shaping or justifying environmental politics (Dove 1983; Watts 1985; Showers 1989; Denevan 1992; Kummer et al. 1994; Fairhead and Leach 1996, 1998; Leach and Mearns 1996; McCann 1997; Perevolotsky and Seligman 1998; Batterbury and Bebbington 1999; Bassett and Koli Bi 2000). Fairhead and Leach (1996, 1998), for example, demonstrate that colonial-era assumptions about the forest history of West Africa were incorrect. They expose problems with widely cited statistics of forest decline and describe mistaken assumptions in forest history analyses. For instance, forest islands in savanna zones were previously seen as relics of ancient forests, when they are actually anthropogenic creations. The above studies also discuss the ideologies

Environment and History **6** (2000): 423–450.

and political-economic contexts which permit the creation and persistence of these stories and assumptions. Dove (1983), for example, shows how assumptions about slash-and-burn cultivation have facilitated the expansion of state control and market exploitation into new territories. By bringing these concerns to Madagascar, I hope to help reshape our understanding of environmental transformations in the island's past and present.

A DOMINANT NARRATIVE

A tropical island, Madagascar has a variety of different ecological zones (Figure 1). The humid east coast is separated from the highlands by an abrupt escarpment. The temperate and mountainous highlands have a winter dry season which lasts four to eight months. The west is characterised by gentler, sedimentary relief and a longer dry season. Currently, dense endemic rainforest covers only 10 per cent of the island, while total forest cover is about 23 per cent (DEF 1996). A band of rainforest hugs the eastern escarpment and continues up to the north, while patches of xerophytic forest and spiny brush exist respectively in the west and south. Large expanses of grazed grassland dominate the interior and highlands. Irrigated rice fields, dryland crop fields and orchards account for about five per cent of the surface area.

Early explorers enthusiastically described the island's highly diverse and endemic flora and fauna, a result of 175 million years of tectonic isolation and a wide range of climates and soils. Of the island's approximately 200,000 plant and animal species, three-quarters are found nowhere else, including 97 per cent of non-avian wildlife (Bradt et al. 1996). Missionaries and explorers in the nineteenth century marveled at the biological wonders, but noted with disdain the slash-and-burn cultivation in the eastern region (e.g. Baron 1890, 1891; Elliot 1892) and lamented the treeless 'barren hills' or 'desert' of the highlands (Sibree 1870; Price 1989; see Figure 2). The French conquered the island in 1896 and throughout their 64-year rule criticised Malagasy vegetation burning practices.

It is out of this context that French naturalists Henri Humbert and Henri Perrier de la Bâthie developed the hypothesis that came to dominate discourse about the island's natural history (Perrier de la Bâthie 1921, 1927, 1936; Humbert 1927, 1949, 1955; see also Burney 1997). Their central assertions, summarised and paraphrased from Humbert (1927, 77-8), are as follows.

> Forests covered nearly all of Madagascar before human settlement. Now only a few natural forests remain; by studying these one can see the alarming progress of deforestation, which is caused by *tavy* (shifting cultivation), logging and grassland fires. Instead of establishing permanent plots and irrigated rice fields, the natives cut the remaining forest for *tavy* in order to cultivate temporary crops. Short-sighted commercial logging is no less harmful, impeding regeneration just like *tavy*. Deforestation was begun by *tavy* in the humid regions and advanced

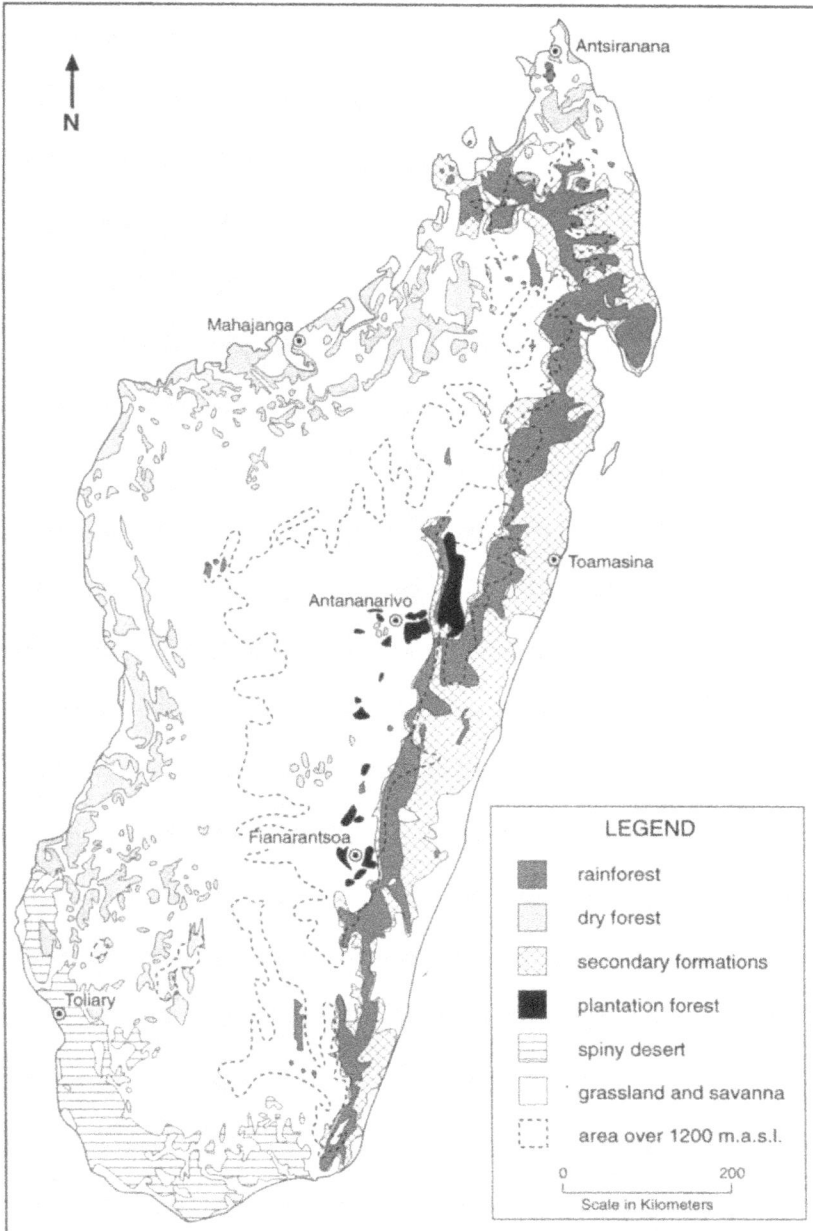

FIGURE 1. Current vegetation zones of Madagascar. Based on AGM (1969), Green and
Sussman (1990), and Conservation International et al. (ca. 1995).

Christian A. Kull

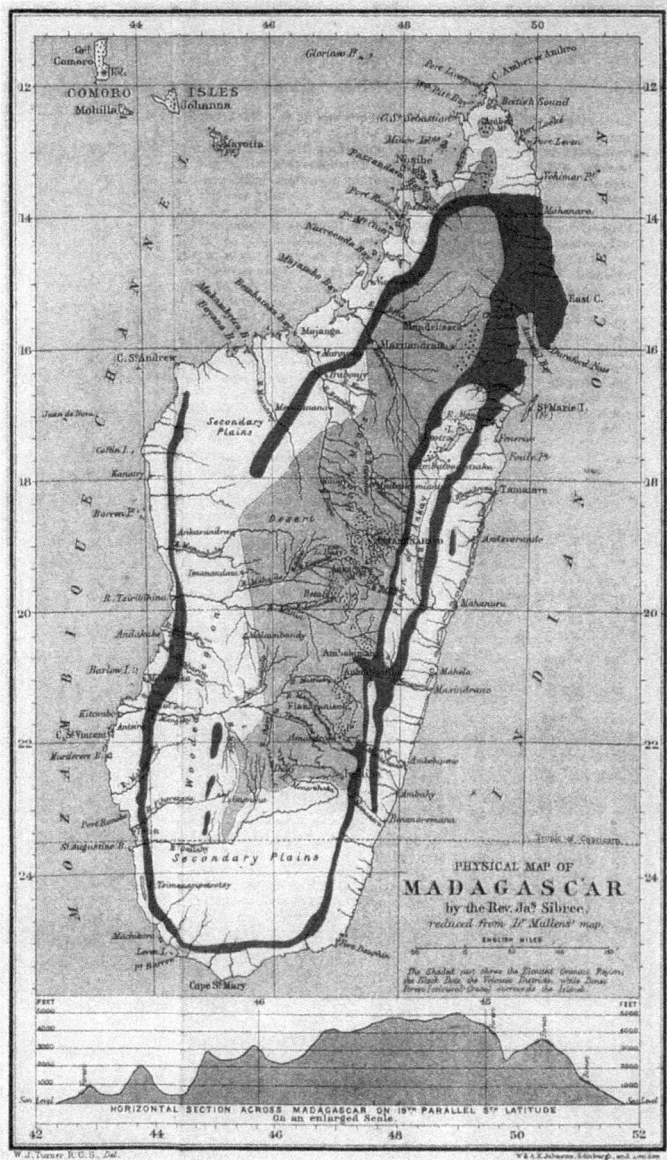

FIGURE 2. James Sibree's (1879) physical map of Madagascar emphasizes the extent of barren hills encountered by early visitors ('desert' and 'high moors') and gives a potentially reasonable, albeit rough, representation of the distribution of 'dense forests' (dark shading) at the time. Light shading denotes the 'elevated granitic regions'; black dots are the 'volcanic districts'.

by rapid-spread fires into the easily flammable forests of the drier highlands and west. The destroyed forest was replaced by grasses which the natives burn annually to renew pastures for their cattle. These fires cause a retreat of forest edges and lead inescapably to the sterilisation of immense areas. The lateritic soils are slowly impoverished, especially in the interior, where the scouring of erosion is rapid after deforestation. The degradation of old pastures pushes the natives to create new ones at the expense of the remaining forest. To avoid the disappearance of the native flora and fauna we must create nature reserves. The reserves will have scientific interest as well as economic interest, since they are refuges of precious seeds and species for industry and pharmaceutics.

Perrier and Humbert's story had a great influence on contemporary writing; Madagascar became considered a type locality for the destruction of indigenous flora by fire and shifting agriculture (Bartlett 1955). The narrative continues to be repeated nearly word-for-word, with only a few modifications, in popular publications (e.g. Murphy 1985; Swaney and Wilcox 1994; Bradt et al. 1996; Holmes 1997; Morell 1999) and by development and environmental organisations (e.g. World Bank 1988; Falloux and Talbot 1993; ONE/Instat 1994; USAID 1997a, 1997b). Meanwhile, scientific understanding of environmental change in Madagascar has continued to evolve. In press and agency documents, however, the Perrier-Humbert story has become a dominant narrative, one that shapes the discursive field within which policy occurs (Ferguson 1990; Roe 1994; Escobar 1995). The story has evolved, of course, since Humbert's day. First, commercial logging is no longer seen as a major threat to forests; instead, attention is focused on *tavy*. Second, the narrative no longer implicitly blames farmer ignorance for *tavy* and grassland fires; instead it sees farmers as squeezed by population growth and poverty, which forces them to make short-term resource-use decisions and ignore long-term sustainability. Finally, scientific evidence from the 1970s that climate desiccation probably played a role in Malagasy vegetation change has recently entered the popular story (e.g. Morell 1999).

The story reaches the public through the media, where journalists use artistic license to dramatise Madagascar's environmental degradation. A particularly effective image is the blood-red, iron oxide-laden rivers and estuaries that 'bleed' into the ocean, visible even from space (e.g. Helfert and Wood 1986; Apt 1996; Holmes 1997; Gallegos 1997; Morell 1999). Another recurrent image is the 'gangrenous wounds' of *lavaka*, the island's erosion gullies (Murphy 1985: 68), also described as 'gaping amphitheatres gouged from barren hills once draped with lush soil-preserving vegetation' (Swaney and Wilcox 1994: 10). These culminate in statements such as 'ravaged by fire, overgrazing, and erosion, the mountains are often entirely devoid of vegetation' (Allen 1995: 4) and 'more than a millennium of slash-and-burn agriculture has produced 100,000 square miles of virtually useless space where people once encountered forest' (ibid.: 13). Such popular accounts of environmental change in Madagascar are full of errors, exaggerations and unquestioned assumptions. The following section seeks to reinterpret the environmental history of the island, addressing both factual discrepancies and explanatory debates, in an attempt to push the dominant narrative towards more defensible claims.

THE MYTHS AND THE DEBATES

The island-wide forest

The narrative asserts that forests once covered the entire island, except for the spiny bush of the desertic southwest, the heathlands of the highest peaks, and the granitic inselbergs of the highlands. Humbert and Perrier's island-wide forest hypothesis is the basis for statements that Madagascar is 90 per cent deforested (see Perrier 1921: 3) and the foundation for much of the dominant narrative. Recent evidence, however, shows convincingly that while humans have dramatically altered the island's vegetation, Holocene forests never covered the entire island.

The Perrier-Humbert island-wide forest hypothesis rested on both empirical observations and inherent assumptions. In the east, the naturalists saw *tavy* eating away at the rainforest, leaving behind fire-climax grasslands. As a result, they concluded that the fire-prone grasslands of the highlands and west must also once have been forested, but had already disappeared due to their drier, more vulnerable characteristics. This logic draws from a Clementsian ecological succession model and a European bias towards forests as the climax vegetation. Maps of climatically-determined 'potential' vegetation zones dominated ecological descriptions of Madagascar; in describing the island's land cover history, it was assumed that each zone used to carry its potential climax vegetation (see Fairhead and Leach 1998: 165).

Empirical observations supporting the Perrier-Humbert hypothesis included forest islands in grassland zones, the testimony of farmers, place names which refer to no-longer extant forests, and the fact that the typically brick- or adobe-built houses in the highlands were constructed of wood until the last century (Parrot 1924; Deschamps 1965; Battistini and Vérin 1967, 1972; Raison 1972). Biogeographical evidence, e.g. from insect distributions across the island (Battistini 1976), also suggested an island-wide forest. Other authors cited the legend of a 'great fire' which seared the highland forests (Savaron 1928; Battistini and Vérin 1967, 1972; Jolly 1980) and statements in the oral history of the central highlands, the *Tantaran'ny Andriana*, which imply that highland forests were more abundant (Berg 1981). Perrier's strongest evidence came from the stratigraphic analysis of highland fossil beds. At Ampasambazimba, near Lake Itasy, burned wood was found among subfossil skeletons, while west of Antsirabe, the Marotampona marsh included a buried layer of tree trunks, branches and fruit that matched the closest existing forests, 80 km to the east (Perrier 1917; Deschamps 1965; MacPhee et al. 1985).

The narrative asserts furthermore that continued burning and a paucity of colonising species prevent the island-wide forest from re-establishing itself. Ecologists theorised that the island's native forest flora evolved in a non-competitive environment, and would thus struggle to regenerate in deforested areas in the face of non-native competitors (Perrier 1936; Koechlin 1972; Koechlin et al. 1974; Gade 1996a). However, Malagasy farmers have names for several pioneering endemic tree species (Rakoto Ramiarantsoa 1995a), and fire exclusion does lead to the growth of woody species (Parrot 1924; Koechlin et al. 1974).

The Perrier-Humbert hypothesis did not convince all contemporaries. Grandidier (1898) argued that the highlands had always been prairie, as did naturalist Baron (see Deschamps 1965) and geographer Gautier (1902). Grandidier based his argument on the fact that highland soils lack a humus horizon typical of forest soils, on the vastness of the grasslands, and on Mayeur's 1777 account of barren highland hills. The debate continued for years, as isolated challenges to various parts of the thesis were raised. Bartlett (1955, 1956) reviewed the literature and decided that there must have been some bushland and savanna before humans arrived – he argued that the extinct three meter high Aepyornis birds probably lived in prairies or open savannas – and that climatic aridification played a role in deforestation. Dez (1970) reached the same conclusion from an alternative reading of the *Tantaran'ny Andriana*, suggesting that the highland landscape in the 13th century consisted of vast grasslands, occasional hillside forests and riparian forests and marshes.

At a widely-attended 1970 conference on conservation in Madagascar, Battistini (1972) posited that the small human population – while perhaps the principal factor for deforestation – could not be held completely responsible, suggesting that climate variations could have contributed. Battistini later suggested that the primeval forest of the highlands was 'perhaps very heterogeneous with vast natural clearings' (Battistini and Vérin 1972: 324). Around the same time, Bourgeat and Aubert (1972) argued from pedologic evidence that highland and western soils are not necessarily relict forest soils, and that the vegetation has changed in the past due to climatic fluctuations. Koechlin (1972) acknowledged that the evidence for an island-wide forest was inconclusive. Soon thereafter Koechlin et al. (1974) argued that the diversity of the flora suggests that forest patches had existed in relative isolation from each other for a long time. They also noted that in the savannas of the southwest, for example, the distribution of vegetation zones including grasslands were closely related to the underlying soils and geology. More recently, Dewar (1984) contended that the central highlands and western regions were originally covered by a mosaic of woodland and savanna, not by a continuous forest. He defended this statement by describing the diverse nature of the highland sub-fossil fauna, including terrestrial herbivores, which suggests a variety of habitats, not just forest.

The evidence that finally overturned the island-wide forest hypothesis appeared in the late 1980s. Palynologist David Burney took cores from several highland lakes and bogs and analysed pollen and charcoal deposits in the sediments. Burney's team has now cored two dozen sites in central, western and northern Madagascar, and similar research has been undertaken by a team of French scientists (Gasse et al. 1994) and by pioneering pollen expert Herbert Straka (1993, 1996). The research shows unequivocally that the highlands and west were never all forest, and instead were a spatial and temporal mosaic of riparian forest, woodlands, heath and grassland. The island's vegetation cover has always been changing; areas covered today by montane rainforest were once heathlands during the last ice age. Charred grass cuticles and woody materials in the sediment cores show that fire was common on the island long before humans arrived around 1500 years ago. However, the arrival of humans is clearly marked in the sedimentary record with a dramatic increase in fire frequency and a significant spread of grasslands (Burney 1987a, 1987b, 1987c, 1993, 1996, 1997; Burney and MacPhee 1988; Dewar and Burney 1994; MacPhee et al. 1985; Matsumoto and Burney 1994).

Where does that leave us? The anthropogenic transformation of Madagascar's environment follows different trajectories in different regions. In the *eastern humid zone*, satellite data show that 66 per cent of the 'original' rainforest has been logged or converted to agriculture and not allowed to re-establish itself, and that at 1980s rates, this rainforest will remain only on the steepest slopes by the year 2025 (Green and Sussman 1990; Sussman et al. 1994). Deforestation is proceeding most rapidly from the east; the western forest boundary near Antsirabe has only retreated 10 km in the past 150 years (Ramamonjisoa 1995), and has remained essentially unchanged from 1950 to 1985 (Green and Sussman 1990: 215n). In the *highlands*, the trajectory of change over the last centuries is from monotonous grasslands towards increasingly imbricated cultural landscapes of irrigated rice, orchards, woodlots and rainfed crop fields. Riparian zones, often the only place where trees were found in the nineteenth century highlands, have been increasingly converted to irrigated rice fields. In the *west*, the boundaries of many forests have remained stable, contradicting assertions that pasture fires eat away at the forest edge (e.g. Chauvet 1972; Roffet 1995). Philippe Morat studied 400 km of forest edge in the west, near Ankazoabo, between 1949 and 1970, and found only three cases of forest decline, the largest of which covered a mere 32 ha (Koechlin et al. 1974). Bertrand and Sourdat (1998: 137) report a study in the Menabe region which documents only minor forest reductions in the past half century. A more recent satellite-based study in the northwest also documents forest stability (Andrianarivo 1990); on the contrary, however, the Ambohitantely forest of the northwestern highlands has clearly shrunk in the face of fire. In the *north*, Gezon and Freed (1999) show that the forests in the Ankarana region have seen pockets of both expansion and decline between 1949 and 1990.

While understandings vary in the environmental community, old ideas based on the Perrier-Humbert hypothesis persist. As a result of the above revisions of the island-wide forest idea, however, one need no longer tolerate the frequent statements that 80 or even 90 per cent of Madagascar's once island-wide forest has been destroyed (e.g. Helfert and Wood 1986; WWF 1992; Bradt et al. 1996; Gade 1996a, 1996b; USAID 1997a). It is simply a fallacy to conclude from the statistic 'Madagascar is 11% forested' that the island is therefore 89% deforested. Similarly, the evidence contradicts assertions that grasslands could not have predated human arrival (e.g. Battistini 1996; Gade 1996a; Lowry et al. 1997). Unfortunately, however, reliable statistics of whole-island deforestation rates since 1900 (or other specified moments in history) will remain unavailable until a baseline is established through careful triangulation between historical maps, explorer's accounts, archival sources and additional paleoecological research.

Causes and causers of deforestation

Deforestation is a critical issue in eastern Madagascar. The popular narrative's analysis of the causes and of the actors in this process contains several over-simplifications and misunderstandings. It asserts that the forest is being decimated by the rural Malagasy, pushed by population growth and poverty to expand and intensify *tavy* agriculture further into the forest (e.g. World Bank 1988; WWF 1992; Allen 1995; Gallegos 1997; USAID 1997b; Webster 1997). The rural Malagasy are portrayed schizophrenically as both ignorant, backward farmers without a care for biodiversity and as potentially wise indigenous resource managers. In this section, I seek to complicate the story of deforestation causality and to contribute to a more realistic picture of the actors.

Population growth is clearly an important factor in pushing farmers to expand *tavy* further into the forest and to reduce fallow periods between *tavy* cycles. Sussman et al. (1994), for example, show that deforestation rates from 1950 to 1985 and population density are positively correlated in the eastern rainforest. Care should be exercised, however, when applying a population-based model of deforestation, for it is the context, not population growth *in itself*, which determines the trajectory of change. In a context of abundant, unregulated resources, population growth will likely cause expanding use of those resources, as it has in eastern Madagascar. However, where population growth is associated with limited resources – whether physically finite or constrained by access and tenure laws – it can, under the right conditions, facilitate landscape enhancements, like aforestation or soil conservation (Tiffen et al. 1994; Fotsing 1992; Fairhead and Leach 1996; Kull 1998; Gray 1999). In the Malagasy highlands, a densely-settled region, vast lands once maintained as grassland now sport imbricated crop fields, rice terraces, woodlots and fruit trees (Ramamonjisoa 1995; Bertrand 1995; Rakoto-Ramiarantsoa 1995a, b; Kull 1998). In these cases, the impact of population pressure was shaped by limited resources, market opportunities, tenure incentives and government policies, resulting in constructive environmental transformations. Therefore, one should be careful in placing blame on population growth; the story can be more complicated.

Poverty is also blamed for forcing rural Malagasy to sacrifice nature for short-term needs. This logic does not hold for two reasons. First, wealth can also produce degradation. Give the average Malagasy *tavy* farmer more money, and deforestation may just as well *increase* as they utilise better tools and pay for additional labour. Second, poverty or no poverty, in their own perspective, Malagasy farmers are not sacrificing nature for short-term needs, they are instead transforming nature to be of more use to them. It is a matter of perspective. Much of Madagascar is a lightly-populated resource frontier, where extensive land-use strategies such as *tavy* and pasture-burning may be both economically and agronomically logical (Boserup 1965; Dove 1983). Poverty does not drive the system, farmer rationality does.

The causes of deforestation, while currently heavily related to population growth near an open resource frontier, deserve additional complication: the story of deforestation should be historically and regionally contextualised. Many specific cases of deforestation involve not just *tavy* farmers, but also commercial interests or political factors. Several examples follow. In the first 30 years of colonial rule, population was stable, and deforestation was largely caused by logging concessions and agricultural displacement for cash-crop cultivation (Jarosz 1993). In this period, between one and seven million ha of primary eastern rain forest were logged, out of an estimated eleven million ha (AOM 1922–30; Humbert 1927, 1949; Heim 1935; Guillermin 1947; Chauvet 1972; Jarosz 1993). Unfortunately, reliable statistics on the nature and extent of the forest cutting, as well as on forest regeneration since that period, do not exist. Today, intensive logging does not threaten Malagasy forests as it does in southeast Asia, yet it does have local impacts. In the 1970s and early 1980s, eleven concessionaires and hundreds of porters operated in the town of Ranomafana, logging the hardwood timber *Dalbergia baroni* (Peters 1994). Other regional forest impacts include commercial cash crop production and timber harvesting in the Ankarana region of the northwest, export maize production in the forests near Tulear, and 1970s famine alleviation policies legalising additional *tavy* in the southeast (Hardenbergh et al. 1995). The Amoronkay region, southeast of the capital, suffered serious deforestation before colonisation due to the charcoal needs of the iron industry (AOM 1900; Dez 1970). In the 1940s, 40,000 ha of forest near Morondava were sacrificed to cultivate corn for World War II; during the 1947 rebellion, rebels hid and farmed in the forests at Betampona and Zahamena (Humbert 1949). Finally, significant deforestation in the Mahafaly spiny forest occurred due to government angora goat raising and missionary successes in fighting traditional beliefs including superstitions which protected the forest (Esoavelomandroso 1986).

The actors in this story of deforestation also need to be better understood, both for their environmentally constructive potential and the constraints placed upon them by the socio-political and economic context. The environmental history of Madagascar stars the Malagasy people as its principal characters. In this story, however, the *tantsaha* (agriculturalists) are both the antagonists, the primary enemies of the environment (Hardenbergh et al. 1995; Hanson 1996), and the protagonists, the wise indigenous people with whom conservationists must work in order to safeguard nature. Morell's (1999) piece is a good example of the two-faced picture of the Malagasy: she describes their 'poor agricultural practices' and features miners who strip and scour the land, but also highlights the potential of working with rural Malagasy to protect forests through traditional *dina* agreements, ecotourism and butterfly farming. Both the antagonist and the protagonist roles are flawed. The antagonist role implies a mistrust and denigration of the *tantsaha* as trespassers and pyromaniac destroyers in an ecological paradise. The *tantsaha* are like farmers and herders the world over, making use of available technologies and resources to gain their livelihoods. Instead of seeing the *tantsaha* as ignorant or desperate, we should recognise their ability to manage resources to their advantage (Esoavelomandroso 1989), whether in expanding *tavy* cultivation, managing their soils for nutrients and erosion (Rakoto Ramiarantsoa 1995a), or maintaining their woodlands for forest products (Gautier 1902: 262; Parrot 1924; Koechlin et al. 1974; Kull 1999b). Sometimes these resource uses can be short-sighted or contrary to environmental goals, yet often they are productive and constructive transformations, at least from the perspective of the *tantsaha*. Rarely would *tantsaha* continue practices clearly detrimental to their own livelihoods! The protagonist role, on the other hand, may excessively romanticise the local people. While the *tantsaha* are generally very experienced in the local conditions of resource management, this does not imply that all can exercise this knowledge equally, nor that a community is homogenous and will make conflict-free decisions. Land distribution, class splits and ethnic or gender differences affect the ability of people to control and manage resources (Schroeder 1993; Suryanata 1994; Peluso 1996), or how market incentives and national politics shape local patterns of resource use (Blaikie 1985; Hecht 1985; Grossman 1993).

The *tantsaha* are experienced resource managers, constrained by socio-political struggles over resource control, by market demands, and by government policies, not ignorant destroyers pushed by hunger and poverty. Their contributions to deforestation vary from place to place and time to time. While population growth fuels much of the phenomenon in the second half of the Twentieth Century, the story of deforestation should include regional exceptions or complimentary factors, as described above, in order to be more accurate.

From incendiarism to fire ecology

In the dominant narrative, fire is the primary tool of deforestation and biological impoverishment and an evocative symbol of destruction. Fires of all types, from contained *tavy* burns in the forest to vast pasture fires in the grasslands, are often lumped together. In this section, I emphasise grassland burning and argue that these fires should be seen as an integral part of Malagasy ecological and agropastoral systems, not as a 'national scourge' (RDM 1980). There has always been, and should always be, fire in Madagascar.

The popular narrative decries the fires which, it writes, 'blacken the hills, sterilise the soils, erode the land, and decimate the forests'. The basic assumption, based on old Western ecological thought, is that fire is external to normal ecosystem processes (Pyne 1995). Burney (1987b, 1996, 1997), however, has demonstrated that fires frequently burned in Madagascar long before humans arrived, ignited by lightning and volcanism. As a result, several endemic species exist that demonstrate adaptations to fire, including the trees *Agauria salicifolia*, *Uapaca bojeri*, *Uapaca densifolia* and *Ziziphus mucronata* (Koechlin et al. 1974; Gade 1985; Rakotoarisetra 1997). Unfortunately, most literature on Madagascar has not come to terms with the huge body of ecological work on fire's role in ecosystems, especially from Australia, South Africa and the United States (see e.g. Goldammer 1990; Braithwaite 1996; Pyne et al. 1996). This literature demonstrates the importance of fire in habitat and species diversity, nutrient cycling and plant reproduction in a wide variety of ecosystems, including tropical grasslands, savannas and woodlands. For example, fires and grazing in the alti-montane prairie of Madagascar's Andringitra mountains have been shown to increase the diversity of terrestrial orchids (Rabetaliana et al. 1999).

The natural role of fire in the highland and western ecosystems of Madagascar was modified when humans arrived. Since settlement, the Malagasy have controlled fire and used this powerful tool to their advantage, as have humans across the globe. Whether to simplify hunting, to increase habitat for favoured species, or to increase grass production for domesticated ungulates, periodic burning is a well-established practice throughout history, from British moorland herders to Australian aborigines, to Maasai pastoralists, and to modern ranchers in Kansas (Homewood and Rodgers 1991; Pyne 1995). In California, up to 13 per cent of all non-desert lands – including forests – may have burned yearly before white settlement (Martin and Sapsis 1991); in Australia's Northern Territories, two-thirds of the land is burned annually (Braithwaite 1991). In Madagascar, one-fourth to one-third of the grasslands are burnt each year (Jolly 1980). These fires primarily serve to renew pastures for cattle grazing, by removing unpalatable lignified grasses and encouraging early sprouting. Many authors call these fires irrational (Bosser 1954; Murphy 1985; Neuvy 1986), yet prescribed burning every one to three years is an accepted management technique in some grasslands, and pasture fires do not necessarily increase erosion (Pyne et al. 1996). The *tantsaha* in the highlands and west also use fire to clear fallow fields or grasslands for plowing, to encourage fertilisation of downslope paddies by erosion, and to control pest populations such as rats and locusts.

Clearly, fires are the main tool by which humans long ago transformed the mosaic vegetation of the highlands into vast grasslands, and by which the *tantsaha* maintain these pastures in their present state. The dominant narrative views these fires as inherently destructive, only rarely allowing for their legitimate role in the agropastoral system (e.g. Jolly 1990; Bloesch 1999). As I described earlier, grassland fires are in general not eating away at forest edges; in the case of the *Uapaca bojeri* woodlands of the highlands, fire may even aid forest maintenance (Kull 1999b). The consequences of *not* burning are even more dramatic: the build-up of fuel could engender destructive wildfires, and the loss of pasture would damage an already fragile rural economy. Fire deserves to be recognised as a key management tool used by the *tantsaha* for hundreds of years in maintaining their landscapes for productive use (Kull 1999a).

Perspectives on erosion

Along with deforestation and fire, soil erosion is a key refrain in the environmental discourse on Madagascar. The World Bank crowned the island 'world champion of erosion', and accounts ceaselessly describe the red, sediment-laden rivers. The consequences of erosion are locally quite severe (Tricart 1953; Le Bourdiec 1972; Rossi 1979; Randrianarijaona 1983; Paulian 1984). The irrigation works at Lake Alaotra, the nation's principle rice-producing area, fill up with 30 to 60 million tons of soil each year. Siltation of the Ikopa River near the capital led to devastating floods in 1982 which left 70,000 people homeless; siltation of the western port of Mahajanga proceeds at almost one metre per year, necessitating costly dredging. In some spots, the deeply incised *lavaka* erosion gullies exceed 30 per km^2 and threaten roadways (Wells and Andriamihaja 1993). In the face of these serious examples of harmful erosion, however, the narrative makes several errors of exaggeration or misinterpretation (Dez 1966: 1227n).

First, the narrative focuses on the visually compelling *lavaka* erosion gullies and thus is prone to over-dramatising erosion. *Lavaka* are vertical-walled gullies in lateritic substrate, reaching 300 m in length and 70 m in width. While *lavaka* are obvious, their contribution to overall erosion often pales compared to the more insidious process of sheet erosion (Stocking 1987, 1996). In addition, *lavaka* in Madagascar are a natural part of this active landscape's evolution in response to tectonism and climatic aridification; they existed before human settlement on the island. Human activities – including pasture burning, cattle grazing, trails, hillfort trenches, canal construction and hillslope cultivation – unequivocally cause only one-quarter of *lavaka* (Wells and Andriamihaja 1993).

Second, authors frequently report shocking statistics of erosion rates without placing them in context. To begin with, statements – such as 'Environmental degradation also results in a rapid loss of an estimated 200 tons per hectare each year' (USAID 1997b: 1) – are misleading, for the figure of 200 t/ha represents extreme cases of erosion and not a widespread situation as implied in

Christian A. Kull

the text. Other authors cite rates varying from 25 to 250 t/ha/yr in the highlands (Randrianarijaona 1983), or 30 t/ha/yr on eastern escarpment *tavy* plots (Roffet 1995). In addition, soil erosion figures are meaningless without a consideration of regional climate, background erosion rates and soil formation rates (Stocking 1986). A soil suffering high rates of erosion in one region may be less degraded than a different soil undergoing light erosion elsewhere, for soils can vary widely in their tolerance to erosion (Hurni 1983).

Third, the sedimentation rates of Madagascar's major rivers are frequently cited as indicators of environmental crisis. The Betsiboka estuary receives 15-50 million t/yr of sediments annually, while the Mangoky River deposits 19-20 million t/yr at its mouth (Randrianarijaona 1983; Paulian 1984). While the rates

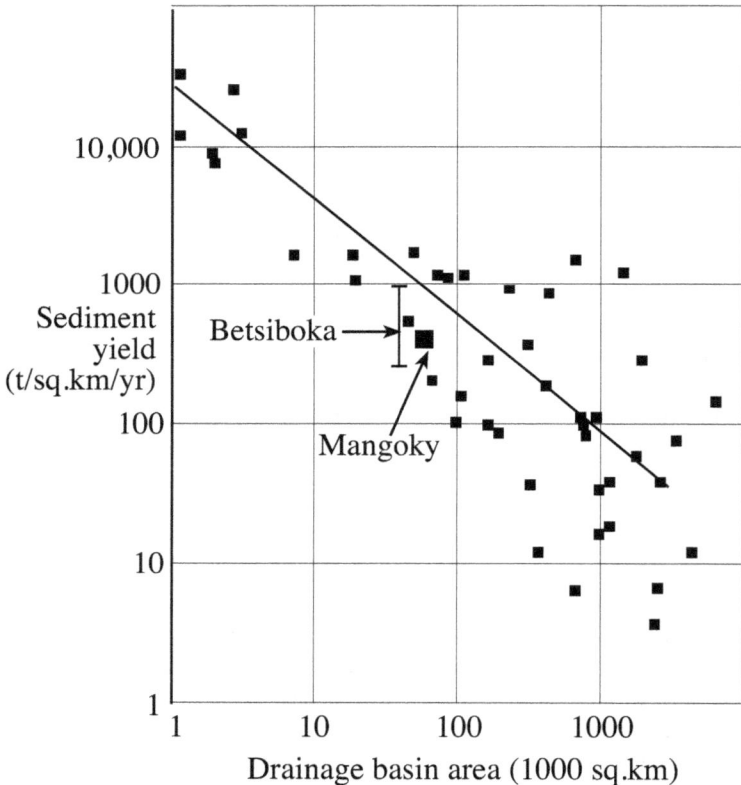

FIGURE 3. A comparison of Madagascar's sediment-clogged rivers with global rivers shows that they are nothing out of the ordinary. Graph plots sediment yields and drainage basin areas for all major sediment-discharging rivers (greater than 10 million t/yr). The Mangoky River drains 55,750 km², and deposits 19-20 million t/yr. The Betsiboka River drains 49,000 km² and deposits between 15-50 million t/yr (Randrianarijaona 1983; Paulian 1984; Ritter 1986: 195).

may appear extreme, they match well with a comparison of sediment yields with drainage basin area of all major sediment-discharging rivers (Ritter 1986; see Figure 3). Based on Figure 3, one concludes that the Malagasy rivers do not represent exceptional soil erosion; rather, their sediment load is comparable to other areas of high sediment yields.

Fourth, the narrative misunderstands the human role in erosion. As mentioned earlier, theories linking dense populations to erosion have recently been contested (Blaikie and Brookfield 1987; Fotsing 1992; Tiffen et al. 1994). More importantly, soil and nutrients lost from one place can either benefit or harm downstream resource users, and this process is well understood by farmers. Farmers in the alluvial *baiboho* of northwestern Madagascar depend on the fertile soils eroded from the uplands (Rabearimanana 1994). Fieldwork in eastern Madagascar demonstrates that *tantsaha* actively manage locally for upslope erosion in order to fertilise rice fields and improve swamp soils (B. Locatelli, pers. comm., 1996), a trend documented elsewhere in Madagascar and the tropics (Wells and Andriamihaja 1993; Rakoto Ramiarantsoa 1995a; Stocking 1996). In sum, the above lacunae in the collective environmental understanding of Madagascar's erosion leads to the over-dramatisation of a phenomenon that is locally destructive yet also 'one of life's certainties' (Stocking 1996: 140).

Woodfuel, agriculture and grazing

The environmental narrative also makes problematic statements and exaggerations with regards to Madagascar's woodfuel supply, agricultural potential and grazing management.

First, woodfuel has been called the most crucial future environmental issue for the island (Vérin 1994; Richard and O'Connor 1997; Gade 1996b). However, the woodfuel crisis expected across the globe during the past 15 years has largely failed to materialise, as prognosticators underestimated alternative fuel supplies, market responses and the ability of trees to grow (Foley 1987; Leach and Mearns 1996; Ribot 1996). In highland Madagascar, the woodfuel 'crisis' has lasted 200 years, spurring the use of alternative fuels, higher prices and tree planting. Nineteenth-century observers noted that highlanders relied on crop residues, cattle dung and dried grass due to wood scarcity (Ellis 1859; Berg 1981; Gade 1996a). Likewise, Sucillon (1897) and Parrot (1924) both commented on the rising and prohibitive cost of fuelwood in the highlands. Instead of a worsening crisis, however, the situation has improved. Today, private eucalyptus woodlots cover 70 per cent of the Manjakandriana region, east of the capital, once described as barren hills of grass (Bertrand 1995; Rakoto Ramiarantsoa 1995a, b), and similar spontaneous zones of woodfuel production are scattered throughout the island.

Second, Madagascar is sometimes portrayed as a nation which cannot feed itself, with poor soils, little agricultural potential and few natural resources (e.g. Neuvy 1986; Gade 1996b). However, only one-tenth of arable lands are cultivated (Allen 1995), including only one-fifth of the *baiboho* alluvial plains of the northwest (Rabearimanana 1994), and a 1961 report from the UN Food and Agriculture Organisation posits that the island has enough resources to support 200 million people (J. Randby pers. comm).

Finally, some authors complain of degradation due to overgrazing (e.g. Allen 1995; Wells and Andriamihaja 1997). Gade (1996b) paints an image of ever-growing cattle herds. Assessments of overgrazing are contested across Africa (Behnke and Scoones 1992) and elsewhere (Perevolotsky and Seligman 1998), and have not specifically been demonstrated in Madagascar. While the human population doubled between 1960 and 1990, the bovine population has hovered around seven million since at least the 1920s (Perrier 1921; Ramaroson and Razafindrakoto 1972/3; ONE/Instat 1994).

Confusion and conflation

Finally, the popular story of Madagascar is often guilty of geographic confusion and conflation. The storytellers forget that the situation in the humid, tropical eastern zone deserves a different analysis than processes in the highlands or west. First, *tavy*, the slash-and-burn cultivation typical of the eastern rainforests, is blamed for consequences elsewhere. Allen (1995) and Berg (1981), for example, blame *tavy* for highland deforestation, when in all likelihood repeated fires related to cattle-raising converted the majority of the highland vegetation mosaic into grasslands (Dewar 1984). *Tavy* is blamed for the siltation of rivers and bays (e.g. Roffet 1995; Apt 1996), yet the rivers experiencing siltation originate in non-*tavy* areas. Others (e.g. Helfert and Wood 1986; Gallegos 1997) imply that *tavy* leads directly to the *lavaka* erosion gullies typical in various parts of the highlands, yet most of these areas are outside the zone of present or historical *tavy*.

Second, the pasture fires of the highlands and west receive blame for deforestation of the eastern rainforest. The 1994 'State of the Environment' report from ONE/Instat, for example, includes an assessment of pasture fires within the section titled 'deforestation'. Ironically, the statistics presented in this assessment show that only 0.1% of the yearly fires, or 2000 ha, occurred in forests! Such confusion and conflation are no doubt often a result of poetic puffery on the part of the authors, yet behind them stands a muddled conception of environmental change in Madagascar shaped by the politics of environmentalism. It is to these politics and to the power of narrative which I turn in the following sections.

NARRATIVES, POWER AND POLICY

Our views of Madagascar as an environmental basket-case are intricately bound up in our cultural imagination, in a colonial and post-colonial history of power and influence, and in real trends of forest loss and extinctions. Many of the errors and misperceptions I highlighted above resulted from the particularities of a Western encounter with an exotic land. Adams and McShane (1992) have shown how our images and preconceptions of Africa – the myths of wildness and of backwards and destructive natives – developed into narratives of degradation that necessitated top-down, reserve-based conservation action. In Madagascar, the story was shaped by outdated ecological notions of climax vegetation and the mistrust of fire, both outgrowths of the ecological context of northern Europe and eastern America (Pyne 1995). The story was developed with a Western bias towards settled, permanent agriculture over swidden techniques. In an era of technocratic modernism and scientific forestry, foreign expertise was valued over local experience (Peluso 1992; Fairhead and Leach 1996; Ribot 1996).

There are a variety of stories possible for Madagascar's environmental history, not all equally valid. The dominance of a certain narrative depends on its correspondence with observed facts, yet also on the power of those telling it. The interpretations of colonial botanists Perrier and Humbert were heard over the stories of the colonised, and this dominance was maintained up through the 1960s. During the early years of socialism, President Ratsiraka used his power to promote an alternative view of resource abundance. Then, as Madagascar suffered in debt and asked for foreign assistance, environmentally-minded donor countries gave the Perrier and Humbert narrative new credence. In the early 1990s, conservation bureaucrats gained unprecedented stature in national politics, especially with the adoption of the Environmental Action Plan. It is telling that the government's *Office National pour l'Environnement* occupies some of the most visible real estate in downtown Antananarivo. The narrative of environmental degradation now serves as a useful means to justify policy interventions as varied as reserve establishment, privatisation of land tenure and agro-technology support, and in fact has become central to the dominant neo-liberal economic development narrative (Bassett and Koli Bi 2000).

The dominant narrative is bound up in the politics, values and structure of the group spreading the story, as well as in the facts of the matter. As Leach and Mearns (1996) point out, received wisdoms are resistant to change not because the facts are not known, but due to political-economic and institutional factors. Received wisdoms are always politically convenient for one interest group or another. Berg notes that deforestation was seen by some as 'a myth invented by colonial administrators intent on agrarian reform' (1981: 295). The more dramatic the story, the more support and conviction can environmentalists bring to their task.

Christian A. Kull

As in the case of eastern Madagascar's deforestation, there are often stark truths embedded within the story and one might ask why a little exaggeration matters. I have been warned several times to be extremely careful in writing this paper, lest it lead to the rethinking of environmental policies in Madagascar. Yet that is my intent, for policies based on myths can be harmful and unproductive. For example, biases about pastoralists led to policies which denied critical resources to the Maasai in the Ngorongoro Conservation Area of Tanzania (Homewood and Rodgers 1991; McCabe et al. 1992). In Madagascar, environmental myths historically contributed to the conclusion that protected areas are the best strategy for conservation, *tavy* and grassland fires should be banned, and land use should be put under more 'rational' management (Dez 1968; Jarosz 1993; Hardenbergh et al. 1995). Humbert (1949) called for increased enforcement power for the Malagasy forest service. The traditional protected areas approach that arose from such ideas has been strongly criticised in Madagascar and elsewhere for leading to resource scarcity, marginalisation of the poor and problems of coercion and resistance. Some projects have the result of decoupling local people from their livelihoods and cultural resource base instead of reinforcing their relations with their environment (Adams and McShane 1992; Hardenbergh 1993; Peluso 1993; Ghimire 1994; Sussman et al. 1994). In the words of Fairhead and Leach (1998: 192):

> ...inhabitants are denied not only their claims and control over valued resources, but also their own understandings of vegetation dynamics and the ecological and social histories with which these are intertwined.

Conservation projects, of course, evolve with the times as they gain experience and take account of new ideas. In Madagascar, the incorporation of local people in conservation actions has been standard policy for ten years now. More recent legislation paves the way to decentralise resource management and transfer to local communities direct control over renewable natural resources such as *tavy* forests and pastures. Finally, conservation programs have moved beyond a strict protected areas focus to an approach based on ecoregions.

However, while agencies adjust their approaches to accommodate new understandings, the stories which appear in the first pages of agency documents, in fundraising letters, and in the popular press continue to replicate the old myths. These are the stories that sell. As Bill McConnell writes (unpub. manus. 1998): 'unfortunately, the declared rationale for continued, and indeed expanding, international efforts at biodiversity conservation stubbornly clings to turn-of-the-[last]-century explanations'. Like Fairhead and Leach (1996) and Roe (1994), I propose the creation of an alternative narrative based in a greater share of 'objective truths'. I aim to accelerate the absorption of this alternative narrative – based on new theories and revised understandings already prevalent in the scientific community – into the world of environmental operators and the popular press.

CONCLUSION

Received wisdoms about the environmental history of Madagascar include much confusion, misunderstanding and misrepresentation. Deforestation and erosion, while very real trends, are exaggerated due to mistaken ideas about pre-settlement forest extent and the eye-catching red soils and erosion gullies. The role of fire, principal tool of landscape change and pasture maintenance, is unnecessarily demonised. Blame is placed on the Malagasy people and problems of poverty and population growth, ignoring economic interests, historical political contexts, community politics and the potential of the people to positively manage their resources. Of course, myths and poetic license are nothing new to the literature on Madagascar, dating back to Marco Polo's roc and the story of the man-eating tree (Dahle 1884). However, as advances in science and painstaking work have increased the scope of our knowledge, we ought to separate myths from fact. New facts, like Burney's paleoecological data points, make a great difference in our understanding of Malagasy environmental history. As these new facts come to light, an alternative story needs to be told, one that is self-reflexive and open to revision. The chapters of the new story are there – in this paper, in the work of Madagascar scholars and in the memory and experience of the *tantsaha* – all that remains is the writing.

ACKNOWLEDGMENTS

I gratefully thank Alison Richard, Bob Dewar, Gerald Berg, Joelisoa Ratsirarson, Deb Paulson, Nabiha Megateli, Bob Sussman, Bill McConnell, Ned Horning, Bruno Ramarorazana, Alain Bertrand and especially David Burney for their help and constructive criticism. I, of course, retain sole responsibility for any errors and the final content of this paper. This research was supported by a graduate fellowship from the US Environmental Protection Agency; the paper was first presented at the 1997 Annual Meeting of the Association of American Geographers.

REFERENCES

Adams, J. S. and T. O. McShane 1992. *The Myth of Wild Africa*. New York: W.W. Norton and Co.

A.G.M. 1969. *Atlas de Madagascar*. Tananarive: Bureau pour le Développement de la Production Agricole and Association des Géographes de Madagascar.

Allen, P. M. 1995. *Madagascar: Conflicts of Authority in the Great Island*. Boulder: Westview Press.

Andrianarivo, J. A. 1990. Analysis of Forest Cover Changes and Estimation of Lemur Population in Northwestern Madagascar Using Satellite Digital Data. Ph.D. Dissertation, Duke University.

AOM 1900. Archives d'Outre Mer, Aix-en-Provence, France. Rapport du garde général des eaux et forêts Lhotelain au Gouv. Gén. de Madagascar, Tananarive, 5. Sept. 1990 (mad-ggm-d/5(18)/1).

AOM 1922-30. Various documents at Archives d'Outre Mer, Aix-en-Provence, France, including Rapport du Service Forestier 1922 (mad-ggm-d/5(18)/5); Direction des Domaines, Etat des baux et concessions forestiers consentis pendant l'année 1922 (mad-ggm-5(4)d13); Rapport général sur le fonctionnement du Service Forestier en 1930 (mad-ggm-d/5(18)/15); Notes de l'inspecteur principal des eaux et forêts Griess (mad-ggm-D/5(18)/9).

Apt, J. 1996. The astronaut's view of home. *National Geographic* **190**(5): 1-30.

Baron, R. 1890. A Malagasy forest. *Antananarivo Annual* **4**(14): 196-211.

Baron, R. 1891. The flora of Madagascar. *Antananarivo Annual* **4**(15): 322-357.

Bartlett, H. H. 1955. *Fire in Relation to Primitive Agriculture and Grazing in the Tropics: Annotated Bibliography.* Ann Arbor: University of Michigan Department of Botany.

Bartlett, H. H. 1956. Fire, primitive agriculture, and grazing in the tropics. In W. L. J. Thomas (ed.) *Man's Role in Changing the Face of the Earth*, pp. 692-720. Chicago: University of Chicago Press.

Bassett, T. J. and Z. Koli Bi. 2000. Environmental discourses and the Ivorian savanna. *Annals of the Association of American Geographers* **90**(1): 67-95.

Batterbury, S. P. J. and A. J. Bebbington 1999. Environmental histories, access to resources, and landscape change: an introduction. *Land Degradation and Development* **10**(4): 279-89.

Battistini, R. 1972. L'homme et l'équilibre de la nature à Madagascar. In *Comptes rendus de la Conférence internationale sur la Conservation de la Nature et de ses Ressources à Madagascar, Tananarive 7-11 Octobre, 1970.* pp. 91-94. Publications UICN Nouvelle Série, Document Supplementaire No. 36. Morges, Switzerland: IUCN.

Battistini, R. 1976. Une preuve de la continuité de l'ancienne couverture forestière à Madagascar: la biogéographie des nymphalides, et specialement des charaxes (lépidoptères). *Madagascar Revue de Géographie* **28**: 123-32.

Battistini, R. 1996. Paleogeographie et variété des milieux naturels à Madagascar et dans les Iles voisines: quelques données de base pour l'étude biogéographique de la <<region Malgache>>. In W. R. Lourenço (ed.) *Biogéographie de Madagascar*, pp. 1-17. Paris: Editions de l'ORSTOM.

Battistini, R. and P. Vérin 1967. Ecologic changes in protohistoric Madagascar. In P. S. Martin and H. E. J. Wright (eds.) *Pleistocene Extinctions: the Search for a Cause*, pp. 407-424. New Haven: Yale University Press.

Battistini, R. and P. Vérin 1972. Man and the environment in Madagascar. In R. Battistini and G. Richard-Vindard (eds.) *Biogeography and Ecology in Madagascar*, pp. 311-338. The Hague: Junk Pub.

Behnke, R. H. and I. Scoones 1992. *Rethinking Range Ecology: Implications for Rangeland Management in Africa.* International Institute for Environment and Development Paper No. 33. Overseas Development Institute.

Berg, G. M. 1981. Riziculture and the Founding of Monarchy in Imerina. *Journal of African History* **22**(3): 289-308.

Bertrand, A. 1995. La sécurisation foncière, condition de la gestion viable de ressources naturelles renouvelables? In F. Ganry and B. Campbell (eds.) *Sustainable land management in African semi-arid and subhumid regions*, pp. 313-327. Montpellier: CIRAD.

Bertrand, A. and M. Sourdat 1998. *Feux et déforestation à Madagascar, revues bibliographiques*. Antananarivo: CIRAD - ORSTOM - CITE.

Blaikie, P. 1985. *The Political Economy of Soil Erosion*. London: Methuen.

Blaikie, P. and H. Brookfield 1987. *Land Degradation and Society*. New York: Methuen.

Bloesch, U. 1999. Fire as a tool in the management of a savanna/dry forest reserve in Madagascar. *Applied Vegetation Science* **2**: 117-124.

Boserup, E. 1965. *The Conditions of Agricultural Growth*. Chicago: Aldine.

Bosser, J. 1954. Les paturages naturels de Madagascar. *Mémoires de l'Institut Scientifique de Madagascar* **Série B, V**: 65-77.

Bourgeat, F. and G. Aubert 1972. Les sols ferralitiques à Madagascar. *Madagascar Revue de Géographie* **20**: 1-23.

Bradt, H., D. Schuurman and N. Garbutt 1996. *Madagascar wildlife: a visitor's guide*. Bucks, UK: Bradt Publications.

Braithwaite, R. W. 1991. Aboriginal fire regimes of monsoonal Australia in the 19th century. *Search* **22**(7): 247-9.

Braithwaite, R. W. 1996. Biodiversity and fire in the savanna landscape. In O. T. Solbrig, E. Medina and J. F. Silva (eds.) *Biodiversity and Savanna Ecosystem Processes*, pp. 121-140. Berlin: Springer Verlag.

Burney, D. A. 1987a. Late Holocene vegetational change in central Madagascar. *Quaternary Research* **20**: 130-143.

Burney, D. A. 1987b. Late Quaternary stratigraphic charcoal records from Madagascar. *Quaternary Research* **28**: 274-280.

Burney, D. A. 1987c. Pre-settlement vegetation changes at Lake Tritrivakely, Madagascar. *Palaeoecology of Africa and the Surrounding Islands* **18**: 357-381.

Burney, D. A. 1993. Late holocene environmental changes in arid southwestern Madagascar. *Quaternary Research* **40**: 98-106.

Burney, D. A. 1996. Climate change and fire ecology as factors in the Quaternary biogeography of Madagascar. In W. R. Lourenço (ed.) *Biogéographie de Madagascar*, pp. 49-58. Paris: Editions de l'ORSTOM.

Burney, D. A. 1997. Theories and facts regarding Holocene environmental change before and after human colonization. In S. M. Goodman and B. D. Patterson (eds.) *Natural and Human-induced Change in Madagascar*, pp. 75-89. Washington: Smithsonian Press.

Burney, D. A. and R. D. E. MacPhee 1988. Mysterious island. *Natural History* **97**(7): 46-55.

Chauvet, B. 1972. The forest of Madagscar. In R. Battistini and G. Richard-Vindard (eds.) *Biogeography and Ecology in Madagascar*, pp. 191-200. The Hague: Junk Pub.

Conservation International, DEF, CNRE, and FTM ca. 1995. *Formations végétales et domaine forestier national de Madagascar*. Map 1:1,000,000. Antananarivo: CI, DEF, CNRE, and FTM.

Dahle, L. 1884. Geographical fictions with regard to Madagascar. *Antananarivo Annual* **2**(8): 403-407.

DEF 1996. Inventaire Ecologique Forestier National. Report, République de Madagascar, Ministère de l'Environnement, Plan d'Actions Environnementales PE1, Direction des Eaux et Forêts, Nov. 1996.

Denevan, W. M. 1992. The pristine myth: the landscape of the Americas in 1492. *Annals of the Association of American Geographers* **82**(3): 369-385.

Deschamps, H. 1965. *Histoire de Madagascar*. Paris: Editions Berger-Levrault.

Dewar, R. E. 1984. Extinctions in Madagascar: the loss of the subfossil fauna. In P. S. Martin and R. G. Klein (eds.) *Quaternary Extinctions: a Prehistoric Revolution*, pp. 574-593. Tucson: University of Arizona Press.

Dewar, R. E. and D. Burney 1994. Recent research in the paleoecology of the highlands of Madagascar and its implications for prehistory. *Taloha* **12**: 79-88.

Dez, J. 1966. Les feux de végétation: aperçus psycho-sociologiques. *Bulletin de Madagascar* **247**: 1211-29.

Dez, J. 1968. La limitation des feux de végétation. *Tany Malagasy/Terre Malgache* **4**: 97-124.

Dez, J. 1970. Elements pour une étude de l'economie agro-sylvie-pastorale de l'Imerina ancienne. *Tany Malagasy/Terre Malgache* **8**: 9-60.

Dove, M. R. 1983. Swidden agriculture and the political economy of ignorance. *Agroforestry Systems* **1**(1): 85-99.

Elliot, G. F. S. 1892. Notes on a botanical trip in Madagascar. *Antananarivo Annual* **4**(16): 394-8.

Ellis, R. W. 1859. *Three Visits to Madagascar During the Years 1853-1854-1856*. New York: Harper and Brothers.

Escobar, A. 1995. *Encountering Development*. Princeton: Princeton University Press.

Esoavelomandroso, M. 1986. La forêt dans le Mahafale aux XIXe et XXe siècle. Séminaire de l'U.E.R. d'Histoire 'Arbres et Plantes à Madagascar', Benasandratra, May 1-4, 1986.

Esoavelomandroso, M. 1989. La destruction de la forêt par l'homme Malgache: un problème mal posé. In M. Maldague, K. Matuka and R. Albignac (eds.) *Environnement et Gestion des Ressources Naturelles dans la Zone Africaine de l'Ocean Indien*, pp. 219-222. Paris: Unesco.

Fairhead, J. and M. Leach 1996. *Misreading the African Landscape*. Cambridge: Cambridge University Press.

Fairhead, J. and M. Leach 1998. *Reframing Deforestation*. London: Routledge.

Falloux, F. and L. M. Talbot 1993. *Crisis and Opportunity*. London: Earthscan.

Ferguson, J. 1990. *The Anti-Politics Machine*. Cambridge: Cambridge University Press.

Foley, G. 1987. Exaggerating the Sahelian woodfuel problem? *Ambio* **16**(6): 367-71.

Fotsing, J. M. 1992. Stratégies paysannes de gestion des terroirs et de lutte antiérosive en pays bamiléké (Ouest Cameroun). *Bulletin Réseau Erosion* **12**: 241-254.

Gade, D. W. 1985. Savanna woodland, fire, protein and silk in highland Madagascar. *Journal of Ethnobiology* **5**(2): 109-122.

Gade, D. W. 1996a. Deforestation and its effects in highland Madagascar. *Mountain Research and Development* **16**(2): 101-116.

Gade, D. W. 1996b. *Madagascar*. American Geographical Society Around the World Program. Blacksburg, Virginia: McDonald and Woodward.

Gallegos, C. M. 1997. Unrealized potential: Madagascar. *Journal of Forestry* **95**(2): 10-15.

Gasse, F., E. Cortijo, J.-R. Disnar, L. Ferry, E. Gibert, C. Kisselet al. 1994. A 36 ka environmental record in the southern tropics: Lake Tritrivakely (Madagascar). *Comptes Rendus de l'Academie Scientifique de Paris* **318**: 1513-1519.

Gautier, E. F. 1902. *Madagascar: Essai de Géographie Physique*. Paris: Augustin Challamel.

Gezon, L. L. and B. Z. Freed 1999. Agroforestry and conservation in northern Madagascar: hopes and hinderances. *African Studies Quarterly* **3**(2): <web.africa.ufl.edu/asq>.

Ghimire, K. B. 1994. Parks and people: livelihood issues in national parks management in Thailand and Madagascar. *Development and Change* **25**(1): 195-229.

Goldammer, J. G., Ed. 1990. *Fire in the Tropical Biota*. Berlin: Springer Verlag.

Grandidier, A. 1898. Le boisement de l'Imerina. *Bulletin du Comité de Madagascar* **4**(2): 83-87.

Granier 1972. A propos du déséquilibre bétail/sol/végétation en savanne. In *Comptes rendus de la Conférence internationale sur la Conservation de la Nature et de ses Ressources à Madagascar, Tananarive 7-11 Octobre, 1970.* pp 132-4. Publications UICN Nouvelle Série, Document Supplementaire No. 36. Morges, Switzerland: IUCN.

Gray, L. 1999. Is land being degraded? A multi-scale investigation of landscape change in southwestern Burkina Faso. *Land Degradation and Development* **10**: 329-43.

Green, G. M. and R. W. Sussman 1990. Deforestation history of the eastern rain forests of Madagascar from satellite images. *Science* **248**: 212-215.

Grossman, L. 1993. The political ecology of banana exports and local food production in St. Vincent, eastern Caribbean. *Annals of the Association of American Geographers* **83**(2): 347-367.

Guillermin, A. 1947. Les forêts. In M. d. Coppet (ed.) *Madagascar*, pp. 23-42. Paris: Encyclopédie de l'Empire Français.

Hanson, P. W. 1996. Coming to terms with the people of Madagascar. Madagascar Cultural Alliance, http://www.mcai.org/people.htm.

Hardenbergh, S. H. B. 1993. Undernutrition, illness, and children's work in an agricultural rain forest community of Madagascar. Ph.D. dissertation, University of Massachusetts Amherst.

Hardenbergh, S. H. B., G. Green and D. Peters 1995. The relationship of human resource use, socioeconomic status, health, and nutrition near Madagascar's protected areas: an assessment of assumptions and solutions. In B. D. Patterson, S. M. Goodman and J. L. Sedlock (eds.) *Environmental Change in Madagascar*, pp. 57-58. Chicago: The Field Museum.

Hecht, S. B. 1985. Environment, development and politics: capital accumulation and the livestock sector in eastern Amazonia. *World Development* **13**(6): 663-684.

Heim 1935. L'état actuel des dévastations forestières à Madagascar. *Revue de Botanique Appliquée at d'Agriculture Tropicale* **15**: 416-8.

Helfert, M. R. and C. A. Wood 1986. Shuttle photos show Madagascar erosion. *Geotimes* **31**(3): 4-5.

Holmes, H. 1997. Mudhopping in Madagascar. *Sierra*. 82: 22-3.

Homewood, K. M. and W. A. Rodgers 1991. *Maasailand Ecology*. Cambridge: Cambridge University Press.

Christian A. Kull

Humbert, H. 1927. Principaux aspects de la végétation à Madagascar. La destruction d'une flore insulaire par le feu. *Memoires de l'Academie Malgache* **Fascicule V**.

Humbert, H. 1949. La dégradation des sols à Madagascar. *Mémoires de l'Institut de Recherche Scientifique de Madagascar* **D1**(1): 33-52.

Humbert, H. 1955. Les territoires phytogéographiques de Madagascar. Leur cartographie. *Année Biologique* **31**(5-6): 439-448.

Hurni, H. 1983. Soil erosion and soil formation in agricultural ecosystems, Ethiopia and northern Thailand. *Mountain Research and Development* **3**(2): 131-142.

Jarosz, L. 1993. Defining and explaining tropical deforestation: shifting cultivation and population growth in colonial Madagascar (1896-1940). *Economic Geography* **69**(4): 366-379.

Jolly, A. 1980. *A World Like Our Own*. New Haven: Yale University Press.

Jolly, A. 1990. On the edge of survival. In F. Lanting, photographer, *Madagascar: A World Out of Time* , pp. 110-21. New York: Aperture.

Koechlin, J. 1972. Flora and vegetation of Madagascar. In R. Battistini and G. Richard-Vindard (eds.) *Biogeography and Ecology in Madagascar*, pp. 145-190. The Hague: Junk.

Koechlin, J., J.-L. Guillaumet and P. Morat 1974. *Flore et Végétation de Madagascar*. Vaduz: J. Cramer.

Kull, C. A. 1996. The evolution of conservation efforts in Madagascar. *International Environmental Affairs* **8**(1): 50-86.

Kull, C. A. 1998. Leimavo revisited: agrarian land-use change in the highlands of Madagascar. *Professional Geographer* **50**(2): 163-76.

Kull, C. A. 1999a. Observations on repressive environmental policies and landscape burning strategies in Madagascar. *African Studies Quarterly* **3**(2): <web.africa.ufl.edu/asq>.

Kull, C. A. 1999b. Woodland stability, non-timber forest products, and fire ecology in the Madagascar highlands. Annual Meeting of the Association of American Geographers, Honolulu, March 24-7, 1999.

Kummer, D., R. Concepcion and B. Canizares 1994. Environmental degradation in the uplands of Cebu. *Geographical Review* **84**(3): 266-276.

Le Bourdiec, P. 1972. Accelerated erosion and soil degradation. In R. Battistini and G. Richard-Vindard (eds.) *Biogeography and Ecology in Madagascar*, pp. 227-259.

Leach, M. and R. Mearns, Eds. 1996. *The Lie of the Land*. Portsmouth, NH: Heinemann.

Lowry, P. P. I., G. E. Schatz and P. B. Phillipson 1997. The classification of natural and anthropogenic vegetation in Madagascar. In S. M. Goodman and B. D. Patterson (eds.) *Natural Change and Human Impact in Madagascar*, pp. 93-123. Washington: Smithsonian Institution Press.

MacPhee, R. D., D. A. Burney and N. A. Wells 1985. Early Holocene chronology and environment of Ampasambazimba, a Malagasy subfossil lemur site. *International Journal of Primatology* **6**(5): 463-489.

Martin, R. E. and D. B. Sapsis 1991. Fire as agents of biodiversity: pyrodiversity promotes biodiversity. Proceedings of the Symposium on Biodiversity of Northwestern California, Santa Rosa, California, Oct. 28-30 1991.

Matsumoto, K. and D. A. Burney 1994. Late Holocene environments at Lake Mitsinjo, northwestern Madagascar. *The Holocene* **4**(1): 16-24.

McCabe, J. T., S. Perkin and C. Schofield 1992. Can conservation and development be coupled among pastoral people? An examination of the Maasai of the Ngorongoro Conservation Area, Tanzania. *Human Organization* **51**(4): 353-366.

McCann, J. C. 1997. The plow and the forest: narratives of deforestation in Ethiopia, 1840-1992. *Environmental History* **2**(2): 138-159.

Morell, V. 1999. Restoring Madagascar. *National Geographic*. 60-69.

Mullens, J. 1875. *Twelve Months in Madagascar*. London: James Nisbet.

Murphy, D. 1985. *Muddling through in Madagascar*. Woodstock, NY: The Overlook Press.

Neuvy, G. 1986. Facteurs inhibiteurs de la production rizicole à Madagascar. In P. Vennetier (ed.) *Crise Agricole et Crise Alimentaire dans le Pays Tropicaux*, Bordeaux: Centre National de la Recherche Scientifique.

ONE/Instat 1994. *Rapport sur l'Etat de l'Environnement à Madagascar*. Antananarivo: PNUD/Banque Mondiale.

Parrot, A. 1924. Déboisement et reboisement à Madagascar. *Bulletin Economique* **4ème trimestre**: 192-195.

Paulian, R. 1984. Madagascar: a microcontinent between Africa and Asia. In A. Jolly, P. Oberlé and R. Albignac (eds.) *Key Environments: Madagascar*, pp. 1-26. Oxford: Pergamon Press.

Peluso, N. L. 1992. *Rich Forests, Poor People*. Berkeley: University of California Press.

Peluso, N. L. 1993. Coercing conservation? The politics of state resource control. *Global Environmental Change* **3**(1): 119-218.

Peluso, N. L. 1996. Fruit trees and family trees in an anthropogenic forest: ethics of access, property zones, and environmental change in Indonesia. *Comparative Studies in Society and History* **38**(3): 510-548.

Perevolotsky, A. and N. G. Seligman 1998. Role of grazing in Mediterranean rangeland ecosystems. *BioScience* **48**(12): 1007-17.

Perrier de la Bâthie, H. 1917. Au sujet des tourbières de Marotampona. *Bulletin de l'Academie Malgache* **Nouv. Ser., 1**: 137-138.

Perrier de la Bâthie, H. 1921. La végétation Malgache. *Annals du Musée Colonial de Marseille* **Sér. 3, v. 9**: 1-266.

Perrier de la Bâthie, H. 1927. Le Tsaratanana, l'Ankaratra, et l'Andringitra. *Mémoires de l'Académie Malgache* **3**.

Perrier de la Bâthie, H. 1936. *Biogéographie des Plantes de Madagascar*. Paris: Société d'Éditions Géographiques, Maritimes et Coloniales.

Peters, W. J. 1994. Attempting to integrate conservation and development among resident peoples of the Ranomafana National Park, Madagascar. Ph.D. dissertation, North Carolina State University.

Price, C. T. 1989. *Missionary to the Malagasy*. New York: Peter Lang.

Pyne, S. J. 1995. *World Fire*. New York: Henry Holt and Co.

Pyne, S. J., P. L. Andrews and R. D. Laven 1996. *Introduction to Wildland Fire*. New York: Wiley.

Rabearimanana, G. 1994. Le Boina. In J.-P. Raison (ed.) *Paysanneries Malgaches dans la Crise*, pp. 1-153. Paris: Karthala.

Rabetaliana, H., M. Randriambololona and P. Schachenmann 1999. The Andringitra National Park in Madagascar. *Unasylva* **50**(196): 25-30.

182

Christian A. Kull

Raison, J.-P. 1972. Utilisation du sol et organisation de l'espace en Imerina ancienne. *Tany Malagasy/Terre Malgache* **13**: 97-121.

Rakoto Ramiarantsoa, H. 1995a. *Chair de la Terre, Oeil de l'Eau.* À travers champs. Paris: Éditions de l'Orstom.

Rakoto Ramiarantsoa, H. 1995b. Les boisements d'eucalyptus dans l'est de l'Imerina (Madagascar). In C. Blanc-Pamard and L. Cambrézy (eds.) *Terre, Terroir, Territoire*, pp. 83-103. Paris: ORSTOM.

Rakotoarisetra, F. N. 1997. Monographie de l'Uapaca densifolia dans la forêt d'Ambohitantely. Mémoire de fin d'études, Université d'Antananarivo, Ecole Supérieur des Sciences Agronomiques.

Ramamonjisoa, J. R. 1995. Le Processus de Développement dans le Vakinankaratra, Hautes Terres Malgaches. Thèse de doctorat d'Etat, Université de Paris I - Panthéon/ Sorbonne.

Ramaroson, S. and D. Razafindrakoto 1972-3. L'élevage à Madagascar: situation actuelle et perspectives d'avenir. *Tany Malagasy/Terre Malgache* **14**: 1-22.

Randrianarijaona, P. 1983. The erosion of Madagascar. *Ambio* **12**: 308-311.

RDM 1980. Rapport du Groupe Interministeriel d'Etudes sur les Feux de Brousse. Repoblika Demokratika Malagasy, Novembre 1980,

Ribot, J. C. 1996. Participation without representation: chiefs, councils and forestry law in the West African Sahel. *Cultural Survival Quarterly* **Fall**: 40-44.

Richard, A. F. and S. O'Connor 1997. Degradation, transformation, and conservation: the past, present, and possible future of Madagascar's environment. In S. M. Goodman and B. D. Patterson (eds.) *Natural change and human impacts in Madagascar*, pp. 406-18. Washington: Smithsonian Institution Press.

Ritter, D. F. 1986. *Process Geomorphology.* Dubuque: Wm. C. Brown Pubs.

Roe, E. 1994. *Narrative Policy Analysis.* Durham: Duke University Press.

Roffet, C. 1995. Madagascar, entre conservation et développement. *Habbanae* **36**: 7-9.

Rossi, G. 1979. L'érosion à Madagascar: l'importance des facteurs humaines. *Cahiers d'Outre Mer* **32**(128): 355-370.

Savaron, C. 1928. Contribution à l'histoire de l'Imerina. *Bulletin de l'Academie Malgache* **Nouv. Ser. Vol. 11**: 61-81.

Schroeder, R. A. 1993. Shady practice: gender and the political ecology of resource stabilization in Gambian garden/orchards. *Economic Geography* **69**(4): 349-365.

Showers, K. B. 1989. Soil erosion in the Kingdom of Lesotho: origins and colonial response, 1830s-1950s. *Journal of Southern African Studies* **15**(2): 263-286.

Sibree, J. 1870. *Madagascar and its People.* London: The Religious Tract Society.

Sibree, J. 1879. History and present condition of our geographical knowledge of Madagascar. *Proceedings of the Royal Geographical Society* **NS 1**(10): 646-65.

Stocking, M. 1987. Measuring land degradation. In P. Blaikie and H. Brookfield (eds.) *Land Degradation and Society*, pp. 49-63. London: Methuen.

Stocking, M. 1996. Soil erosion: breaking new ground. In M. Leach and R. Mearns (eds.) *The Lie of the Land*, pp. 140-154. Portsmouth, NH: Heinemann.

Straka, H. 1993. Beiträge zur Kenntnis der Vegetationsgeschichte von Madagaskar (Vorläufige Mitteilung). *Festschrift Zoller, Dissertationes Botanicae* **196**: 439-449.

Straka, H. 1996. Histoire de la végétation de Madagascar oriental dans les derniers 100 millenaires. In W. R. Lourenço (ed.) *Biogéographie de Madagascar*, pp. 37-47. Paris: Editions de l'ORSTOM.

Sucillon, L.-C. 1897. Notes sur le cercle de Tsiafahy. *Colonie de Madagascar. Notes, Reconnaissances et Explorations* 2(8): 118-127.

Suryanata, K. 1994. Fruit trees under contract: tenure and land use change in upland Java, Indonesia. *World Development* 22(10): 1567-1578.

Sussman, R. W., G. M. Green and L. K. Sussman 1994. Satellite imagery, human ecology, anthropology, and deforestation in Madagascar. *Human Ecology* 22(3): 333-354.

Swaney, D. and R. Wilcox 1994. *Madagascar and Comoros: a Travel Survival Kit.* Hawthorn, Australia: Lonely Planet.

Tiffen, M., M. Mortimore and F. Gichuki 1994. *More People, Less Erosion.* Chichester: John Wiley and Sons.

Tricart, J. 1953. Erosion naturelle et erosion anthropogène a Madagascar. *Revue de Geomorphologie Dynamique* 4 suppl.: 225-230.

USAID 1997a. Congressional Presentation FY 1997 on Madagascar.

USAID 1997b. Strategic objectives agreement, USAID and Government of Madagascar. http://www.info.USAID.gov/countries/mg/sntzsoag.htm.

Vérin, P. 1994. *Madagascar.* Paris: Karthala.

Watts, M. J. 1985. Social theory and environmental degradation. In Y. Gradus (ed.) *Desert Development*, pp. 14-32. Dordrecht: D. Reidel Pub. Co.

Webster, D. 1997. The looting and smuggling and fencing and hoarding of impossibly precious, feathered and scaly wild things. *The New York Times Magazine*. 26-33; 48-61.

Wells, N. A. and B. Andriamihaja 1993. The initiation and growth of gullies in Madagascar: are humans to blame? *Geomorphology* 8: 1-46.

Wells, N. A. and B. R. Andriamihaja 1997. Extreme gully erosion in Madagascar and its natural and anthropogenic causes. In S. M. Goodman and B. D. Patterson (ed.) *Natural Change and Human Impact in Madagascar*, pp. 44-74. Washington: Smithsonian Institution Press.

World Bank 1988. *Madagascar Environmental Action Plan, preliminary version.* World Bank, USAID, Coop. Suisse, Unesco, UNDP, WWF.

WWF 1992. *WWF-International Conservation Programme Packet 1992-1993.* Gland, Switzerland: WWF.

Renarrating a Biological Invasion:
Historical Memory, Local Communities and Ecologists

Karen Middleton

INTRODUCTION

My paternal grandmother, my ancestor, told me the story of how Malagasy Cactus was once our food.

When *vazaha* [French colonisers] arrived in this land, they could see no people because they were all hiding in the thickets of prickly pear. This intrigued the *vazaha*. 'What are you doing in the woods?' they asked.

'Oh, we're just in the woods', came the reply.

'Ok', said the *vazaha*, 'explain properly. How do you survive in the woods? What do you eat?'

'We get by on prickly pear cactus', replied the *gasy*. 'That's our staple food'.[1]

You see people in those days weren't farmers. They didn't grow crops, didn't wear clothes. All they did was eat prickly pears in the woods.

The French were surprised to hear this! It was something they'd never encountered before.

Our food, Malagasy Cactus, was a wonderful food for us *gasy*. It made us podgy, plump. The colonisers killed it because they couldn't find workers. We didn't *choose* to give up Malagasy Cactus. It was killed by the state because people hid in the thickets.

The French at Toliara fetched *poizy* to kill Malagasy Cactus. So then there was no food. And the French ordered them to stop fighting, to stop the killing. All people did in those days was quarrel and murder one another in amongst the prickly pears.

And after Malagasy Cactus died the *vazaha* said 'Here are hoes, here are axes, grow crops, cultivate fields. Here's money. You'll be *gasy* no longer. You'll become *vazaha*.'

And they distributed *raketambazaha* [lit. foreign, coloniser's or white man's cactus]. *Raketambazaha* was already around. But when Malagasy Cactus died, the French ordered [people] to plant it so 'you can eat the leaves'.[2]

Interview, Sikina, Befeha, 3/12/2002.

Environment and History **18** (2012): 61–95.
 doi: 10.3197/096734012X13225062753624

Karen Middleton

Invasion by alien species has attracted a great deal of academic attention over recent decades.[3] A rising concern with the economic and environmental impacts of introduced species has also stimulated interest in methods of control.[4] The historical case study has become a key tool in developing understandings of these phenomena. Natural scientists look regularly to the past to provide a broader evidential base for theory about invasive species and their management than can be provided by the standard scientific methodologies.[5] Scholars in the humanist disciplines explore the importance of interests, values, needs and aspirations in determining attitudes to exotic species by revisiting contestations in the past.[6] By excavating historical narratives about species introductions, they highlight the deeply cultural significances that underpin the very concepts of 'useful plants' and 'weeds'.[7]

For scholars of all disciplines, however, written documents can be less than satisfactory points of departure. A natural science enquiry into alien species based on the searching of historical records often encounters unreliable species identifications and a paucity of quantitative data that limit the value of the case study in building predictive models,[8] while marked asymmetries in the opinions recorded for posterity restrict insights into historical debates. It is generally accepted that the cost–benefits of introduced species are seldom distributed evenly among 'stakeholders'. Moreover, many major incidents of species spread and control took place in European colonies. People at the margins of state, the argument might go, are usually less well represented in written sources than are elites, governments, and colonial rulers. Thus, the histories of introduced species are likely to be retold more from the perspective of scientists and government officials than of the local communities who actually lived with the plants and were impacted on by their control.

The problems of imbalance in written sources can be overstated. Marginal 'voices' are often better represented in written records than is at first apparent and/or can be recovered by reading such records for the 'silences' or 'against the grain'. Even so, it may still be thought useful to supplement search of the written records with the collection of oral evidence from local people, particularly in instances of species spread and eradication that occurred in living memory. The precedents for such an undertaking appear encouraging. An extensive literature in environmental history now claims to draw on local knowledge and social memory to provide alternative insights into past landscapes, often challenging received wisdoms.[9]

There is, however, hardly a consensus on the place of memory in academic history-writing.[10] Many historians value oral testimony, seeing it as a source specific to subordinate and marginalised historical experience and thus a way of producing alternative (and in colonial contexts genuinely non-Western) histories. Some even question the preference given by 'scientific history' to written documentation as historical evidence.[11] But others view popular narratives as a qualitatively different and less reliable resource than written material.

Highlighting a ceaseless dialectic of remembering and forgetting, they point to the role of narrated pasts in identity construction and to the way memory is co-opted by contemporary political projects.[12] Some note that 'memory' is often characterised by epistemologies and mythical temporalities that are at odds with the linear perspectives of time that inform modern history. In its strongest version, this approach reduces memory to a form of cultural production that constantly reconstructs the past in line with present needs. From this perspective, there is no true or false memory. People are in effect 'remembering the present' (to deploy Johannes Fabian's turn of phrase) rather than the past.[13]

This paper addresses some of these issues by exploring community memories of a biological control programme in French colonial Madagascar involving insect predation on a prickly pear. I shall refer to this prickly pear as 'Malagasy Cactus' in translation of *raketa gasy*, the name my Malagasy informants gave to the plant.[14] A variety of Opuntia (Cactaceae) native to South America, it had been introduced into Madagascar in the late eighteenth century, via Tôlañaro (Fort Dauphin) in the southeast. By the late nineteenth century, when France annexed Madagascar as a colony, the plant was found throughout the island but had become especially dominant, some say invasive, in its southernmost parts. Boom subsequently turned to bust in the 1920s when the plant succumbed to an introduced cochineal insect within the space of a few years.

In its day the circumstances and consequences of this event caused intense controversy in French colonial circles. While commentators agreed that an extraordinary landscape transformation had been wrought in southern Madagascar, they agreed on little else. One faction emphasised the socio-economic benefits that Malagasy Cactus had brought to vulnerable people trying to survive in a dryland while the opposing faction saw dense, thorny thickets that grew up to 4–5 metres high and 7–8 metres deep as an obstacle to 'progressive' forms of land and labour use.[15] In the following decades relatively little was published on the incident. On the whole, opinion within the colonial administration came to view the intervention as ill-advised. Today, there has been a resurgence of interest in the story, partly in the context of modern environmental concerns.[16] Identified early as a 'hotspot' for megabiodiversity, Madagascar has long been the focus for international conservation efforts but it was only in the mid-1990s that greater recognition of biodiversity in the island's 'spiny dry forest' began to focus global attention on the natural history of the Malagasy Deep South.[17]

Given the enduring impasse over key issues in Malagasy Cactus history, it is tempting to turn to social memory in an attempt to address all kinds of questions on which we seem unable to agree on the basis of available written evidence. Was Malagasy Cactus invasive? How important, economically speaking, was the plant? To what extent were local people dependent on Malagasy Cactus? Did they welcome its spread? Was Malagasy Cactus killed accidentally or intentionally? And, one of the most contested points in colonial literature, did people die in large numbers as a result?

At first blush local people would seem ideally placed to settle at least some of these questions. No one, surely, is better able to remember Malagasy Cactus and what happened when it died than those who lived with the plant and who directly experienced its loss? Yet longstanding debates over the significance of memory for historical reconstructions, even of quite recent events, suggest possible difficulties with such an approach.

This essay has three objectives. First, by comparing and contrasting two sets of narrative collected in the early 1980s and the early 2000s, it charts the radical revisions that local memories of Malagasy Cactus have undergone over these two decades. Secondly, it shows how stories told about the eradication of Malagasy Cactus have become a powerful rhetorical tool in the context of a present-day controversy over another prickly pear. Experience of biological invasion in the present has been reshaping historical memory, while reinterpreted narrative of past biological control is informing current debates. Thirdly, it relates these narrative shifts to broader political and social developments in Madagascar, highlighting the influence of new environmental discourse and new forms of *vazaha* [stranger] experience – in the shape of green governmentality and humanitarian assistance – in mediating historical narrative. Altogether these data *seem* to support a view of memory as cultural production of present time and as a less than reliable source for verifying the past. And yet, by exploring the complexity of memory production, including the interplay between personally acquired autobiographical memory and socially transmitted memory, the article also points to the possibility of a more nuanced approach to local knowledge of the past.

CONTEXT OF STUDY

The interview data reported in this essay were collected in the Karembola region of dryland Madagascar (Tsihombe and Beloha Districts) in 1981–1983 and in 2002–2003. The first data set was collected almost incidentally during extended fieldwork for a doctorate in social anthropology. The second data set – a more diverse set of narratives – was collected twenty years later in the course of environmental history research. From the perspective of studying memory, it is important to register not only the twenty years that had elapsed between the two field studies but also the fact that both sets of interview data relate to an event that took place in the late 1920s – that is, several decades before. Thus, even in the first study in the early 1980s 'eyewitness' informants were already drawing on distant memories of Malagasy Cactus and its eradication while younger informants were 'remembering' events they had not themselves seen.[18] Secondly, it should be noted that narrators in the second study (2002–2003) did not have the benefit of written materials and tape-recordings I had made in the early 1980s and so were generally unaware of the ways in which local narrative, including their own narrative, had changed over the intervening years. In a non-literate society such as rural Karembola, where transmitted memory is oral rather than inscribed, important narrational shifts can occur without the narrators themselves being necessarily conscious of the fact. Thirdly, it is relevant to note that, in addition to collecting oral histories, as a social anthropologist I have always employed participant-observation techniques aimed at eliciting in-depth qualitative understandings of local knowledge and practice and that in both field studies this approach enabled me to go beyond formal interview data to explore the ways in which Malagasy Cactus narration was used in everyday life. Fourthly, it is worth recording that, while I draw occasionally on archival research to comment on interview material in this article, detailed discussion of the written records through confrontation of multiple data sources for the purposes of historical reconstruction is not my objective here.

Turning to the broader economic and political contexts of the comparative study, in some respects rural Karembola changed little between the early 1980s and the early 2000s. In both fieldwork periods almost all villagers were subsistence farmers, cropping manioc, maize, sweet potatoes, millet, pumpkins, melons, squash and various legumes, with restricted market participation. With harvests entirely dependent on hand hoe technology and low, poorly distributed and unreliable rainfall, agricultural productivity was low. Many households also reared zebu cattle and small ruminants that they sold to buy in food when field crops failed. However, wealth in livestock was very unequally distributed and in both periods there were households in all communities with inadequate holdings to meet their economic needs. Some cash was derived from selling extractive plant products (ricin, periwinkle) and lobster via middlemen to global markets or from trading fish and contraband tobacco. But local opportunities for income diversification were limited and in many households cultivated foods were necessarily supplemented in the dry season and extended droughts by trapped and gathered foods. Some Karembola migrated to other regions of Madagascar where they typically found work as agricultural labourers, night watchmen and rickshaw pullers.

In the early 1980s, as in the early twenty-first century, prickly pears made a crucial contribution to local food security, enabling farmers to bridge food and fodder shortfalls. The fruits and cladodes (leaf pads) have been harvested for human consumption, the succulent cladodes exploited to feed and water livestock, and the spinier varieties used to make stout hedging for vegetable gardens and cattle pens.[19] It might even be the case that prickly pears became more important during the 1990s as the introduction of free market economics and the growing appropriation of land for nature conservation, coupled possibly with environmental change, contributed to the growing pauperisation of many households in this already disadvantaged region. Indeed, while Karembola to date has been less affected by external mining and conservation interventions than districts further east, the experiences of economic liberalisation, green governmentality, and humanitarian assistance have all reshaped Malagasy Cactus narrative in significant ways.

MALAGASY CACTUS: THE 'NATIONALIST' NARRATIVE

In the early 1980s, when I first conducted fieldwork in Karembola, there was a shared narrative about Malagasy Cactus. Briefly, the story went like this: Malagasy Cactus was a plant native to Karembola, a plant the ancestors found growing in profusion when they came into the land. Its fruit and leaf pads became staples for the ancestors and their cattle, bringing them prosperity in an otherwise thirsty land. The French colonial administration, however, hated Malagasy Cactus because people hid among the thickets and survived on the fruit. The plant made it difficult to pacify the region, to recruit labour for public works and settler plantations and to get the taxes in. So the v*azaha* introduced an insect to kill the plant. The consequences were devastating as people and cattle, suddenly deprived of food and water, either died in vast numbers or fled the region for other parts of the colony, many never to return. 'The time of the ancestors' was over, and 'foreign time' began. Karembola was left a 'broken land' where 'people could only nod their heads in agreement whenever *vazaha* spoke'.[20]

For present purposes, I shall dub this the 'nationalist' version of Malagasy Cactus history because it overlaps with narrative that appeared in the Malagasy nationalist press soon after the event. (The most obvious difference between the two versions is that while the nationalist press in the early 1930s offered an essentially secular account of the colonial intervention, my informants embedded it in a more 'religious' account.)

In the early 1980s this narrative was a communal narrative in the sense that it was known and reproduced by all Karembola I met. In fact, the story formed a normative component of our encounter, following shortly on after the greeting 'Hail *vazaha*! What brings you to this thirsty land?' A defining statement of the *gasy–vazaha* relationship, narrative around 'the moment the cactus died' at this time provided the foundational myth of Karembola society and the cornerstone of Karembola identity.

That the demise of Malagasy Cactus had led to such a strong and consistent nationalist narrative can, I think, be taken as indicative of the plant's erstwhile importance to Karembola culture and economy. It would be wrong to suppose, however, that narrators were solely concerned to remember the defining moment in Karembola history.

Another reason why rural Karembola in the 1980s continued to renarrate the decades-old tragedy was because Malagasy Cactus narrative offered an oblique way of commenting on more recent political developments. Such an interpretation may at first seem counter-intuitive in that the narrative of French complicity, which forms the core of the nationalist narrative, seems wholly in keeping with the anti-imperialist rhetoric of the early Ratsiraka state. Coming to power in the wake of the Malagasy Socialist Revolution, this was a state which actively encouraged its citizens to recover memories of colonial suffering as a way of contrasting itself to its immediate predecessor, Tsiranana's so-called 'neo-colonial' regime (1960–1972).[21] In fact, by the early 1980s, deteriorating economic conditions and a top-down socialism meant that few Karembola still bought into the idea of a 'second independence'. Rather, they described themselves as living in a land still controlled by *vazaha*, where Karembola could only nod their heads in agreement to whatever *vazaha* proposed. In this context, the master narrative of Malagasy Cactus with its reflections on state power and Karembola impotence remained as apposite a commentary on governance and economy under the Second Republic as it had been under colonial rule.[22]

NEW CENTURY, NEW NARRATIVE

I have described the nationalist version of Malagasy Cactus history that was current in early 1980s Karembola. Returning to the subject in the early twenty-first century (2002–03), I was genuinely startled to discover that the communal narrative I had internalised twenty years earlier was now attracting lively, often acrimonious, debate. While every Karembola I spoke to still attributed the death of Malagasy Cactus to *vazaha* intervention, they now disputed almost every other detail of the story: whether Malagasy Cactus had been invasive; whether the French had been right to intervene; whether local people had also desired its eradication; and, perhaps most startling of all, given the evidence, whether any local people had actually died or suffered as a result. While the basic storyline from the 1980s persisted, its meaning, the social and political lessons that individual narrators drew from it, had in many cases been dramatically reworked. In less than a generation, widely shared memory and interpretation had given way to surprisingly diverse narrative about the past.

In some quarters local narrative had become sympathetic to the French colonial administration. While the motives ascribed to the intervention had remained constant, viz., to pacify the region, to secure labour for settlers in more fertile parts of the colony by creating famine and to facilitate tax collection, for many narrators these were now the perfectly justified motives of a 'civilising' state.

I first became aware of significant disparity between narrative versions when informants described the method by which Malagasy Cactus was killed. In the 1980s Karembola had been unanimous in stating that 'Malagasy Cactus had been eaten (predated) by insects *(biby)*', an opinion that was consistent with the scientific facts.[23] By 2002–03 there was widespread disagreement among narrators on whether the French had conducted chemical or biological warfare (or both) against the plant. To be sure, the narrative of biological control still had many adherents, especially among surviving eyewitnesses. In the following, for example, an elderly speaker recalls how

> The *posy* that killed it [Malagasy Cactus] was very red, white outside, red when dead [...]. We were afraid of the blood, there was blood. 'What's this *biby*?' we wondered. We didn't eat it because we thought it might kill [us too].[24]

But many informants now had in mind some form of chemical warfare, stating explicitly that 'It was the French from Tuléar who killed it, with *poizy* ... the plant was killed by *poizy*, not insects' (*poizy, tsy biby*).[25] Some elaborated this distinction by comparing the agent deployed against Malagasy Cactus with the chemical products that the government uses to control pests such as locusts, cockroaches and rats.[26]

The idea of a *double* killing, a combined package of targeted measures, involving both chemical and insect agents, staggered over an extended period of time to ensure full control of the plant, had also become popular. Most commonly, the French were described as deploying chemicals in the first instance to drive people from the thickets before turning to living predators to finish the job

> What killed Malagasy Cactus? When the French arrived, they couldn't see the population ... So they applied *poizy*, applied a medicine (*isiañe fanafoly*), Malagasy Cactus died, and everybody left the woods. Then [there was] a creature, a living thing (*biby, raha veloñe)* which ate [the plant]. You can still see the insect on surviving stands. But actually I'm not sure who killed it and how because it's only what I've been told. The population went hungry, the cattle too. Everything perished. After which this land became dark.[27]

Another distinctive feature of early twenty-first century narrative, and one that seems to be directly correlated to the rising conviction about chemical control, was the frequent references to an airborne campaign. Thus, according to the respondent just quoted, 'plane[s] flew past carrying the *poizy* [chemicals], spraying it, dropping it on the plants, whereupon [Malagasy Cactus] died'.[28]

In theory, the idea of a two-pronged campaign against Malagasy Cactus could appear plausible. Multiple strategies involving chemical products and various insect species, supplemented in some instances by mechanical measures such as felling, stacking and burning the plants, were deployed in parallel campaigns against Opuntia in South Africa and Australia (though never to my knowledge planes). The aim almost invariably was to improve efficacy when the application of single measures failed to achieve desired levels of control.[29] There is no evidence, however, that chemicals (let alone planes) were used against Malagasy Cactus. Cochineal predation on this plant was too rapid to require such help.

Another, to my mind particularly disconcerting, narrative development was that, by 2002–03, famine denial had become rife. The very core of 1980s narrative had been that the loss of Malagasy Cactus had resulted in a killing famine that decimated the cattle herds, left large numbers of Karembola dead and drove a mass exodus of survivors from the land. This in turn had fractured the social fabric, depriving Karembola of its spiritual and political leadership, leaving only 'children' behind.

By 2002–03 famine narrative, while evidently still current, no longer monopolised the imagination because significant numbers of Karembola now insisted that food had not been a problem when Malagasy Cactus died. Some claimed that the fruit had only ever been a mainstay for vagabonds, slaves and bandits, people living on the margins of Karembola society. The implication was not only that eradication had fairly restricted economic impacts but that indolent, quasi-criminal elements got their just deserts. Others still portrayed Malagasy Cactus as the ancestral tree of providence that 'made all things flourish in this thirsty land', but undermined this received wisdom by claiming either that 'people fell back on wild food (*tindroke*), they found *sonjo*, *bazaha* [other cacti], when Malagasy Cactus died', or that people had made an immediate transition from Malagasy Cactus dependency to agriculture by taking up hoes and cropping the now vacant land. These early twenty-first century renarrations departed dramatically from popular memory of twenty years earlier but resonated powerfully with narratives produced by colonial protagonists of eradication seventy odd years before. Following a visit to the neighbouring Mahafale region in 1929, for instance, Georges Petit related how he had observed its residents eagerly planting manioc and maize in land liberated from prickly pear.[30] He and like-minded commentators also insisted that Malagasy Cactus fruit had never been a human staple and that the spineless cactus (*raketambazaha*) would provide a superior fodder plant.[31] Such views would have been heretical in 1980s Karembola. Famine narrative was deeply entrenched.

Initially I was inclined to attribute these representational shifts to generation and the passage of time. I reasoned that, in the early 1980s, many Karembola who had witnessed Malagasy Cactus were still alive. They had eaten the fruit, they had lived its landscapes and they had witnessed its death. Perhaps more importantly, at the time of that first fieldwork, Malagasy Cactus survivors monopolised leadership positions in Karembola communities. As village elders, priests, orators and family heads they were in a position to approve and propagate their 'eyewitness' version of events. In those arguably more deferential days, it was predominantly their discourse that shaped public discourse. By 2002 surviving eyewitnesses were naturally far fewer and most had been little more than young children at the time Malagasy Cactus died. Moreover, given their advanced age, they no longer controlled public memory: effective power had passed to men of middle age.

I intuitively supposed that younger generations, who had only heard about Malagasy Cactus third- or fourth-hand, might be more likely to map current perceptions of *vazaha* practice upon 'memories' of the past. Take, for instance, the confusion over chemical control. Although to my knowledge chemicals have *never* been deployed against Opuntia in Madagascar, their routine use to control all manner of other pests and disease in the island, including, most notably, the widespread application of DDT in the 1950s, had made chemical control more familiar to most Karembola by the late twentieth century than biological control.[32] As for revisionist narrative of an airborne attack on Malagasy Cactus, its most likely inspiration was the Locust Control Service, which for decades has conducted flights in the Malagasy Deep South.[33] Other possible sources for misremembered history, which may have been compounded to make aerial bombardment a symbol of *vazaha* power, include wartime bombing by British planes during the campaign to take Madagascar from Vichy France; flights in the 1950s aimed at seeding artificial rain; and, further afield, aerial bombardment of East Coast Madagascar during the repression of the 1947–1948 anti-colonial rebellion.

On closer inspection, age-related explanation proved too simple, in that the production of anachronistic narrative was by no means confined to younger generations. In some instances even those who had witnessed the original event had changed their story, become less certain of the details, over twenty years past. One elderly woman whom I shall name Nirisoa narrated how the French had brought chemical-spraying planes from Tuléar, but readily admitted when questioned that she 'didn't actually see the plane[s] [but] only saw Malagasy Cactus die'. She went on to describe a '*biby* [that] was white, white but red inside when cut. The red got on your clothes', unmistakable eyewitness description of the cochineal insect.[34] In this instance, we appear to be dealing with quite complicated layers of disjunctive memory, where a detail held for decades in autobiographical memory now sat alongside a 'false' or distorted memory. This is by no means unusual. Studies have repeatedly demonstrated that so-called 'eye witness' testimony is often composed of ocular evidence combined with, and reinterpreted in the light of, narrative heard from others, at the time or subsequently.[35]

It might be useful in this instance to adopt an intersubjective approach to memory (narrative) production. Lambek and Antze, for example, propose to distinguish among author, narrator, character, reader, elicitor and censor when analysing narrative change.[36] While there is not the scope to pursue such a fine-grained approach in this article, it is tempting to speculate that Nirisoa's narrative represents an attempt to combine her *personal* experience of insects that stained one's clothing red (what we might term local knowledge grounded in empirical observation) with other narrative versions of the same history she had heard. It may even be that 'memory' of a two-pronged attack on Malagasy Cactus has become popular in Karembola partly as an way of reconciling competing versions authored by persons with competency and authority in different domains. Many elderly narrators appeared to lack the confidence to assert the veracity of their childhood memory. Although they could have documented their narrative of biological control by visual inspection of surviving stands of Malagasy Cactus, they seemed vulnerable to suggestion, if not outright censorship, by younger audiences more familiar with *vazaha* technology and to whom a modernising, albeit secondhand, narrative perhaps originating beyond Karembola offered a seemingly more credible explanation of the event. By combining both narratives into a two-part temporal sequence, the narrator (young or old) avoids having to make a judgment on which version is correct.

RAKETAMENA AS CONTEXT OF RENARRATION

I have made some preliminary suggestions for analysing certain imaginative retellings of Malagasy Cactus narrative using an intersubjective model that pays attention to the way memory is informed by interactions between various actors accessing both external and local knowledge, including the ideas Karembola themselves hold about generation and authority. A more immediate context for Malagasy Cactus renarration in the early 2000s, however, was the controversy that had erupted over a contemporary prickly pear.

Raketamena or 'red prickly pear' (so-named on account of the colour of its fruit) is a fairly recent arrival in Karembola that has been spreading aggressively since (at least) the 1970s and has now claimed vast tracts of land. Some reports describe it as *Opuntia stricta* or *Opuntia stricta* (Haw.),[37] a species ranked among 'One Hundred of the World's Worst Invasive Alien Species'.[38]

Despite the growing ascendancy of biodiversity discourse during the latter years of Ratsiraka's Second Republic, environmental reviews published in the 1980s had tended to gloss over the issues raised by naturalised Opuntia in the Malagasy Deep South.[39] Informally, I was told that conservationists were very much against the introductions but were keeping quiet because they knew that local peoples relied upon the plants. However, the Rio Earth Summit in 1992 and the subsequent Convention on Biological Diversity (CBD), to which Madagascar became a signatory, had changed public discourse by defining invasion by alien species as the most serious threat to global indigenous biodiversity after habitat destruction. Directing signatories to take practical action to 'prevent the introduction of, control or eradicate those alien species which threaten ecosystems, habitats or species', this Convention emboldened conservationists to articulate their opposition to Opuntia in southern Madagascar on the grounds of their alleged impacts on the highly specialised '*Didiereaceae-Euphorbia* bush'.[40]

By the turn of the century the World Wide Fund for Nature (WWF), increasingly active in the 'Spiny Forest Androy Sub-Region' of its recently designated 'Dry Forest Ecoregion' had become involved. However, rather than publish indiscriminate critiques of all Opuntia introductions, WWF was targeting *raketamena* – a focus which benefited from some local community support. Two WWF-commissioned reports had described the negative impacts of *raketamena* infestations on biodiversity and subsistence farming in Marovato and Tranovaho

Communes (Tsihombe and Beloha Sub-Prefectures) and claimed that there was a consensus within local communities on the need for eradication.[41]

Following extended criticism of an approach that was overly focused on the conservation of plants and animals to the detriment of local people's needs, WWF, like other key stakeholders in Madagascar, had reformulated its policy in the late 1980s to embrace a rhetoric that stressed the need to integrate development and environmental concerns. In so far as there was, or at least appeared to be, a coincidence between its biodiversity objectives and local socio-economic interests, a campaign against *raketamena* appeared to promise WWF a project that would deliver on this rhetoric. Identifying such projects had become pressing following the 2002 elections. Pro-capitalist Marc Ravalomanana, the newly elected President of Madagascar, was underscoring his expectation that environmental organisations would make an important contribution, alongside foreign investment and the private sector, to achieving millennium goals of poverty eradication and economic growth.

The WWF-sponsored reports had overstated the matter, however, in claiming that an initiative against *raketamena* had unanimous local support. Certainly all the Karembola I interviewed in Marovato and Tranovaho Communes in 2002–03 viewed *raketamena* as a prolific spreader that was infesting fields and grazing lands. However, local opinion was deeply divided over what should happen to the plant. Many villagers did call for its swift and total eradication. But others protested vehemently that the fruit was keeping them alive and that without *raketamena* they and their families would die. In effect, *raketamena* had differential cost–benefits for individual households according to their resource endowments. No one suggested that *raketamena* was an ideal food or made light of its impacts on agro-pastoral productivity; but for poorer community members, struggling to survive a series of bad harvests without the resources to access food at market, its value as a famine food outweighed its costs. By contrast, wealthier families (in 2002–03) not only had little need for *raketamena* – those short of food were raising cash by selling cows and goats at market – but many blamed staggering livestock losses on the plant.[42] Critically, *raketamena* spread was viewed as undermining elite power, which depends on producing food and livestock surpluses for use in patronage and exchange networks.

These different valuations of *raketamena* would probably have remained a source of grievance within local communities had external agencies not become involved. A meeting during which 'community leaders' had discussed *raketamena* with *vazaha* (the term *vazaha* here designates a mixture of expatriates and Malagasy nationals, notably government agronomists and WWF personnel) had greatly inflamed the issue when, or so it was widely reported, these *vazaha* had promised to introduce insects to eradicate the plant. By November 2002, when I arrived in the field, popular belief in the imminent prospect of eradication had ratcheted up emotions and focused everybody's attention on the *raketamena* debate.

This controversy over *raketamena* had had discernible impacts on Malagasy Cactus narrative. First and most obviously, by 2002–03 it had become almost impossible for Karembola to narrate the story of Malagasy Cactus without working in an extended metacommentary on *raketamena*. This involved comparing or contrasting the two species on a host of characteristics from the botanical (spininess, modes of reproduction and propensity to invasiveness) to the utilitarian (comparative cost-benefits as hedging, fodder and human food). Comparative exercises extended to what might be predicted about the consequences of *raketamena* eradication for livelihoods and food security on the basis of what happened when Malagasy Cactus died. Even the very elderly who had experienced Malagasy Cactus in person had revised their narration in the light of contemporary debates. Recent experience of *raketamena* was reshaping 'memories' of Malagasy Cactus for all informants regardless of age.

Consider, for instance, Zomana, a sprightly septuagenarian who had eaten Malagasy Cactus as a small boy. In 1981, when I first made his acquaintance, he had subscribed apparently without reservation to the nationalist narrative of the death of Malagasy Cactus as the event that put an end to *gasy* times. Twenty years later, he had reoriented his narration to address the issue of *raketamena*, a plant that was visibly encroaching on land all about his house. According to Zomana now, Malagasy Cactus had

> spread exactly like *raketamena*. If its fruits dropped, they all rooted too. [Malagasy Cactus] just grew without being planted. Nowadays [people] are enlightened, were people in those days wise? .... *Raketamena* is giving us a hard time, it's changing the land.[43]

Reflecting on *raketamena* in the present, Zomana here recovers a 'memory' of Malagasy Cactus invasiveness he had never articulated in my presence before. We know that landscape often presents deeply evocative cues for remembrance.[44] Yet the central event in Karembola history – the landscape transformation of the 1920s – had always been invisible precisely because Malagasy Cactus had disappeared. Indeed, the endless retellings of the story in the early 1980s were driven partly by the desire to reveal Malagasy Cactus as a latent presence in the contemporary landscape, to make explicit a concealed history. Now in the early twenty-first century, seventy odd years after the event, the sight of the *raketamena* infestations seemed to be awakening dormant memories, acting as a prompt to surviving eyewitnesses to look beyond communal narrative and remember how things really were.

Zomana here not only articulates memories of plant invasiveness he had not mentioned in the 1980s. He also pushes the analogy between Malagasy Cactus and *raketamena* to the point of appearing to question the wisdom of the ancestors. 'Nowadays [people] are enlightened', he says; 'Were people back then wise?' Zomana's temporal scales have also shifted: it is the incursion of *raketamena* and the suffering it is causing, rather than the death of Malagasy

Cactus, that he now highlights as the sign of altered, more difficult times. At the very least, experience of *raketamena* in the present has prompted Zomana to produce historical narrative of greater ambiguity than before. Nonetheless, unlike many younger, non-eyewitness narrators, Zomana holds back from portraying the eradication of Malagasy Cactus as progress. Despite its putative invasiveness, he still insists, as in the 1980s, on the excellence of Malagasy Cactus as a food. 'It was one of the Opuntia that truly nourish', he says, 'a proper meal, people's true food'.[45]

It is clear that, with the WWF seeking to push forward a programme of *raketamena* eradication or control, the Malagasy Cactus story had acquired critical meaning in the context of contemporary debates. *Raketamena* was presenting serious existential choices for rural Karembola and renarrated Malagasy Cactus history offered villagers on all sides of the debate an opportunity to voice situated commentaries on the benefits and the risks.

As a general rule, poorer Karembola, those for whom *raketamena* is a vital resource and who feared for their own survival in the event of its eradication, tended to reproduce the long-standing narrative of the great famine when Malagasy Cactus died. They found deep parallels between their own dependency on *raketamena* at times of hardship and the erstwhile importance of Malagasy Cactus as a food. By emphasising how intensely the ancestors had suffered when *their* staple was taken from them, they hoped to underscore the food security risks that *raketamena* eradication would pose. Listen, for instance, to Celestine, one of *raketamena*'s most impassioned defenders. She had heard, she told me, that *vazaha* were planning to kill *raketamena* and wanted to 'notify' them that

> We, the poor [*ondate tsy manan-draha*, lit. 'people who have nothing'], we don't want *raketamena* killed. We don't have cattle, goats, sheep to sell to buy food. ... We don't have poultry ... [*Raketamena*] is our food from dawn to dusk . ...We couldn't manage to hoe [our fields] if we didn't have *raketamena* to eat.

Without prompting, Celestine proceeded to draw analogies between the fate that awaited her family and other needy households were *raketamena* to be eradicated and what happened when Malagasy Cactus died:

> They went hungry. It was their only food. So when the *vazaha* killed it, everybody fled, died on the way. That's how the many came to go north ... If you killed *raketamena*, we couldn't even go north; we don't have the fare. We'd just die on the way. When *raketamena* dies, we die [too]. Only those with money made it [when Malagasy Cactus died].[46]

From this, one might suppose that it was generally more affluent Karembola who were producing and promoting new, historically problematic, readings of the story (notably, famine denial) in support of a pro-eradication stance. Certainly, many supporters of *raketamena* eradication did make light of the possible consequences by denying that anyone had gone hungry when Malagasy

Cactus died. Just as *gasy* in the 1930s had turned to other wild foods or set to work with hoes, so, they argued, people would do the same if *raketamena* were killed. Similarly, those who transposed the memory of recent food relief back onto Malagasy Cactus history now projected this memory forward onto *raketamena* eradication, insisting that

> Of course hunger is to be feared [if *raketamena* were to die] but you *vazaha* have always taken care of us when we've been hungry since ancestral times … For two years [1992–3] we received rations (*hanem-bode*, lit. 'orphan's food') because we couldn't plant crops. Butter-beans, maize, you gave food. If *vazaha* do kill *raketamena*, we shan't go without.[47]

Others, taking a more robust line, argued that, just as only good-for-nothings had relied on Malagasy Cactus, so too only the feckless relied on *raketamena* today. And, since these *raketamena* eaters elected not to work, unlike 'worthy' villagers, it would be entirely 'their own fault if they suffered, starved when *raketamena* dies'.[48]

Generally, however, the use of Malagasy Cactus narrative was less homogeneous on the pro-eradication side. Not everyone who wanted action to rid 'Madagasikara' of *raketamena* automatically engaged in famine denial, cited Malagasy Cactus as a precedent, portrayed it as invasive or thought its eradication had been good. Many informants, especially but not exclusively the elderly, argued the case against *raketamena* by highlighting all the ways in which modern pest *differed* from ancestral plant. Thus, Vontana, an elderly man who had known the pre-cochineal landscapes around Tranovaho and hated *raketamena* with a vengeance, noted that Malagasy Cactus and *raketamena* were rather similar in that neither plant 'stops fruiting all year round' (in contrast to the shorter seasons of other locally present Opuntia varieties) but that otherwise their fruit bore no comparison as a food. He remembered Malagasy Cactus as

> A splendid food, rather tart but it fattened you up, a superb food, nourished people, cattle, goats. … whereas *raketamena* is a bad food, nothing but bones and skin. It's killing people, livestock, killing the land.[49]

This style of argument was particularly common among the powerful and the wealthy. These men and their wives had both the most at risk from *raketamena* invasion and the most invested in tradition. They owed their power and authority partly to their ability to deliver above-average levels of agro-pastoral productivity and partly to their manipulation of ancestor-focused rhetoric, of which conventional Malagasy Cactus history constituted a key component. Such people generally condemned *raketamena* and upheld traditional Malagasy Cactus narrative by enumerating a series of botanical and economic contrasts between the two plants:

> Malagasy Cactus was awaiting the arrival of people. People found Malagasy Cactus when they came into the land. But this [*raketamena*] is a modern thing.

... *Raketamena* didn't nourish people in this land in ancestral times ... When *vazaha* obliterated Malagasy Cactus, we were annihilated because [we] ate it. But if *raketamena* died, we'd [be able to] cultivate our fields.[50]

Community leaders were generally reluctant to deny the great famine because it meant contradicting the core assumption of the nationalist story, namely, that Malagasy Cactus had been the ancestors' food. They preferred to deny the relevance of history to current policy decisions by invoking critical contrasts between then and now. Thus, they generally argued that *raketamena* eradication would not have the same terrible consequences for Karembola society because 'in those days there were no markets whereas nowadays markets at Soamanitse and Marovato mean you can buy food, there's money, and oxen carts to bring in food'.

In short, it was not possible to generalise about uses of Malagasy Cactus narrative in the context of contemporary struggles over *raketamena*. While many Karembola did make use of a re-narrated memory of Malagasy Cactus to argue the case for *raketamena* eradication, there was no necessary correlation between the position an individual took on *raketamena* and the type of Malagasy Cactus narrative he or she produced. Only in the broadest sense could one say that the Malagasy Cactus story had been appropriated and re-narrated to serve the conflicting interests of rich and poor.

STORYTELLING IN A 'GOVERNANCE STATE'

To analyse Malagasy Cactus narrative as argument embedded in present-day conflict over *raketamena* implies a certain instrumentality to the storytelling. It suggests that people were tailoring the story they told about the past in order to defend their interests in the present. Or as one theorist of memory puts it, 'people turn to the past to find what they need to support present interests; they find the past they want'. Instrumentality was certainly an important aspect of the 2002–03 renarrations. One reason why recollections of Malagasy Cactus came spilling unprompted into everyday conversation was because as a *vazaha* I was perceived by many Karembola as someone who might carry influence with the authorities if I could be persuaded to put their side in the *raketamena* debate. But renarration was also more complex.[51] While the *raketamena* controversy had undoubtedly stimulated interest in and appropriation of Malagasy Cactus narrative, not all the narrative shifts reported in this paper can be explained in such terms. Rhetorical uses of the past were also informed by broader cultural and intellectual developments in Madagascar. These included the influence of new environmental knowledge seeping into rural communities (and memory created through this).

Take, for instance, discourse on prickly pear invasiveness, an issue that was conspicuously absent from 1980s commentary but had become pivotal to twenty-first century renarrated pasts. All Karembola interviewed in 2002–03, whether they were for or against its eradication, described *raketamena* as invasive.[52] But while all villagers (regardless of eyewitness status) conjured up imagery of an ancestral landscape populated by Malagasy Cactus thickets, only some believed that it had been invasive too. For many informants the very idea of a Malagasy Cactus infestation was a symbolic contradiction because it implied disorder in the ancestral landscape, and the basic tenet of conventional local history was that Malagasy Cactus had been a 'good', 'ancestral' thing. Viewed as a native plant, its remembered profligacy was more often attributed by such informants to natural distribution. It was an aspect of the bounty the ancestors enjoyed. Divine Providence had created Karembola as a dryland but had endowed it with plenty of cactus to ensure that everything thrived. Consequently, where conservationists might be inclined to draw analogies between two weed species, many Karembola insisted on profoundly different interpretations of Malagasy Cactus proliferation and *raketamena* spread:

> Malagasy Cactus was awaiting the arrival of people. People found Malagasy Cactus when they came into the land. ... This newcomer is different. Malagasy Cactus was there from the beginning, growing. It wasn't invasive; it was something good, there from the start. There's land where it grows, land it liked, unlike *raketamena* which spreads ...[53]

It should be noted that when informants said that Malagasy Cactus 'wasn't planted but simply grew' (*tsy amboleañe fa nitiry avao*), they were not necessarily vocalising the idea that Malagasy Cactus had been a weed. In so far as informants saw Malagasy Cactus as a native species, they could hold the idea of a luxuriant monospecies *raketa* forest growing spontaneously in favoured habitats, much as a British naturalist might extoll the beech woods of the Chilterns.[54]

But it was equally clear that other Karembola now 'remembered' Malagasy Cactus as an indigenous weed, a plant that spread without human assistance, a plant that in its day had been as troublesome as *raketamena* now was

> I don't know where Malagasy Cactus originated. It was something the ancestors found here. It killed the land just like *raketamena*. There's none today; it's all gone. If it [starts to] grow, it soon dies. But in the old days it was invasive, a disaster, without parallel.[55]

Again, there was no necessary correlation between the position an individual took on *raketamena* and the type of Malagasy Cactus narrative he or she produced. While some informants made use of a re-narrated memory of Malagasy Cactus invasiveness to bolster arguments against *raketamena*, advocates of *raketamena* eradication could equally well contrast their patterns of spread.

Whatever the narrator's position, Malagasy Cactus retellings had been impacted by exposure to conservationist discourse. For instance, where informants in the early 1980s would have simply asserted that Malagasy Cactus had 'accompanied the ancestors from time immemorial', many now volunteered to detail the natural econiches (*toerana*) they imagined it had once preferred, much as they described typical econiches for other (truly) indigenous species of the southern bush such as *fantiolotse* (*Alluaudia procera*) or the various Euphorbia (such as *famata* and *ametse*). Constantly pressed by western or western-trained botanists for this kind of local knowledge regarding key 'hotspot' native plants, Karembola appear to have responded by incorporating the language into a traditional belief system that remembers Malagasy Cactus as a dominant member of an *ur*-form botanical Karembola.

In substituting a nominally more 'scientific' discourse for the mythic language that inflected previous renarrations, some informants portrayed Malagasy Cactus as what in equilibrium biology would be termed an old-fashioned 'climax vegetation' – that is, a vegetation that has evolved naturally for certain Karembola habitats. Listen again to Vontana:

> Malagasy Cactus was only plentiful in the places where it grew. It just grew; it wasn't planted. There was lots of it, but it didn't spread. It was only found where it was found. It wasn't invasive, though there was certain terrain where it abounded, particular spots where it was plentiful.[56]

There was something double-edged about these seeming borrowings from equilibrium biology. Hybridised modes of discourse could be both simultaneously supportive and subversive of conservationist positions. While renarrated memory of Malagasy Cactus as a *raketamena*-like invader endorsed conservationist discourse on the dangers of alien species (in effect conceding two instances of dramatic invasion of Karembola by prickly pears), the misappropriation of other typically conservationist idioms of econiches and balance-of-nature paradigms to sustain memory of Malagasy Cactus indigeneity was more subversive in that it placed a globally ubiquitous exotic on a par with rare, highly valued endemic species. One might almost suspect such informants of parodying the biological nativism still favoured by conservationists in Madagascar from a non-equilibrial or 'New Ecology' perspective, were they not so obviously convinced that Malagasy Cactus had been an ancient native plant. While WWF had found local allies over its plans for *raketamena* eradication, those allies did not always deploy the potential of Malagasy Cactus history as it would.[57]

It will be apparent that a social history of Malagasy Cactus remembering in Karembola between the early 1980s and the early 2000s cannot be undertaken without reference to the growing interactions that took place between local people and conservationists over these years. Karembola were not only producing diverse renarrated pasts partly in response to rumours of a pending *vazaha* intervention against *raketamena*; environmental discourse was also hybridising with local knowledge more generally to produce new language about the natural world. In this respect, the approach to knowledge advanced in this article departs fundamentally from the stark dichotomisations that currently predominate in studies of interactions between conservationists and local peoples in Madagascar. Existing approaches emphasise conflict and opposition, whereas in the case I am describing local people have also been adopting, adapting and subverting transnational environmental rhetoric, thereby lending new cadences to remembrance of Malagasy Cactus and to commentary on *raketamena*. This 'rhetorical traffic' was particularly marked in villages located on the periphery of Cap Sainte Marie Special Reserve but was evident throughout Karembola, where conservationist discourse had been disseminated into the remotest village by radio, government directives and, to a lesser extent, primary schools.[58] Nor was this a one-way process. There is evidence that WWF in 2002 was partly inhibited in its discourse on *raketamena* (at least in local contexts) by the continuing authority of traditional Malagasy Cactus narrative (see below).[59]

WWF was not, however, the only or even the most significant international organisation active in Karembola in 2002–03. Madagascar's transition from state socialism to neo-liberalism from the mid-1980s through the 1990s had seen the island become increasingly subject to governance by external actors (for example, World Bank and IMF), while a grave famine in 1991–92 brought international humanitarian assistance to the Malagasy Deep South on an unprecedented scale. As a result, WWF's 'Spiny Forest Androy Sub-Region' overlapped with a 'Zone of Food Insecurity' claimed by the United Nations World Food Programme (WFP). Despite a shared rhetoric on needing to integrate environment and development, these two agencies conceptualised the region and its problems in very different ways. In 2002–03 such differences spilled over into attitudes to prickly pears.[60]

In contrast to WWF personnel, personnel of agencies broadly concerned with humanitarian assistance were opposed to *raketamena* eradication on account of the plant's importance in local livelihoods as an emergency resource.[61] This was unsurprising since prickly pear consumption was among the multiple indicators of food security that Projet SAP reporters (*Système Alerte Précoce/ Early Warning System*) were expected to monitor on the ground. Secondly, and less predictably, such personnel were also committed to the nationalist narrative about the past. Interviewed on 29 November 2002, when he was busy organising urgent supplies for thirteen communes in difficulty, Randranjafizanaka Achilson of WFP-Fort Dauphin explained how

> We [WFP and WWF] once met at Tsihombe [a small town some 40 kilometres northeast of Cap Sainte Marie and administrative centre for Tsihombe District], because we were all staying at *Hotel Paradis du Sud*. They [WWF] approached us, not officially but informally, they approached WFP informally, as lead donor, about the possibility of a project for the management of *raketamena* in the context of our 'Food-for-Work' schemes. ... But WFP doesn't want to get involved in *raketa* eradication because people eat *raketa*. *Raketa* are beneficial. We knew the history of the cochineal, how people didn't want to work in plantations, how the French introduced the cochineal to destroy and kill the plant. ... how it ended in famine.[62]

There was a marked asymmetry in the use agency personnel made of history. In their conversations about *raketamena*, staff of agencies concerned with humanitarian assistance made explicit and unprompted reference to the great famine that followed the eradication of Malagasy Cactus. Personnel of conservationist agencies, by contrast, seldom mentioned the event. It was not (as far as I could judge from my interactions with them) that the latter actively disavowed public memory of the great famine. It was rather that their narrations focused on *raketamena* in the present and made limited use of narrative around Malagasy Cactus as a context for or evidence in the *raketamena* debate. It may be that they felt unable to develop a historiography powerful enough to challenge narrative of the killing famine, a narrative to which so many local gatekeepers still subscribed. Similarly, they may have ventured to deploy renarrated memory of Malagasy Cactus invasiveness to comment on *raketamena* only to encounter the same fierce rebuttals as I faced whenever I questioned memorialisation of Malagasy Cactus as a native plant. In such circumstances, conservationists may have decided that their case against *raketamena* was better served (at least when engaging with Karembola villagers) by avoiding historical analogies altogether rather than risking alienating key local allies. To this extent, local memory of Malagasy Cactus had shaped WWF discourse on invasive species just as conservationist discourse had shaped local discourse.

WWF had, however, underestimated both the extent to which the nationalist narrative of Malagasy Cactus also existed as public memory within parallel institutions (governmental and non-governmental) where it appears to circulate largely without texts, and the degree to which these parallel bodies would use story-telling around Malagasy Cactus as a rhetorical tool in policy debates. Thus, at the United Nations Food and Agricultural Organisation (FAO) offices in Antananarivo (the capital of Madagascar), I was given to understand that the deeply entrenched association between *raketa* eradication and famine (which retellings of the nationalist story reproduce again and again) would probably act as a brake on its approving a funding application for action against *raketamena* that WWF had recently submitted to FAO.[63] 'We know that people eat cactus at Fort Dauphin', one Malagasy staff member commented, 'Yes we know, we know all about Malagasy Cactus ...'[64] She cited the WWF submission on *raketamena* as an example of how 'researchers from outside Madagascar submit proposals and make arguments that we Malagasy just know are erroneous without needing to do any research'.[65]

It would be mistaken to suppose from this asymmetry in storytelling that an alliance between WWF and rural elites was balanced by an alliance between WFP and the rural poor. Certainly, in terms of formal procedures, WFP was more sensitive to the need to consult all sections of local communities than were conservationist agencies at this time. But in practice the delivery of food aid to Karembola villages in 2002–03 was deeply embedded in local power structures. Not only did many Projet SAP reporters charged with collecting primary socio-economic data find it quicker and easier to consult community leaders than to conduct house-to-house enquiries, but community leaders subsequently controlled local participation in (and benefit from) Food-for-Work programmes. In a region where state infrastructure has declined sharply and private enterprise has not filled the gap, WWF and WFP and their respective satellite agencies offered valuable resources to community leaders, a means of building followers and of advancing their own political and economic interests over those of the poor.

One outcome of the different positions WFP and WWF personnel took on *raketamena* and their failure to agree a version of the past was that local élites were left relative autonomy in the history they produced.

Production of evolutionary narrative

I have suggested that by 2002–03 commentary on the present as well as memory of the past had become imbricated in encounters between local communities and new external agents of change. Just as local discourse on Malagasy Cactus had incorporated idioms from green governmentality, so too it had drawn on experience of food aid. Traditional Malagasy Cactus narrative had bled into WFP policy on *raketamena* while the experience of humanitarian assistance had led some Karembola to revise their Malagasy Cactus history.[66]

One of the most striking developments was the greater number of Karembola who produced evolutionary narrative that represented the eradication of Malagasy Cactus as a civilising event that transformed primitive, anarchic forest-dwellers, ignorant of money and agriculture, into clean, industrious farmers settled in an open, domesticated landscape. Incorporating basic colonial tropes and metaphors about the perfectibility of Malagasy people and the improvement of Malagasy landscapes, these linear narratives, with their implicit acceptance of the 'civilising' mission of French colonialism, were deeply reminiscent of the modernising discourse that had inspired the original 1920s campaign against Malagasy Cactus. Not that I should like to draw an over-neat dichotomy between 1980s and early twenty-first century narrative. Even in the 1980s Malagasy Cactus had been embedded in a transformative discourse on the ancestors.[67] But no one in that first study had ventured the opinion that its eradication had been a wholly positive event.

Important political and economic changes in the intervening years had transformed Karembola discourse on economy and personhood. I was particularly struck by the way conflicting Malagasy Cactus renarrations resonated with the kind of debates over state socialism and free market solutions that figured in the elections that had brought Ravalomanana to power earlier in the year.[68] A continuing electoral process kept such debates topical during fieldwork. That narrative around Malagasy Cactus should be a popular idiom for such reflections in Karembola is perhaps hardly surprising, given the way alternative visions of economy and development – progressive and traditional – had fired 1920s colonial debates.[69]

From this perspective it is interesting to compare narrational positions on *vazaha* over time. I indicated that the way *vazaha* figured in early 1980s narrative was quite complex. A story, which appeared to chime with the isolationism of the socialist era, had been an oblique way of critiquing the Antananarivo-based Marxist regime. By contrast, early twenty-first century rhetoric seemed to be more about co-opting *vazaha* than expelling, hating or blaming them, the term *vazaha* here designating the myriad international para-statal bodies that had descended on the Malagasy Deep South. New, generally more positive, attitudes to foreigners and to foreign interventions were another factor that had made the nationalist narrative of Malagasy Cactus less appealing to many Karembola than before. Yet informants were by no means agreed on what greater *vazaha* involvement meant. Some affirmed their faith in the power of a market economy and foreign investment coupled with a (largely imagined) indigenous Karembola capitalism to enrich local people and redress food insecurity. Others saw *vazaha* more as kindly providers of food aid. Reinterpreted Malagasy Cactus narrative could be made to resonate with either position, if ambivalently.

All these elements of Malagasy Cactus renarration interacted with commentary on *raketamena* but were by no means reducible to it. Or, to put it another way, the *raketamena* question, as a particular dilemma requiring choices to be made, often tapped into other great issues of the day but did not explain these broader existential concerns.

CONCLUSION

I began this paper by noting sharply contrasting opinions within the academic community on the value of oral memory as a historical resource. With respect to Malagasy Cactus, I have shown how, in contrast to the homogenous, totalising narrative I collected in the 1980s, Karembola in the early 2000s were producing a range of diverse texts. I have attributed these developments to changing political contexts, to a new invasive prickly pear, to internal social differentiation, to the spread of external agencies and conservationist discourses and to reconceptualisations of *vazaha* and the past. I have suggested that Malagasy Cactus renarrations are to be understood in terms of complex influences, working to different time scales, deploying varying narrative styles. In particular, while contemporary interest led people to remember differently, narrative was far from tailored neatly to suit present needs and certain renarrations were problematic for conservationists seeking local support.

One question might be whether differing research interests and styles of fieldwork between the early 1980s and the early 2000s affected these findings. In the first study, I relied on participant-observation to research Karembola kinship and ritual. Malagasy Cactus history was not something I actively elicited; it was narrative that everybody produced. At the time I did not probe this narrative, partly because I met no dissenters, partly because many of those recounting the story had witnessed the events it purported to describe and partly because their version was consistent with ethnography I had read before coming to the field. Moreover, the story seemed persuasive. After all Karembola is a dryland and we were eating lots of prickly pear. In 2002–03, a shorter time in the field and a more narrowly focused research topic necessarily brought greater reliance on semi-structured interviews. More pertinently, as my research interests shifted to environmental issues and my knowledge of invasive species expanded, I asked questions that would not have occurred to me before. It is therefore very possible that my questioning may have contributed to subjects remembering differently. On the other hand, many of the renarrations cited in this article were voiced spontaneously by villagers and can be seen as expressing their own agenda and concerns. My own sense is that, while interactivity is a fundamental and intractable dimension of fieldwork, the key point is that, owing to the temporal coincidence between the spread of raketamena in Karembola and the highlighting of invasive species in global discourse, both I and my informants had new ideas and experiences to work with.

Here precisely the data point to an interesting question. Does the adoption of new knowledge, or new ways of framing old knowledge to address new interests, make memory *less* reliable (because the past is renarrated or 'distorted' in the light of current concerns) or *more* reliable (because serious contemporary challenges stimulate narrators to look more critically at 'official' narrative they have inherited from the past)?

The temptation when faced with patently ongoing revisions of popular narrative of the kind documented in this article is to fall back on well-worn clichés about remembering and forgetting being 'thus locked together in a complicated web' as one version of the past competes with another[70], or about memory telling us more about people's values and objectives in the present than anything useful about the past.

I would take a different position. I agree that variable and discordant stories of Malagasy Cactus need to be located within social histories of remembering. Indeed, I hope I have shown how memories of Malagasy Cactus have been (partly) shaped by politico-economic interests in the present and why it is important to ask how people use particular views of the past in interactions today. But it also seems to me possible to recognise the reconstructive nature of memory without either reducing memory to current use or accepting the modern dictum that all versions of the past are or should be seen as equally valid retellings of a complex event.

'Truth' is of course a difficult word that carries highly moral connotations. Something may be 'true' in many senses, even if the details reported do not correspond exactly with all of the facts. There is much to be gained by following the approach adopted, for instance, by Luise White who, in her book *Speaking with Vampires*, argues that seemingly fanciful memories about vampires can in fact be more truthful than academic histories because they tell us about how Africans themselves perceived colonialism. She suggests further that such beliefs became factually critical in that they impacted materially on colonialism.[71] Proceeding along similar lines, I suggested that anachronistic stories about chemical-spraying planes express truths about Karembola perceptions of *vazaha*. (In a region where a ploughed field is still a rare sight and where many households rely for at least part of the year on small cavities in the limestone rock for their water supply, planes remain even now iconic of *vazaha* modernity, a subject of endless speculation in the field.) We can also see these imaginative retellings as capturing something of the scale of devastation, the extraordinary nature of the event, the overwhelming sense of colonial power Karembola felt at this onslaught on their traditional way of life. Better understandings of past subjectivities can in turn inform current debates on invasive species by giving some sense of what it feels to be an ordinary person whose familiar landscape is transformed by chemical or biological control. But the fact remains that, as a report of the method deployed to eradicate Malagasy Cactus, this narrative element is factually incorrect.

What is clear is that one cannot say that overall that one or other version bears a closer relation to what, after Janice Haaken, we might term 'the concrete facticity of events'.[72] On some points the earlier 1980s narrative is more reliable. Malagasy Cactus *was* predated by an insect. Chemicals and planes were not involved. A serious famine in Karembola *did* follow shortly after its eradication, though – revisionists have a point – it is possible to argue that the loss of Malagasy Cactus was by no means its only cause.[73] On other points, the early twenty-first century renarrations offer fresh and important insights. They open up prickly pear invasiveness for popular debate within local communities. They also disclose the fact of Malagasy Cactus survival (at infinitely reduced levels), bringing the case study into line with scientific models of biological invasion. Both these issues remained screened off in 1980s narrative.[74]

At the same time there are issues in Malagasy Cactus historiography that neither the traditional 'nationalist' narrative nor the revisionist narratives address. Most obviously, the 1980s narrative glosses over important social differentiation in how Karembola in the 1920s and 1930s experienced the death of Malagasy Cactus by narrating it as a collective trauma. While it is beyond the scope of this present essay to report the archival evidence, it is clear that the plant's eradication had uneven impacts both within communities and between communities, depending partly on access to other resources, itself a function of locality, rank, gender and age. This is not to say that the narrative of flight, death and pauperisation I collected in the 1980s was 'false' memory. Archival records confirm the historical authenticity of such narrative as *generic* statements. But the idea of a *collective* trauma, in the sense that all Karembola suffered and suffered in equal measure, *is* open to challenge. If differential impacts were not revealed in narrative produced in the early 1980s, it was partly because the narrators were the survivors (or their descendants) or, more precisely, since those who escaped to more fertile parts of the island were also survivors, those who survived and subsequently prospered in their homeland. We might say that, by the early 1980s, Karembola had forged a collective identity around a partly appropriated and essentially depersonalised trauma.[75] Those who died or who 'went away, never to return' were mostly nameless. The story had few individuated characters. It was organised around a collective 'we' in which Karembola past and present were merged.

Early twenty-first century narrative does little to address this lacuna. While discord around *raketamena* in the present seems to have opened up the intellectual space to question the long-standing idea of the death of Malagasy Cactus as a collective trauma, and to speculate on the kind of intracommunity struggles that might (but only might) have once attended Malagasy Cactus, there was still a striking lack of personal detail in the accounts. Even narrators like Celestine who expressed their fears by drawing links between their present reliance on *raketamena* and what happened when Malagasy Cactus died, spoke in broad historical sweeps and were unable to provide particularised histories for named victims of the famine. They may be the destitute of Karembola society now but they too are the descendants of those who survived then.

Some of the data suggest interesting connections between individual autobiographical memory and collective memory. Anthropologist Benedict Anderson proposes that the presence of a narrative is an index of people having forgotten the original formative experience. Having to 'have already forgotten' tragedies of which one needs unceasingly to be 'reminded', he argues, is a characteristic device in the construction of nationalism.[76] Looking at communal narrative about Malagasy Cactus from the 1980s, the highly impersonal narration is suggestive of Anderson's argument. This really was a stylised, homogenised, dare I say fossilised, social or collective memory that had lost the immediacy of original experience. Yet the extremely diverse and highly individualistic memory of twenty years later suggests a more complicated trajectory between memory and narrative than that Anderson outlines, as narrators like Zomana zigzagged back from communal narrative to recover original experience, finding all kinds of fresh connections between the present and the past. It is now possible to see that, while collective storytelling may have dominated public narrative in the 1980s, individual autobiographical memory of original experience survived, latent, private, unvoiced. As anthropologist Maurice Bloch has argued, it is important not to overlook the distinction between collective or social memory and individual autobiographical memory. The latter may retain elements that collective memory has 'forgotten', making them available decades later as a resource for local knowledge to 'recover and reuse'.[77]

I suggested that the sight of *raketamena* prompted Zomana to remember (articulate) things about Malagasy Cactus he had never articulated before. Another instant of active re-remembering occurred when Vontana, reminiscing in his little timber house, observed:

> You know, there was something unusual about Malagasy Cactus. The fruits grew one on top of the other, at times leaves also appeared in the chains.[78]

Eighteen months earlier, while working in the Paris Muséum National d'Histoire Naturelle, I had found an unpublished sketch of Malagasy Cactus recording precisely this pattern of growth. Made by colonial botanist Perrier de la Bâthie, an accompanying note in his hand describes how these curious multi-headed chains eventually grew so heavy that they caused the branch to droop.[79] This example (where a detail retained by individual memory is corroborated by independent evidence) suggests that by careful triangulation it might be possible to move beyond a view that draws on memory to explore contested values and contested places to one that asks objective questions about plant biology and landscape history.[80]

Such a methodology stands in stark opposition to approaches which advocate the collection of oral memory primarily to gain insights into sociocultural perceptions and values, a 'sense', a 'feel', a 'flavour' of what events meant to the people involved. Historian Pier Larson expressly warns that 'mining social memory for nuggets of evidentiary "raw material", once proposed as the proper treatment of oral tradition, is fraught with problems and contradictions'.[81] Of course social memory embodies its own interpretations and meanings that historians must take seriously in their professional reconstructions. But there is equal danger in an overly holistic approach. Evidence presented here suggests that it is precisely the odd detail that has *not* been integrated into coherent narrative that may be more reliable whereas the 'feel', the 'sense' to the story – precisely because it is a *story* – may be the least reliable.

In a context where more powerful actors are beginning to use Malagasy Cactus history with a view to framing conservationist policy for practical interventions in alien species in Madagascar, it would be particularly untimely to pass over local oral evidence in favour of written documents.[82] The challenge rather is to elaborate the relationship between history and memory by developing critical methodologies that allow a systematic inclusion of oral memory (and forgetting) as a source for the past.[83] This means placing memory in relation to present contexts and to the culturally determined forms in which narrative is expressed, including local ideas about testimony and truth. It also means paying closer attention to the articulation of social and individual memory than historians and anthropologists are generally minded to pay.

ACKNOWLEDGEMENTS.

Fieldwork in Madagascar has been generously supported by the Economic and Social Research Council (UK) in 1981–1983 and by the Nuffield Foundation and the Rhodes Trust in 2002–03. For assistance in Madagascar I am indebted to the University at To-liara, to government officials in Beloha and Tsihombe and to staff of Cap Sainte Marie Special Reserve. Participants in the European Association for Environmental History (UK) conference on Invasions and Transformations, St Antony's College, Oxford, 15 September 2009 provided useful comments on an earlier draft as did Sam Middleton. I also thank two anonymous reviewers for thoughtful suggestions and advice. As ever, my greatest debt is to the peoples of Karembola for their extraordinary kindness over so many years. It remains to stress that the interpretations presented in this paper are my own, and that I alone am responsible for any errors.

NOTES

[1] *homan-raketa avao* (lit. all we eat).

[2] A more or less spineless Opuntia, probably a cultivar of *Opuntia ficus-indica*, introduced into the region by French military officers in the early 1900s and promoted well into the 1950s by the colonial state.

[3] J. Drake, H. Mooney, F. di Castri, R. Groves, F. Kruger, M. Rejmánek and M. Williamson (eds.) *Biological Invasions: A Global Perspective* (Chichester: John Wiley & Sons, 1989); Quentin Cronk and Janice Fuller, *Plant Invaders: The Threat to Natural Ecosystems* (London: Chapman and Hall, 1995); O. Sandlund, P. Schei and Å. Viken (eds.) *Invasive Species and Biodiversity Management* (Dordrecht: Kluwer Academic, 1999); H. Mooney and R. Hobbs (eds.) *Invasive Species in a Changing World* (Washington DC: Island Press, 2000); Daniel Simberloff, 'A Rising Tide of Species and Literature: A Review of Some Recent Books on Biological Invasions', *Bioscience* **54** (2004): 247–54; Harold A. Mooney, Richard N. Mack, Jeffrey A. McNeely, Laurie E. Neville, Peter J. Schei and Jeffrey K. Waage (eds.) *Invasive Alien Species – A New Synthesis* (Washington: Island Press, 2005); Charles Perrings, Mark Williamson and Silvana Dalmazzone (eds.) *The Economics of Biological Invasions* (Cheltenham: Edward Elgar, 2000).

[4] See e.g. Gregory Ruiz and James Carlton (eds.) *Invasive Species: Vectors and Management Strategies* (Washington: Island Press, 2003); L. Child, J. Brock, G. Brundu, K. Prach, Petr Pyšek, P. Wade and Mark Williamson (eds.) *Plant Invasions: Ecological Threats and Management Solutions* (Leiden: Backhuys, 2003); and many of the works cited in note 2.

[5] See, for example, in addition to many of the works listed in notes 2 and 3, Mark Williamson, *Biological Invasions* (London: Chapman & Hall, 1996); and, on prickly pear, Helmuth Zimmermann and V.C. Moran, 'Ecology and Management of Cactus Weeds in South Africa', *South African Journal of Science* **78** (1982): 314–320.

[6] Humanist scholarship on exotic species is vast and growing but see, for example, Peter Coates, *American Perceptions of Immigrant and Alien Species; Strangers on the Land* (Berkeley, Ca. and London: University of California Press, 2006); Iftekhar Iqbal, 'Fighting with a Weed: Water Hyacinth and the State in Colonial Bengal, c. 1910–1947', *Environ-*

ment and History 15 (2009): 35–59; William Beinart, *The Rise of Conservation in South Africa: Settlers, Livestock, and the Environment 1770–1950* (Oxford: Oxford University Press, 2003), Ch. 8. For an overview that makes a strong argument for interdisciplinary approaches to the subject, see William Beinart and Karen Middleton, 'Plant Transfers in Historical Perspectives: A Review Article', *Environment and History* 10 (2004): 3–29

[7] The quest for an 'objective' technical language free of value judgements has spurred extensive literature – see e.g. Charles S. Elton, *The Ecology of Invasions by Animals and Plants* (London: Methuen, 1958); Pierre Binggeli, 'Misuses of Terminology and Anthropomorphic Concepts in the Description of Introduced Species', *Bulletin (British Ecology Society)* **25** (1994): 10–13; James Perrin, Mark Williamson and Alastair Fitter, 'A Survey of Differing Views of Weed Classification: Implications for Regulation of Introductions', *Biological Conservation* **60** (1992): 47–56; David Richardson, Petr Pyšek, M. Rejmanek, Michael Barbour, F. Panetta and Carol West, 'Naturalization and Invasion of Alien Plants: Concepts and Definitions', *Diversity and Distributions* **6** (2000): 93–107. For the purposes of this essay, I use the term 'invasive' to mean a plant that has the capacity to spread exponentially without human assistance and that poses problems for locally available methods of control.

[8] Cf. Richard N. Mack, 'Assessing the Extent, Status, and Dynamism of Plant Invasions: Current and Emerging Approaches', in Mooney and Hobbs (eds.) *Invasive Species*, pp. 141–68; M. Sagoff, 'Do Non-Native Species Threaten the Natural Environment?' *Journal of Agricultural and Environmental Ethics* **18**, 3 (2005): 215–36.

[9] James Fairhead and Melissa Leach, *Misreading the African Landscape* (Cambridge: Cambridge University Press, 1996); Terence Ranger, *Voices from the Rocks: nature, culture and history in the Matapos Hills, Zimbabwe* (Oxford: James Currey, 1999); Ramachandra Guha, *The Unquiet Woods: Ecological Change and Peasant Resistance in the Himalaya* (New Delhi: Oxford University Press, 1989); K. Sivaramakrishnan, *Modern Forests: Statemaking and Environmental Change in Colonial Eastern India* (Stanford, Ca.: Stanford University Press, 1999); Ann Grodzins Gold and Bhoju Ram Gujar, *In the Time of Trees and Sorrows: Nature, Power and Memory in Rajasthan* (Durham, NC: Duke University Press, 2002); Partha Chatterjee and Anjan Ghosh, *History and the Present* (New Delhi: Permanent Black, 2002). For an exploration of prickly pear in South Africa that supplements written sources with oral evidence see William Beinart and Luvuyo Wotshela, 'Prickly Pear in the Eastern Cape since the 1950s – Perspectives from Interviews', *Kronos: Journal of Cape History*, **29** (2003): 191–209. To my knowledge, the study of introduced Opuntia in India and Queensland, Australia has not yet attracted this kind of approach.

[10] On history and memory, see Maurice Halbwachs, *The Collective Memory* (New York: Harper and Row, 1980 [1950]; Pierre Nora, 'Between Memory and History: les lieux de mémoire', *Representations* **26** (1989): 7–23.

[11] Ajay Skaria, *Hybrid Histories: Forests, Frontiers, and Wildness in Western India* (New Delhi: Oxford University Press, 1999); Paul Thompson, *The Voice of the Past* (Oxford: University of Oxford Press, 1988).

[12] Michel-Rolph Trouillot, *Silencing the Past: Power and the Production of History* (Boston: Beacon, 1995); Paul Antze and Michael Lambek (eds.) *Tense Past. Cultural Essays in Trauma and Memory* (London: Routledge, 1996); Elizabeth Tonkin, *Narrating Our Pasts: The Social Construction of Oral History* (Cambridge: Cambridge University Press, 1992).

[13] Johannes Fabian, *Remembering the Present: Painting and Popular History in Zaire* (Berkeley and Los Angeles: University of California Press, 1996).

[14] There is an emerging consensus that the correct identification for Malagasy Cactus is *Opuntia monacantha*. However, while I recognise that it would be useful for readers to have the scientific names for prickly pears discussed in this article, I am reluctant to make definitive pronouncements given the inconsistences in a literature too often authored by non-specialists in *Opuntia* taxonomy.

[15] See e.g. Georges Petit, 'Introduction à Madagascar de la cochenille du Figuier d'Inde (*Dactylopius coccus,* Costa) et ses conséquences inattendues', *Revue d'Histoire Naturelle,* **10**, 5 (1929): 160–173; Edmund François, 'De l'emploi de l'Herbe Kikuyu à Madagascar', *Revue de Botanique appliquée* **105** (1930): 287–292; Henri Perrier de la Bâthie, 'Les famines du Sud-Ouest de Madagascar, causes et remèdes', *Revue de Botanique appliquée,* **14**, 151 (1934): 173–186; and, on the pro-Malagasy Cactus side, Raymond Decary, 'La question des raiketa dans l'Extrême-Sud de Madagascar', *Bulletin Économique de Madagascar* **1** (1927): 92–96; 'À propos de l'Opuntia épineux de Madagascar', *Revue de Botanique appliquée* **8**, 77 (1928): 43–46. An account of these positions is given in Karen Middleton, 'Who Killed "Malagasy Cactus"? Science, Environment and Colonialism in Southern Madagascar (1924–1930)', *Journal of Southern African Studies* **25** (1999): 215–248.

[16] See e.g. Middleton, '"Who Killed 'Malagasy Cactus'"?'; Jeffrey Kaufmann, 'Forget the Numbers: The Case of a Madagascar Famine', *History in Africa* **27** (2000): 143–157; Pierre Bingelli, 'Cactaceae, *Opuntia* spp., Prickly Pear. *Raiketa, Rakaita, Raketa*', in S. Goodman and J. Benstead (eds.) *The Natural History of Madagascar* (Chicago: University of Chicago Press, 2004), pp. 335–8.

[17] On 'spiny dry forest' see R. Rabesandratana, 'Flora of the Malagasy Southwest', in A. Jolly, P. Oberlé and R. Albignac (eds.) *Key Environments: Madagascar* (Oxford: Pergamon Press, 1984), pp. 55–74; Werner Rauh, *Succulent and Xerophytic Plants of Madagascar* (Mill Valley, California: Strawberry Press, 1995 and 1998). On Madagascar's emergence as a 'mega[bio]diversity country', see Christian Kull, 'The Evolution of Conservation Efforts in Madagascar', *International Environmental Affairs* **8** (1996): 50–86.

[18] Unfortunately it is not possible to compare these interview data with earlier memory because local Malagasy opinion was not recorded in any detail at the time.

[19] On the utility of prickly pears in semi-arid zones see G. Barbera, P. Inglese and E. Pimienta-Barrios (eds.) *Agro-Ecology, Cultivation and Uses of Cactus Pear* (Rome: FAO, 1995); Henri Le Houérou, 'The Role of Cacti (*Opuntia* spp.) in Erosion Control, Land Reclamation, Rehabilitation and Agricultural Development in the Mediterranean Basin', *Journal of Arid Environments* **33** (1996): 135–59.

[20] Cf. Karen Middleton, 'Circumcision, Death, and Strangers', *Journal of Religion in Africa,* **27** (1997): 341–73.

[21] Cf. Françoise Raison-Jourde, 'Une rébellion en quête de statut: 1947 à Madagascar', *Revue de la Bibliothèque nationale* **34** (1989): 24–32.

[22] Cf. Karen Middleton, 'From Ratsiraka to Ravalomanana: changing narratives of prickly pears in dryland Madagascar', *Études Océan Indian,* **42–43** (2009): 47–83. On Malagasy Socialism see Maureen Covell, *Madagascar: Politics, Economics and Society* (London and New York: Frances Pinter, 1987). On popular memory as a privileged domain of resistance to hegemonic narratives see Rubie Watson (ed.) *Memory, History and Opposition under State Socialism* (Santa Fe: School of American Research Press, 1994); Gerald

Karen Middleton

Sider and Gavin Smith (eds.) *Between History and Histories: the Making of Silences and Commemorations* (Toronto Buffalo London: University of Toronto Press, 1997).

[23] Almost all published accounts, whatever their other antagonisms, concur that Malagasy Cactus was killed by a cochineal insect or *Dactylopius* species (see e.g. Henri Perrier de la Bâthie, 'Introduction à Tananarive du *Coccus cacti* ou Cochenille du Figuier d'Inde', *Bulletin économique de Madagascar et dépendances* **21**, 3–4 (1924): 222; Petit, 'Introduction à Madagascar': 163; Cl. Frappa, 'Sur *Dactylopius tomentosus* Lam. et son acclimatement à Madagascar', *Revue de pathologie végétale et d'entomologie agricole* **19** (1932): 48–55; J. Mann, 'Cactus-Feeding Insects and Mites', *Bulletin (U.S. National Museum)* **256** (1969): 139; Raymond Decary, *L'Androy (Extrême Sud de Madagascar). Essai de monographie régionale*, Vol. II (Paris: Société d'Éditions Géographiques, Maritimes et Coloniales, 1933), pp. v–vi).

[24] Vontana, Tranovaho, 15/1/2003.

[25] Sambo, Marobey, 5/12/2002.

[26] There is significant semantic instability around key vernacular in these narratives, with some speakers using the terms *biby* and *posy/poizy* as synonyms for 'insect' (as for example when a speaker glossed *posy* as '*biby*, living creatures') while others drew a clear distinction between *biby* meaning 'insect' and *posy/poizy* meaning 'chemicals' (as when they likened 'this *posy*' to 'medicines used against cockroaches' *(fanafoly bararaoke))*. These open-ended meanings seem to be both related to and productive of the now widespread confusion over the agent of control.

[27] Betaimboroke, 15/12/2003.

[28] Betaimboroke, 15/12/2003. 'Memory' of an airborne chemical campaign is not a wholly recent production. Though I recorded no such narrative in Karembola in the 1980s, there is newspaper evidence that such stories were already circulating in the 1960s, at least amongst Tandroy emigrées in Anosy (D.R. 'A l'écoute d'un migrant antandroy', *Lumière* 1821, 11/4/1971). More recently, Kaufmann reports a narrative from Androka, Mahafale that has aeroplanes dropping *cochineal-infested cladodes*, rather than chemical products, on Malagasy Cactus (Jeffrey Kaufmann, 'Cactus Pastoralism on Madagascar', Ph.D dissertation, University of Wisconsin-Madison, 2001, p. 205). Given the polysemy and ambiguity of terms such as *poizy/posy* noted in note 26, it would have been useful to have this testimony as original, unredacted text. In any case, the narrator is a gendarme 'whose homeland was several hundred kilometers north'. It is interesting that the two earliest modernising narratives on record involve narrators who are not only 'outsiders' to local society but also in their different ways 'progressive', civilising agents in the south: the gendarme because he is educated and closely associated with state power; the narrator of the text published by *Lumière* because he has left his native Androy to become a rice farmer elsewhere. While external narrative of this kind may have been a source for Karembola retellings, neither the gendarme nor the emigrant, it should be noted, posits the double-sequenced hit involving chemicals and insects that characterised many peasant narratives in 2002–2003.

[29] See e.g. for South Africa D. P. Annecke and V. C. Moran, 'Critical Reviews of Biological Pest Control in South Africa. 2. The Prickly Pear, *Opuntia ficus-indica* (L.) Miller', *Journal of the Entomological Society of Southern Africa* **41** (1978): 161–188; W. Pettey, 'The Biological Control of Prickly Pears in South Africa', *Scientific Bulletin of the Department of Agriculture and Forestry Union of South Africa* **271** (1947–48): 1–163. And for Australia J. Mann, *Cacti Naturalized in Australia and their Control* (Brisbane:

Government Printer, 1970).

[30] Petit, 'Introduction à Madagascar': 168.

[31] Perrier de la Bâthie, 'Les famines du Sud-Ouest'; François, 'De l'emploi'. Critics argued that nowhere near enough of this variety had been planted or that neither humans nor livestock wanted to eat it. *Saonjo* is a later introduction.

[32] Although a Laboratory of Agricultural Entomology was created in 1931, chemical interventions remained the preferred government option in Madagascar until at least the 1970s. Cf. J. Appert, M. Betbeder-Matibet and H. Ranaivosoa, 'Vingt années de lutte biologique à Madagascar – Twenty years of biological control in Madagascar', *L'Agronomie Tropicale* **24** (1969): 555–85.

[33] Although this service was created in 1928, its deployment of planes in the Malagasy Deep South began much later. Indeed, the first aircraft of any kind to fly over the region are said to date to 1931 (Decary, *L'Androy*, p. 246). For the record, the cochineal had reached Karembola by 1928.

[34] Nirisoa, Befeha, 3/12/2002. Not that plane narrative is necessarily incompatible, imaginatively speaking, with biological control (see note 28) but none of my narrators made such a link.

[35] On suggestion/post-event misinformation and false memory see especially the pioneering studies by Elizabeth Loftus and colleagues (Elizabeth F. Loftus and J. E. Palmer,'Reconstruction of automobile destruction: an example of the interaction between language and memory', *Journal of Verbal Learning and Verbal Behavior* **13** (1974): 585–9; Elizabeth F. Loftus, D. G. Miller and H. J. Burns, 'Semantic integration of verbal information into a visual memory', *Journal of Experimental Psychology: Human Learning and Memory* **4** (1978): 19–31); Elizabeth F. Loftus, 'Leading questions and the eyewitness report', *Cognitive Psychology* **7** (1975): 560–72; Daniel L. Schacter (ed.) *Memory Distortion* (Cambridge, Mass. and London: Harvard University Press, 1995). For more critical positions see D. A. Bekerian and J. M. Bowers, 'Eyewitness testimony: were we misled?' *Journal of Experimental Psychology: Learning, Memory and Cognition* **9** (1983): 139–45; R.E. Christiaansen and K. Ochalek, 'Editing misleading information from memory: evidence for the coexistence of original and postevent information', *Memory and Cognition* **11** (1983): 467–75; Michael McCloskey and Maria Zaragoza, 'Misleading postevent information and memory for events; arguments and evidence against memory impairment hypotheses',*Journal of Experimental Psychology: General* **117**, 1 (1985): 1–16.

[36] Michael Lambek and Paul Antze, 'Introduction', in Antze and Lambek (eds.) *Tense Past*, p. xviii.

[37] ANGAP-FOFIFA-WWF (n.d.) 'Rapport de Mission à Cap Ste Marie (du 03 au 06 mai 2001) et Propositions des Actions Futures', Unpublished Report; J. A. Randriamampia- nina, Solosieva, and J. Rajaonarison, 'Rapport de Mission dans le Sud (synthèse) (du 26 novembre au 6 décembre 2001 et du 21 août au 4 septembre 2002)', FOFIFA DRA Antananarivo/FOFIFA centre régional du Sud et Sud-Ouest Toliara/WWF Fort Dauphin/ ANGAP Direction Régionale de Toliara, September 2002. Without actually rejecting these identifications, I indicate certain inconsistencies elsewhere (Karen Middleton, 'Red Prickly Pear and the World Wide Fund for Nature: Rural Poverty and Invasive Species in Dryland Madagascar', unpublished report, 2003).

[38] Global Invasive Species Database, 2011. Opuntia stricta (http://www.issg.org/database/ species/ecology.asp?si=19&fr=1&sts=sss. Accessed 20/1/2011.)

[39] M. Jenkins (ed.) *Madagascar, an Environmental Profile* (Gland and Cambridge: IUCN/UNEP/WWF, 1987); M. Nicoll and O. Langrand, *Madagascar: Revue de la Conservation et des aires protégées* (Gland: World Wide Fund for Nature, 1989), p. 124, 178; Jean-Louis Guillaumet, 'The vegetation; an extraordinary diversity', in A. Jolly, P. Oberlé and R. Albignac (eds.) *Madagascar: Key Environments* (Oxford: Pergamon Press, 1984), pp. 32–3.

[40] Rauh, *Succulent and Xerophytic Plants* [1995], pp. 55, 68; [1998], pp. xi, 66; V. Soarimalala and M. Raherilalao, 'Pression et menaces dans la région forestière sèche malgache', *Malagasy Nature* 1 (2008): 159.

[41] ANGAP-FOFIFA-WWF, 'Rapport'; Randriamampianina et al.'Rapport'. ANGAP (Association Nationale pour la Gestion des Aires Protegées) was a parastatal agency charged with managing natural parks and reserves in Madagascar while FOFIFA (*Foibe Fikarohana momban'y Famboleana*) is a government department concerned with agronomic research.

[42] Further detail on stakeholder perceptions and livelihood strategies is given in Middleton, 'Red Prickly Pear'. Because the *raketamena* issue was so controversial in Karembola villages, the names of informants mentioned in this article have been changed.

[43] Zomana, Tranovaho, 8/1/2003.

[44] See eg Simon Schama, *Landscape and Memory* (New York: A.A. Knopf/Random House, 1995); David Lowenthal, *The Past is a Foreign Country* (Cambridge: Cambridge University Press, 1985); Edward Casey, *Remembering: A Phenomenological Study* (Bloomington: Indiana University Press, 1987); Renato Rosaldo, *Olongot Headhunting* (Palo Alto: Stanford University Press, 1980); Stephen Feld and Keith Basso, *Senses of Place* (Santa Fe, N.M.: School of American Research, 1996).

[45] Zomana, Tranovaho, 8/1/2003.

[46] Celestine, Befeha, 12/12/2002. In fact, some were transported by labour recruiters while many more walked a thousand kilometres and more to find work. Poor people such as Celestine did not necessarily reproduce the nationalist storyline unadulterated. It was quite possible for them to introduce anachronistic narrative (airborne chemical campaigns) and even new views of *vazaha* power (see below) while retaining the core element of the great famine that rang so true with their own experience and fears.

[47] Lahibe, Bevazoa, 4/12/2002.

[48] *Komitim-pokontany,* Marobey, 5/12/2002

[49] Vontana, Tranovaho, 15/1/2003.

[50] President, Soamañitse, interviewed Barabay, 23/1/2003.

[51] On the limitations to interest-based models of memory cf. Carolyn Hamilton, *Terrific Majesty: The Powers of Shaka Zulu and the Limits of Hisorical Invention* (Cambridge: Harvard University Press, 1998).

[52] The Karembola verb I translate as 'to be invasive' is *mandakake*. It means 'to spread', 'to cover ground' and in some contexts the English gloss 'invasive' might be overstated. But such a gloss is fully justified when, as with respect to *raketamena*, informants deploy the verb coupled with other descriptive terms that convey the idea of a plant with 'bad habits' (*fomba raty*), 'a pest' (*biby manahirañe*) they 'can't remove' (*tsy afake*). I have no record of any instance from the 1980s in which such vocabulary was applied to Malagasy Cactus.

[53] President, Soamañitse, interviewed Barabay, 23/1/2003.

[54] Published interpretations of Malagasy Cactus spread range from broadly biological models that see the plant as an aggressive invader (Perrier de la Bâthie, 'Introduction à Tananarive') to a 'social planting' thesis at the other extreme (Kaufmann, 'Cactus Pastoralism'), with certain authors arguing for a more complex model that recognises the interplay of natural and social factors in plant spread. Karembola narrative offers a third interpretation: Malagasy Cactus was neither spread by people nor self-propagating alien but a divinely-appointed cornucopia in an otherwise thirsty land.

[55] Tsyambone, Ngarata, 5/12/2003.

[56] Vontana, Tranovaho, 15/1/2003.

[57] For some key critiques of Clementsian climax from 'New Ecology' perspectives see R. McIntosh, 'Pluralism in Ecology', *Annual Review of Ecological Systematics* **18** (1987): 321–41; D. Botkin, *Discordant Harmonies: A 'New Ecology' for the Twenty-First Century* (New York: Oxford University Press, 1990); D. Sprugel, 'Disturbance, Equilibrium, and Environmental Variability: What is "Natural" Vegetation in a Changing Environment?', *Biological Conservation* **58** (1991): 1–18. On nativist trends in conservationist biology, see Jonah H. Peretti, 'Nativism and Nature: Rethinking Biological Invasion', *Environmental Values* **7** (1998): 183–92.

[58] Although in recent decades conservation of Madagascar's unique biodiversity had figured prominently in state education, exposing younger Karembola to new ideas, school attendance and adult literacy in Karembola remained low. Wealthier rural Karembola were not necessarily more educated or better travelled: they were more likely to have appropriated conservationist discourse through the radio and their dealings with government and agency personnel than through formal schooling. The term 'elite' is probably a misnomer in the Karembola context if it leads reader to think of an educated, transnational elite such as is found in other parts of Madagascar.

[59] The dichotomising perspective is pervasive in anthropological studies of conservation in Madagascar but some recent examples include Lisa L. Gezon, *Global Visions, Local Landscapes. A Political Ecology of Conservation, Conflict, and Control in Northern Madagascar* (Plymouth, UK: AltaMira Press, 2006); Janice Harper, 'Memories of Ancestry in the Forests of Madagascar', in P. Stewart and A. Strathern (eds.) *Landscape, Memory and History: Anthropological Perspectives* (London, Pluto, 2003), pp. 89–107; Eva Keller, 'The Banana Plant and the Moon: Conservation and the Malagasy Ethos of Life in Masoala, Madagascar', *American Ethnologist* **35** (2008): 650–64. For some recent non-Madagascar ethnographies that seek to get grips with the complexity of knowledge interfaces between conservationists and local people, see James G. Carrier, 'Biography, Ecology, Political Economy: Seascape and Conflict in Jamaica', in Stewart and Strathern (eds.) *Landscape, Memory and History*, pp. 210–28; Dan Brockington, 'The Politics and Ethnography of Environmentalisms in Tanzania', *African Affairs* **105**, 418 (2006): 97–116; Anna Tsing, *Friction: an Ethnography of Global Connection* (Princeton, NJ: Princeton University Press, 2005); and J. Peter Brosius, 'Endangered Forest, Endangered People: Environmentalist Representations of Indigenous Knowledge', *Human Ecology* **25** (1997): 47–69, from whom the expression 'rhetorical traffic' is borrowed.

[60] On 'governance states' and the rise of what have become known as BINGOs (Big International Non-Government Organisations) see Graham Harrison, *The World Bank and Africa; the Construction of Governance States* (London: Routledge, 2004); J. Boli and G. Thomas (eds.) *Constructing World Culture: International Nongovernmental Organisations since 1875* (Stanford: Stanford UP, 1999); Paige West, *Conservation is*

Our Government Now: The Politics of Ecology in Papua New Guinea (Durham and London: Duke University Press, 2006).

[61] This article reports on statements made by individual staff-members at FAO, WFP, and SAP and interviewed by the author in Beloha, Ambovombe, Fort Dauphin and Antananarivo in 2002/2003. I cannot comment on official or emerging policy regarding *raketamena* within these organisations as I had no access to internally circulated reports.

[62] Randranjafizanaka Achilson, Assistant de Programme PAM [WFP], interviewed at WFP regional office Fort Dauphin, 29/11/2002.

[63] Mark Fenn and Namie Ratsifandihamanana, 'Impacts sociaux et écologiques de l'expansion de la plante envahissante Opuntia stricta ou *Raketamena* dans le Sud. Demande financement', n.d. [2002].

[64] Interview, FAO Headquarters, Antananarivo, 7/2/2003.

[65] The WWF application had been forwarded to FAO HQ in Rome, where local history (even if it was known) was unlikely to carry influence with an organisation already favourably disposed to biological control.

[66] This paper reports the situation in 2003. WFP more recently has been sponsoring *Opuntia* control in the south (pers. comm. Dorothee Klaus, Director UNICEF-Madagascar 25/11/2010). It is indicative of WFP sensitivity about these projects that repeated requests for further information have met with no response.

[67] On the multilayered symbolism of Malagasy Cactus see Middleton, 'From Ratsiraka to Ravalomanana'.

[68] Cf. Middleton, 'From Ratsiraka to Ravalomanana'.

[69] Middleton, 'Who Killed Malagasy Cactus?'

[70] Rubie S. Watson, 'Introduction', *Memory, History and Opposition*, p. 18.

[71] Luise White, *Speaking with Vampires: Rumor and History in Colonial Africa* (Berkeley: University of California Press, 2000).

[72] Janice Haaken, *Pillar of Salt: Gender, Memory, and the Perils of Looking Back* (London: Free Association, 1998), p. 118.

[73] See e.g. Georges Petit, 'Quelques aspects de la géographie végétale et des cultures à Madagascar', *Bulletin Association Géographes Français* 77 (1934): 37–9.

[74] Informants in the 1980s so invariably described Malagasy Cactus as 'dead' (*mate*) that I had believed the plant to be extinct, a misconception that much scientific literature does little to correct: see e.g. M. Julien and M. Griffiths, *Biological Control of Weeds. A World Catalogue of Agents and Their Target Weeds* (Wallingford: CAB International, 1999), p. 49.

[75] On the homogenising processes involved in the construction of an 'imagined community, imagined self', see Lambek and Antze, 'Introduction', p. xx.

[76] Benedict Anderson, *Imagined Communities: reflections on the origins and spread of nationalism* (London: Verso, 1991), p. 203.

[77] Maurice Bloch, 'Autobiographical memory and the historical memory of the more distant past', *How We Think They Think: Anthropological Approaches to Cognition, Memory, and Literacy* (Boulder, Colorado: Westview Press, 1998), pp. 114–27.

[78] Vontana, Tranovaho, 15/1/2003.

[79] Herbarium, Muséum national d'histoire naturelle, Paris, Madagascar Opuntia 2714.

[80] See e.g. Fairhead and Leach, *Misreading the African Landscape.*

[81] Pier Larson, *History and Memory in the Age of Enslavement, Becoming Merina in Highland Madagascar, 1770–1822* (Oxford: James Currey, 2000), p. 287. For similar strictures, see Mark Hobart (ed.) *An Anthropological Critique of Development: the Growth of Ignorance* (London: Routledge, 1993); Roy Ellen and Holly Harris, 'Introduction', in R. Ellen, P. Parkes and A. Bicker (eds.) *Indigenous Environmental Knowledge and its Transformations. Critical Anthropological Perspectives* (London & New York: Routledge, 2000), pp. 1–33.

[82] See e.g. Bingelli, 'Cactaceae', which attempts to develop a Malagasy Cactus historiography for conservationist use from a limited number of secondary sources of varying quality.

[83] A. Appadurai, 'The past as a scarce resource', *Man* **6**, 2 (1981): 10–19; Philippe Joutard, *Ces voix qui nous viennent du passé* (Paris: Hachette, 1983); Luisa Passerini, *Fascism in Popular Memory: the Cultural Experience of the Turin Working Class* (Cambridge: Cambridge University Press, 1987).

Environment, Ethnicity and History in Chotanagpur, India, 1850–1970

Vinita Damodaran

Recent studies following the pioneering work of writers such as Eric Hobsbawm, Terence Ranger, and Benedict Anderson in Europe and Leroy Vail in Africa have effectively argued that ethnicity and ethnic ideologies are historically contingent creations.[1] This line of research challenges the ahistorical and primordial assumptions underlying nationalist and ethnic ideologies. In the words of Leroy Vail 'If ethnic consciousness was a product of historical experience, then its creation and elaboration would be a proper subject of enquiry for historians.'[2] However, while there has been a great deal written on ethnicity and ethnic identities as constructed through class, race and gender ideologies,[3] the ways in which ethnic identities have been constructed around images of the land and the changing meanings of the landscape is neglected.[4] In particular, there is very limited work on the ways in which images of the landscape have fuelled cultural and political resistance against a dominant and aggrandising state.[5] Moreover, most studies of the landscape have concentrated on the arable parts of the landscape while the non-arable forest, pastoral, mountain, marsh and other areas is ignored. The growing domain of environmental history has recently attempted to document the forest and non-forest parts of the non arable environment and its relationship with indigenous peoples in the colonial context.[6] However there is a need to understand the links between images of the landscape, the construction of identities and cultural resistance, especially in South Asia. In particular, scholars will need to discuss the complexities of the relationship between the spiritual memory of remembered landscapes and modern politics. This study explores the important contemporary issues of environmental degradation, ethnic and regional dissidence through an analysis of the ways in which identities have been constructed around images of the land and forest in Chotanagpur. The narrative extends over the colonial and post-colonial period thus avoiding the confusing division of labour between historians and political scientists, and is conducted in the light of recent theoretical advances made in the historiography of India.

In the southern part of Bihar state, the region of Chotanagpur is now the site of a widespread movement for statehood for its 30 million inhabitants, many of whom belong to indigenous (*adivasi*) groups. The demand for Jharkhand also includes other districts in the neighbouring states of Orissa, Bengal and Madhya Pradesh, a total of 23 districts are involved. The movement centres mainly on an explicit assertion of the 'rights' of the indigenous people of the region to take charge of their own territory and resources and to revive a diverse and rich

Environment and History **3** (1997): 273–298.

culture which has been long suppressed by a state controlled by the dominant non-*adivasi* Hindu elite.[7] The Jharkhand movement regards itself as being linked to earlier nineteenth century tribal protest movements against the dispossession of ancestral lands and forests and other interventions into a traditional way of life by the colonial state. Their historical consciousness enables the Chotanagpuris to construct their history as one that evokes the memory of past exploitation which has helped to attract a wider constituency to the political platform of the Jharkhand parties and has enabled the construction of a pan-tribal identity. The discourse of marginality forcefully articulates the history of the region as one in which the indigenous inhabitants of Chotanagpur were alienated from their lands and forests, gradually peripheralised in regional politics and subjected to the whims of the colonial state. In this context, the institutions of the courts and the police are seen as bolstering the interventions of a predominantly Hindu elite in local society. [8] Much of this discourse has hinged on claims that local communities are the best stewards of the landscape and by reference to a global environmentalism. Following independence, the Jharkhand parties received a new impetus in the context of renewed exploitation. The situation of the *adivasis* steadily worsened as the nationalist state, on the grounds of a new state ideology of 'tribal assimilation' increased its incursions into Chotanagpuri society, with harsh consequences for its inhabitants.

Given a sustained attack on the material position and identity of the indigenous people in Chotanagpur, their minority discourse in the present period has emerged as the outcome of damage systematically inflicted on their cultures under both colonial and post-colonial rule. When I use the term culture, I do not use it in a homogenous or static form. Rather, I would like to invoke James Clifford when he chooses to see culture as constructed and disputed and constantly reshaped through displacement and interaction. There is therefore a need to locate Chotanagpuri culture in a history of cultural and inter-regional relationships.

The question that should be asked here is whether Chotanagpuri cultural identity was shaped in the colonial theatre? This would lead us to an argument made popular by the Comaroffs in the context of southern Africa based on the notion that Tswana ethnicity was a profoundly historical creation and that the Tswana sense of difference from the 'other' was to emerge as a central trope in Tswana historical consciousness only through their interaction with white culture.[9] This assertion is problematic, for while it can be argued that the interaction with colonial culture was a dynamic one and resulted in new cultural forms, this does not mean there was no strong sense of identity in Chotanagpur previously. The identity of different Chotanagpuri communities was one that had developed over the centuries through their migration to the hills away from the plains, through their encounters with different groups in their journey into the hills and finally through the multiplicity of their engagements with the forests and the land in the place of their final settlement.[10] In the later period, in the seventeenth and eighteenth century the encounter of various Chotanagpur communities with Hindu moneylenders had resulted in the rejection of notions of exchange and property of the plains peoples. To construct the Chotanagpuris as the Comaroffs have done for the Tswana is to construct them as unchanging and timeless before the encounter with colonial rule. The people in Chotanagpur had lived in villages, travelled and encountered other cultures in the past and had a clear sense of historical consciousness.[11] As Clifford writes 'the chronotrope of culture then ... comes to resemble as much a site of travel encounters as of residence.... culture seen as a rooted body that grows, lives and dies is questioned, constructed and disputed. Historicities, sites of displacement, interference and interaction come sharply into view.'[12] Community has to be seen as a constructed entity and a sense of place as one that is 'contingent and negotiated'.

In the nineteenth and twentieth century it can be argued that, as a result of ever increasing interventions by a modernising colonial state, *adivasi* culture was gradually reduced to the status of a 'minority culture' by the dominant culture resulting in a growing *adivasi* consciousness and a sense of an authentic *adivasi* identity. The destruction involved in the process of interaction with a dominant culture was manifold, resulting, in Chotanagpur, in dramatic changes in the relationship between local people and their environment and in the dismantling of a previously functional economic system. In concert with this material destruction, the cultural formations, language and the diverse modes of identity of the people were irreversibly affected and displaced by a single mode of historical development within which tribal cultures were perceived as 'underdeveloped', 'imperfect', 'childlike' or even 'criminal'. [13] There is little doubt that this resulted in a growing sense of *adivasi* identity which was to fuel a long cycle of protest and resistance

The construction of tribals as 'simple and childlike' and in need of progress runs right through much of colonial writing in the period, albeit with some important exceptions.[14] Tribal practices were defined in colonial and later national ethnographies as static, unchanging and primitive. Ecologically, the developmental ideology of the colonial state in the nineteenth century brought havoc in its wake to hitherto remote regions such as Chotanagpur, where rapid deforestation (primarily to meet the timber needs of the new railways) and open cast mines transformed the relationship of the local people with their environment. Migration into the region, which had a long history, increased dramatically in the nineteenth century and resulted in heavy pressure on the land, dispossession of indigenous people's lands and emigration outside the region by *adivasi* communities who found themselves fast becoming minorities in their own districts. A large percentage of the population of Ranchi district were registered as having emigrated to the Assam tea gardens between 1864 and 1880 and many thousands more had left unbeknown to the officials. It was a sad flight of a proud people subordinated to the ranks of 'coolie' labour in mines and plantations. One needs to examine the historical background to this development.

Vinita Damodaran

THE LANDSCAPE OF SERVITUDE

Over the nineteenth century, colonial ecological science had clearly demonstrated the adaptability and the productivity of local agricultural practices. However, generations of agricultural experts still believed that a profitable agricultural system could not be created in this area without mechanisation and its concomitant resettlement. As a result, these experts were contemptuous of local agricultural practices. As the Ranchi settlement report recorded for the area 'the system of cultivation is primitive and the soil is poor. Irrigation is neglected and manuring practised on the uplands. They (the people) are thriftless and indolent and only the pinch of poverty drives them to undertake any sustained employment'[15]We can see here the nature of the discourse that helped to construct Chotanagpur as an object of knowledge and as a subject of intervention.

Colonial constructions of the landscape had a similar tone.The forests were in the early part of the nineteenth century primarily regarded as a resource. The perceived policy of the colonial rulers was to extend cultivation at the expense of forest tracts and to exterminate all wild and dangerous game. The rewards offered by the state to destroy tigers effectively decimated the population of these magnificent beasts. In the Santhal Parganas, E.G. Man reported in the 1860s that where elephants and rhinos were abundant as late as the 1830s and 1840s 'now the latter are extinct and of the former but three are left'[16] In Ranchi, the district gazetteer recorded the unchecked destruction of forests in the district in the nineteenth century.[17]A major cause of the destruction of the jungles in most districts was the sale or the lease of the forest to contractors for supply of railway sleepers. Entire forests were destroyed to supply the timber necessary for railways. The opening of the railway to Ranchi and Lohardagga in the nineteenth century and the improvement of communications by road led to the sale of jungles previously untouched. Coupled with this, landlord encroachments on village jungles and the sale of timber to outsiders further exacerbated the destruction of the jungle.

During these developments the colonial administration continued to argue that traditional forest use seriously exacerbated the destruction of the forest. It was said that peasants were continuing age old practices which had scarcely harmed the forest when it was in abundance, but now only further threatened its destruction. *Jhuming* (shifting cultivation) was also said to threaten jungles in remoter parts of the district. Interestingly the writers of the settlement report in 1906 were forced to concede that the destruction wrought by such methods accounted for only a small percentage of the deforestation in Dalbhum.[18]

In spite of this the government and the forest department still embarked on a wholesale programme of forest reservation and exclusion of indigenous communities from the forest. This state programme had a far-reaching impact on the lives of the local people. In many places the landlord and the state had long battled with each other to secure large areas of jungle land, extinguishing the traditional common rights of the people. Under the Permanent Settlement of Bengal, all such rights had been transferred to the landlord, to enable such lands to be taxed. The landlords then proceeded wherever they could to extinguish peasant rights to common property resources in several parts of the region. As the forests on which they depended on for their food resources dwindled, the region began to suffer from famines and the late nineteenth century saw a 'rash of famines'.

Agricultural marginality

During this period, landlordism came to dominate rural social relations. While the northern districts had begun to be heavily overrun by Hindu immigrants before the advent of the British the pace of change increased under colonial rule and even the southern districts began to feel threatened. The Mundas in Ranchi for example had managed to hold on to their traditional *Khuntkhatti* tenures in the face of outsider landlord encroachment. By the time of the Census operations in 1881, the original indigenous population in the districts of Palamau and Hazaribagh was only 36% and 34% while in the remoter districts of Ranchi and Hazaribagh it was 74% and 75%. In the northern districts of Palamau and Hazaribagh and in the Santhal Parganas increasing subinfeudation and the growing spread of debt bondage were the main grievances of the peasantry. In the Santhal Parganas, the main causes of distress were the grasping and rapacious mahajans, the misery of hereditary bondage, and the unparalleled corruption of the police and the impossibility of redress in the courts. In certain districts, as in Birbhum, the invasion of the pargana by a powerful English company bent on destroying the rights of the Ghatwals or Bhumij *khuntkhattidars* gave rise to much disturbance. These settler rights all across Chotanagpur were challenged by colonial courts and superior land interests. One settlement report recorded 'It is common experience in Chotanagpur that the aboriginals are ruined by their incapacity to state their claims intelligently'. Clearly the discursive and political framework in which they had to operate disadvantaged the indigenous people. The incidence of debt bondage began rapidly to spread and peasants were often forced to borrow money from the landlord or moneylender to meet local needs. The system that developed was known as *kamiaouti* and led to the absolute degradation of the *kamias*. By gradually advancing small sums of money to them in lieu of land many Hindu landlords converted their tenants into slaves and reduced their holdings to tiny homestead lands. Some of the poorest peasantry migrated away from their homeland to work elsewhere. The new coal mines in Ranchi and Dhanbad and the tea gardens of Assam and Bhutan found a cheap and abundant labour force among a dispossessed and impoverished rural population. One Santhali song records this process thus

> Sahrul I could not dance
> My *jhuri* has gone as a coolie

> Only with him can I dance
> My *jhuri* has gone as a coolie [19]

The grave effects of unrestrained interventions into the lives of the indigenous peoples of Chotanagpur were apparent by the middle of the nineteenth century, even to the conservative colonial state. Following the experiments of Augustus Cleveland in the Santhal Parganas in the latter half of the eighteenth century, and his attempts to preserve tribal institutions and concepts of 'justice', the colonial state was forced to recognise certain special areas and the entire province of Chotanagpur was put under the charge of an officer, designated agent to the governor-general in 1831.[20] Under the Government of India Act of 1870, the British Parliament conferred on the Governor-General-in-council the power to approve and sanction as laws, regulations made by the local government for the administration of certain special areas. This rather belated recognition was followed by the Scheduled Districts Act of 1874. This act empowered the local government to declare (for each of the tracts specified in the Act) which enactments were in force and to what modifications or restrictions they were subject. However, apart from giving these areas special status, these changes came too late to have any significant effect on the lives of the indigenous peoples of Chotanagpur.

In the twentieth century, the colonial state continued to experiment with cosmetic reforms. The Chotanagpur Land Alienation Act was strengthened in 1908 to prevent further alienation of tribal lands. Anthropologists such as Grigson, Grierson and Verrier Elwin became critics of colonial state policy towards tribes and vehement advocates of 'isolationist' policies intended to preserve the way of life of the indigenous peoples.[21] These men passionately believed that the tribals of central and north-eastern India had to be protected against modernisation. It has been argued that the British government found in this new ideology a cloak for its reluctance to spend heavily in tribal areas which were not considered to be major contributors to the imperial treasury. The ideology was also seen as useful in preventing the spread of nationalist ideas in these regions. Whatever the motives, this new thinking came too late to rectify the ills resulting from forced modernisation in Chotanagpur.[22]

THE REMEMBERED LANDSCAPE

> *Oh for the days when men knew no cares*
> *and drank their fill of home brewed ale*
> *Woe to the age when men on earth below*
> *Do daily die of famine*

> (Munda song recorded by S.C. Roy in the 1900s)

Identities were transformed in the context of this rapid ecological and cultural change. As Steve Daniels notes (and this is certainly true both of ethnic and national identities), 'Identities are often defined by legends and landscapes, by stories of golden ages, enduring traditions, heroic deeds and dramatic destinies located in ancient or promised homelands with hallowed sites and scenery. The symbolic activation of time and space often drawing on religious sentiment gives shape to the imagined community of the nation.'[23] It must be noted here that the term landscape is a complex concept. As Cosgrove argues the term can be seen as a 'socio-historical construct', a way of seeing projected onto the land which has its own techniques and which articulates a particular way of experiencing a relationship with nature. It can be argued in a similar fashion, that the landscape of Chotanagpur has been reclaimed and reconstituted as a complex symbolic terrain for definitions of Chotanagpuri identity.[24] It is useful here to note, as Sahlins does in the context of Hawaii, that 'the landscape and its legends inscribe a criticism of the existing regime. In the current jargon, the landscape is text. Places and names evoke an alternative society older, truer and more directly related to the people.' In this way were the landscapes of Chotanagpur organised by stories and legends of conquest and defeat and through memories of better times.

The symbolism of the landscape and past readings of it

In the context of Amazonia, Peter Gow shows how the Bayo Urubanka river was lived as a human landscape by local native peoples through a multiplicity of engagements with the forest and the river, with each other in acts of generosity, in narration and in encounter with the dead and with spirits.[25] In a similar fashion were the landscapes of Chotanagpur lived as a human landscape in the past. The original forests were spread out over thousands of square miles, especially in the districts of Hazaribagh, Singhbhum, Palamau and Ranchi, all of which had large forest areas. Indigenous rulers had tended to preserve the forest for military reasons and as Walter Hamilton noted in 1820, in several parts of Chotanagpur the woods had been forested with great care by the rajas as a protection against invasion from without. The trees were mainly either moist deciduous or dry deciduous and the whole division had a very rich growth of *sal* (*Shorea robusta*).

Ranchi district in the centre of the Gangetic tract was one area where the *sal* tree was most dominant. The best *sal* forests were found here in the valleys where in good soil straight trees up to 120 feet in height and with a girth of up to 15 feet were found. In the valleys, especially in sheltered situations, the principal companions of the *sal* were the *asan*, (*Terminalia tomentosa*), *gambhar*, *kend* and *simal*. The *mahua* tree (*Bassia latifolia*) was common throughout Chotanagpur and was very important to the local economy. In the villages better fruit bearing trees grew like the *jamun* (*Eugenia jambolina*) *karanj* (*Ponamia glabra*) *tetar*

(*Tamarindus indica*),*bael* (*Aegle marmegos*),jackfruit (*Autocarpus integrifolia*) *pipal* (*Ficus religiosa*) and *ber* (*Ficus benglensis*).There were many other forest shrubs and trees which yielded fruit and which afforded valuable food supplements in years of scarcity. Slacke, in his report on the settlement operations in Chotanagpur estate in 1882, enumerated 21 species of seeds and the fruits of 45 uncultivated trees which were used as food in addition to 34 trees the leaves of which were used as vegetables, and 18 species of edible roots. Slacke also gives the names of 97 forest products used as medicines, 28 used as oil and gums, 17 used as dyes and 33 creepers or barks of trees used as rope fibres. The length of these lists gives us some indication of the economic value of the jungles to the indigenous inhabitants. Valentine Ball noted in the 1860s that several of the communities were heavily dependent on jungle products. For example the keriahs of the Jolhari hills, who were not settled agriculturists, relied on the jungle for a supply of fruits, leaves and roots. This they supplemented with rice procured from the lowland agricultural communities by trading jungle products such as honey, *lac*, *sal* seeds and leaves and *tusser* cocoons. [26]

Most of the communities, even the settled agricultural ones and particularly the women of these communities had a highly sophisticated technical knowledge of their jungle habitat. The Ho's of Singhbhum, for example, had names for all the common plants and those of economic importance to them and, like the forest Mundas, were well versed in the edible properties of plants. The Birhor, in the extreme east of Singhbhum, were a wandering community who lived by snaring monkeys and by collecting the fibre of the *Bauhinia vahlii* creeper.[27] The forest environment, and a knowledge of it, were thus of critical importance to the local people, particularly in dietary terms. This importance in terms of food was paralleled by an equal significance in systems of belief; and the two were not truly separable. Chotanagpur folk taxonomy was completely embedded in and mediated by the local cultural order.

Munda understandings of the landscape and its productivity seemed to encompass conceptual links between women and forests. Every Munda village, for example, had its own particular spirits whose duty it was to look after the crops.These spirits were known as *bongas,* which was a generic name referring to spirits and the power or force of mountains, hills, forests, trees, rivers, houses and village. One such spirit, known as Desawali, played a large part in Munda festivals which were connected with the cultivation of the land. The home of this presiding deity was the *sarna* or sacred grove, a little path of jungle, that when all else was cleared for cultivation, was left as a refuge for the gods where they might live apart. At all seasons of the year offerings were made in the *sarna*, for on the favour of the Desawali depended the success and failure of the crops. The other communities such as the Oraons also had a festival connected with the sacred grove which was observed with much ceremony in March and April. It occurred when the *sal* tree was in flower and its graceful plume-like blossoms decked the earth. All the villagers assembled in the sarna where the *sarna burhi* or the woman of the grove was said to reside. The sahrul festival was an act of rejoicing in the jungle which had come into flower. It was a fecundity ceremony, a marriage of the earth with the sun. On the assumption that the soil is ready to be quickened, the fertility of the jungle is used to stimulate the fertility of the fields.The fecundity of the forest thus made it in many ways a female space.[28] This was expressed, for example, in certain rituals among Oraons involving a *sal* sapling in which a slit had been made to resemble a female organ, which was used to augment the procreative energy of men. As Roy noted a week before the Phagun festival in March and again before the Sahrul festival in April or May the boys of the dormitory were led forth by the Dhumkuria Mahato to a suitable secluded place some way off from the village where they would smear their bodies with red earth. The boys then proceeded to spit into the slit in the sapling and insert their organ into the slit. All the time the Mahato stood behind the *sal* sapling with the new *sal* wood stick in his hand and as each boy was about to step back after performing the magic operation the mahato struck the boy on his thigh, a little below the groin.[29] Here we find individual trees representing a woman. The Munda and Santhal aesthetic traditions seemed to represent this connection. Women's domestic wall designs in large parts of Chotanagpur incorporate forest and foliage motifs into their active aesthetic traditions to induce fertility for themselves and in after-life.[30] It is important to note here that this symbolic construction of the special relationship of Chotanagpuri women with the natural environment had a basis in the prevailing gender division of labour which made these women primarily responsible for fetching fuelwood, fodder and water.

The propitiating of the female spirits of the grove was, however, mainly done by men with women being excluded. The sacrifice began by sacrificing fowls before a rough image of mud or stone. At night the villagers returned home with *sal* blossoms and marched to the beating of the drums and the blowing of the horns, with much dancing along the way. The following morning the women, gaily decked with *sal* blossoms, carried baskets filled with the same blossoms which they placed over the door of every house for luck. The festivals of the sacred grove were very important for the settled agricultural communities of Chotanagpur and emphasised the importance of the forest and its flowering seasons in the ritual life of the communities.The forest was central to human life, and forest and village together made up a spiritual and moral entity. It was to be protected and preserved and the people must respect it. Clearly a symbiotic relationship prevailed between the people and the landscape in Chotanagpur.[31] This interpretation is clearly different from the perception of the Mende of Sierra Leone in an anthropological study conducted in 1980, in which the forest is regarded as opposing mankind and as an obstacle to human progress; the Mende believing that the bush contained innumerable dangers including evil spirits and that it constantly threatens farms.[32] The Chotanagpuris on the other hand tended to experience the forest and village as ontologically part of each other, the one being the life force for the other's continuing existence. An interesting study among the Nayaka of south India by Nurit Bird David argues for a similar cosmic economy of sharing.[33] The Nayakas converse, dance, sing and even share cigarettes with the spirits of the forest which they invoke with shamanistic experience.

One should point out here that this is not a story of an unchanging pristine landscape, the narrative is not a version of Chotanagpur's idyllic environmental past. Clearly the people's engagement with the forests in multiple ways saw the creation of a human landscape, one that was ever changing with past forests giving way to settlements, shifting agricultural practices and altering boundaries between villages and forests. Nature was not *out there*, it was a lived relationship for the people of Chotanagpur. Only detailed environmental histories of villages will allow us to piece together these rich and complex stories.[34] However, what can be forcefully argued here is that Chotanagpuri understandings of the landscape, their stories of nature, and their lived history were to differ radically from the perceptions of nature and the land of colonial scientists and policy makers and later of a modernising nationalist elite.[35] The destruction of forests that was to occur as a result of colonial intervention in the nineteenth century and later was to change this relationship between the people and their environment drastically and forever. However, the memory of the landscape was to live on and it became a repository of Chotanagpur's nostalgic past to be revived in complex oppositional contexts. Simon Schama has noted that 'Landscapes are culture before they are nature; constructs of the imagination projected on to wood, water and rock.... once a certain idea of the landscape, a myth, a vision establishes itself in an actual place, it has a peculiar way of muddling categories, of making metaphors more real than their referents; of becoming in fact part of the scenery.'[36]

The despoliation of the forested landscape and the transformation of the people's relationship with their environment in Chotanagpur in the nineteenth century, was a powerful memory which was revived in periods of cultural resistance. In the latter half of the twentieth century this resulted in specific cultural images of the landscape being evoked through the ritual festivals of the scared grove and emerging as a factor in protest. Images of the landscape have long played a role in cultural resistance in other regions. For example, as Daniels notes, the 'ethnic nationalism fuelling the dissolution of the Soviet Union was codified by pictorial images of independent homelands'.[37] It can be argued, that resistance may have been fanned by memory of better times in a less despoiled setting. These memories were present in the 'landscape of their current servitude'.[38] In the context of Chotanagpur, the 'remembered landscape' was to fuel a long struggle of protest.

THE POLITICS OF MARGINALITY

All through the nineteenth century the indigenous communities had sought to protest against the growing incursions into their lives. Beginning with the unrest in Tamar in 1816 and the Munda rebellion in 1832 the last decades of the nineteenth century saw unrest in the almost every district of Chotanagpur. The Birsa Munda uprising in the 1890s which began as a protest against colonial forest laws was the culmination of this period of rebellion. British forest reservation laws had long proved irksome to the Mundas and in the context of the degradation of their forest environment, the exploitation by Hindu money lenders and an modernising colonial state, they rose in protest.

By the twentieth century there was thus an established tradition of resistance in the region. In 1915 British reports began to mention the Tana Bhagat disturbances among the Oraons in Ranchi district. Their movement, which had strong religious overtones, aimed to redress local grievances against zamindars and traders. The Oraons were apparently enjoined by divine command to 'give up superstitious practices and animal sacrifices', to 'stop eating meat and drinking liquor', to 'cease ploughing their fields', and to withdraw their field labour from non-*adivasi* landowners. [39] One of the movement's leaders, Sibu Oraon, had been reported as distributing leaflets to the effect that *zamindari* raj had come to an end, that it was no longer necessary to pay rent or chaukidari tax, that the Marwaris were selling cloth very dear and should be turned out and their cloths burnt and that the bones of Muhammadans should be broken because they killed cows. [40] The movement had a large following and was said to resemble Birsa Munda's uprising of 1895. By the 1920s, the Tana Bhagat movement had acquired 'disturbing' links with the Congress movement in the rest of Bihar. Gandhi's non- cooperation struggle resulted in renewed agitation in Chotanagpur and the protesters intensified their demands for low rents, restoration of rights to the jungle and abolition of forced labour. [41] The restriction of access to forests and fresh water fisheries resulted in a wave of protest among Oraons, Mundas and Santhals in the 1920s and 30s in Midnapur, Bankura and Singbhum and the rest of Chotanagpur. Many of these were fuelled by memories of better times, by stories of their fathers' times when all jungles were free and all *bandhs* open to the general public. [42]

When they lost their lands and forest to *dikus* (outsiders) the communities refused to recognise the loss as legitimate. William Archer, a colonial official at the time, records the ceremonial taking possession of villages by Tana Bhagats, the planting of the Tana Bhagat flag in the village and the reallocation of land among followers. He recorded that to the question 'Where are your title deeds?' the Bhagats replied 'The answer is: my spade, my axe, my ploughshare are my title deeds... ploughing is the writing of the golden pen on golden land'. To the argument 'Your lands have been auctioned for arrears of rent and purchased by another', the reply was 'When a man buys a mat he rolls it up and takes it away, similarly unless the purchaser has rolled up my land and taken it away how can he be said to have purchased them?'[43] This is similar to the native readings of the landscape in Amazonia. As Peter Gow notes the native people of the region were implicated in the landscape through moving around in the landscape and leaving traces on it. It was a lived space and no piece of paper in a far off government office which determined who owned what could transform the people's own complex relationship with the land.[44]

The Tana Bhagat movement is interesting because the Oraon followers adopted vegetarianism and an austere lifestyle and incorporated other Gandhian symbols in an attempt to strengthen their hands against the landlords. It can be argued that the movement's appropriation of the symbols of the national movement helped to promote a feeling of solidarity with a wider struggle against an oppressor state. Gandhi was also understood in terms of the people's own religious consciousness. The *adivasi* world was filled with divine and semi-divine beings and Gandhi was considered a divine force of this type with powers to mediate between the *adivasis* and nature.[45] The people of Chotanagpur were producing their own brand of culture and modernity.

Modernity and its discontents

The second decades of the twentieth century saw the development of a modern political idiom which sought to arrest the marginalisation of the indigenous groups. Through their interactions with colonial institutions a new ethnicity began to emerge. A pan-*adivasi* identity was asserted under the Chotanagpur Unnati Samaj, which was started in 1915. It embodied inter-denominational unity of the missions for political purposes.[46] The leaders of this organisation were mainly Christian converts, English educated students belonging to the Munda and Oraon tribes who at times tended towards sectarian behaviour against non-tribal autochthones. The rising *adivasi* consciousness was given impetus by activities of missionaries and by colonial writings which categorised people into essentialised tribal identities with fixed boundaries. One can argue, with some reservation recorded above, as the Comaroffs do in the context of the Tswana, that the 'long conversations with the missionaries had set the terms of the encounter which sought to make Africans through their everyday dress, agriculture, architecture and so on- through formal education ... the various ways in which the culture shown by the churchmen took root in the social terrain of the Tswana to be reinvented or reified into ethnic tradition ... some to be creatively transformed, some to be redeployed to talk back to whites parts of the evangelical message ... giving rise to novel forms of consciousness and action.'[47] However, it must be noted that the growing Christian influence and the activities of missionaries did not result in a revival of tribal traditions; rather there occurred a distancing from tribal ways of life, which began to be seen as primitive. Many of the Christian tribals embraced the language of modernity with zeal.

In 1937, the Unnati Samaj was reorganised as the Adivasi Mahasabha and, for the first time, raised the question of a separate Jharkhand state. The immediate cause of the formation of the Mahasabha was the experience of the first elections held in 1937 under the Government of India act of 1935. The Congress swept the polls and there was a growing realisation among the educated tribals that unless they organised themselves, the Congress would hold sway in Chotanagpur as elsewhere. It was felt that the Congress had little in it to offer the indigenous inhabitants of Chotanagpur and was a party of the *dikus* (foreigners). This provided the impetus for some Christian and non-Christian tribals to join forces under the Adivasi Mahasabha. The Adivasi Mahasabha continued its efforts to forge a pan-tribal identity and also opened up its membership to non-*adivasis* of that region. However, the strict distinction between tribals and non-tribals in the popular mind was to crystallise further after the announcement of the Scheduled Tribes list in 1936. In its manifesto in 1937, the Sabha emphasised unity among the different tribal groups above all. This emphasis on unity was in keeping with a growing understanding that only a broad movement would strengthen the hands of the Sabha vis-à-vis the state. Its expressed objectives were the improvement of the economic and political status of *adivasis* in Chotanagpur.[48] The party was opposed to the Congress in this period and was seen as loyalist by the British. It therefore remained outside the mainstream of nationalist politics.[49] The attitude of the *adivasi* leadership to the Congress in this period was in part because the Congress was seen as a party of the *dikus* which had little respect for tribal tradition and culture. In the 1940s the party gained new support for its movement from the Muslim League which hoped to establish a corridor through Chotanagpur to link it with east Pakistan. The nature of the relationship between the Congress and the Adivasi Mahasabha can be gauged by the violence which erupted during the elections in 1946. In the elections in the region fights broke out between the Mahasabha and the Congress at various polling stations. In Kunti district five *adivasis* were killed and several injured in the violence, generating widespread condemnation.[50] A renewed period of mistrust and hostility between the Congress and the Adivasi Mahasabha followed.

Events in the preceding period had resulted in ethnic arguments based on the notion of a seperate tribal identity losing its force in favour of regionalism, as the Jharkhand party began to embrace the discourse of western modernity. In 1950 the Adivasi Mahasabha was wound up to form the Jharkhand party, which gradually changed its policy vis-à-vis the Congress. The discourse of the new party did not emphasise tribal traditions either ecological or cultural. Instead, there was an abandonment of the old language in favour of a modern one. The party then had to broaden its base and attempted to enlist more non-Scheduled-Tribe members. By the 1950s it was clear that ethnicity alone could not be the basis of a political dialogue in Chotanagpur. The census of 1951 showed that the tribals had become a minority in Chotanagpur. The party was therefore thrown open, at least in principle, (as embodied in its constitution) to all Chotanagpuris. A significant transition from ethnicity to regionalism emerged as the formative factor in the movement. This is not to say that ethnic arguments completely lost their force but only that the Jharkhand party saw it as politically tactful to air more regionally based arguments. Notions of ethnicity had to be reconstituted in different historical moments. The history of cultural contact with the plains peoples, migration both to and from Chotanagpur, and inequality were challenging notions of a pure *adivasi* identity. Many of the more recent migrants into Chotanagpur were in fact poor low caste plains Hindus. It was felt that the new Jharkhand parties had to contend with this change by abandoning the old political language.

The foremost ideologue of the Jharkhand Party in the 1950s was a western-educated Munda, Jaipal Singh. He had been active in the Adivasi Mahasabha and epitomised the new breed of leadership, which was western- educated, Christian and had an urban outlook. As a charismatic leader, he had a large following.[51] Under his leadership, the concept of Jharkhand was enlarged to include all the regions that once formed the Chotanagpur administrative division. The party had decided that it would use constitutional means to achieve its goal. That the policy worked is clear, for the popularity of the party in the 1950s rapidly increased. It swept the polls both in 1952 and 1957, emerging as the major political organisation in the Chotanagpur/ Santhal Parganas area. The 1957 elections then saw it extend its influence into Orissa, where it captured five seats. It displayed remarkable unity, and thousands of people turned up at party meetings to show support. However, despite this show of strength, the States Reorganisation Commission in the mid-1950s turned down the plea for a separate Jharkhand state.

By the late 1950s, the party entered a period of decline. At the leadership level, there was a growing split between the Christian and non-Christian *adivasis* on account of the former controlling high party positions. There was also a growing realisation among the people that the party had failed to deliver the goods. It did not have any concrete agrarian or environmental programme and the leadership was drawn from the high strata of tribal society, that is, mainly from Mundas and Mankis (village headmen) in many parts and from the Manjhi and educated Christians from the Munda and Oraon areas. As agrarian conditions continued to deteriorate new measures were needed to remedy widespread impoverishment, but the party organisation was too weak, and it had no radical programme.[52] Eventually, the search for funds led it into dealing with the hated *diku* class of exploiters. In 1962, the party accepted as member an ex-zamindar of Chotanagpur and appointed a secretary from the money-lending community. In the absence of a clearly articulated programme of a cultural and political revival they were threatened with incorporation by the dominant national party the Congress. By 1962 the Congress, with its programme of *garibi hatao*, actually seemed to be more in tune with mass demands to end poverty. Support for the Congress correspondingly increased with the decline of the Jharkhand party. In the 1962 elections the Jharkhand party was reduced to 20 seats in the Bihar Assembly and it appeared that it could no longer maintain itself as a viable political organisation. The merger of the Jharkhand party with the Congress was thus a natural corollary to these events. Jaipal Singh accepted a portfolio in the Bihar cabinet and many of his supporters never forgot this betrayal.[53] The merger signalled the end of the Jharkhand party as a party of the people and effectively outlawed the radical stream. This showed the success of the dominant ruling party at the centre, the Congress in countering the threat posed by the Jharkhand by affecting a merger, denying *adivasi* claims an authenticity and by reinforcing the claims of the nation state.

A period of dissent followed, with grassroots activists struggling to build a political base. and to articulate a new ethnicity based upon the notion of a separate cultural identity and a remembered past. A radicalisation of politics was inevitable given the increased exploitation of the tribal lands and forests under the Congress regime. Indeed, Congress policy towards tribal areas in the post-independence period had totally alienated the *adivasis*. After 1947, the 'isolationist' thinking of the colonial rulers was heavily criticised by the nationalist state wedded, as it imperial precursors had been, to the ideology of development.[54] The report of the Scheduled-Tribes Commissioner, known as the Dhebar report on the Indian state's policy towards tribals, argued that the British policy of isolating them had resulted in their exploitation.[55] Professor Ghurye, a long time critic of British rule, voiced this change in thinking when he stated that 'the policy of protecting the so called aborigines through the constitutional expedient of excluded areas or partially excluded areas evoked a protest from politically conscious Indians and was resented by many of them.'[56] In a conscious attempt to move away from the British policy towards the tribes, the new policy was unashamedly assimilationist, its professed aim being to draw the tribes into the mainstream of Indian political culture. The aim of the nation state in denying *adivasi* claims to an authenticity recognised the possibilities of such an admission in creating sites of resistance and empowerment and moved swiftly in the post-independence period to crush all such hopes. The consequences of such a policy were predictably disastrous and fuelled more tension in the region.[57] It also paved the way for further violent histories of economic, political and cultural interaction that was to exist alongside with cultures of displacement and transplantation. This was to lead to a new cultural revival and a rearticulation of ancient rhetoric especially that relating to the land.

The government of Bihar pushed ahead with a massive exploitation of the forest and mineral wealth of the region while maintaining in its official 'tribal' policies that the 'tribals' should be allowed to develop according to their own genius. After the 1950s, thousands of acres of *adivasi* land were lost to new industries. The cities of Ranchi, Dhanbad and Jamshedpur continued to grow rapidly through an ever-increasing in-migration of non-*adivasi dikus*. By 1961, there were already half a million migrants in Dhanbad and Singhbhum. There was also an extensive loss of land through sales by *adivasis* to non-*adivasis* not only for business purposes but for erecting residential buildings.[58] The result was an increasing 'de-tribalisation', with communities such as the Bauris becoming descheduled on account of their development as coal miners. The 1971 census disclosed an alarming state of affairs. The percentage of the 'scheduled tribes' in the population of the districts of tribal Bihar had fallen sharply in the decade from 1961: in Ranchi from 61.61 to 58.08, in Singhbhum from 47.31 to 46.12 and in the Santhal Parganas from 38.24 to 36.22. This was not only due to the slow growth rate of the *adivasi* population, which was in fact among the lowest in India, but the influx of people from other parts of Bihar. In this period, struggles to halt these dramatic changes developed under a new leadership and resulted in the creation of political organisations like the Birsa Seva Dal and the Jharkhand Mukti Morcha (JMM). The new political extremism was reflected most clearly in the formation, in 1973, of the JMM, whose object was to form a separate Jharkhand state, end the exploitation of 'tribals' by 'non-tribals', and secure preferential treatment for 'sons of the soil' in the matter of employment. Strategic claims for authentic tribal traditions and histories emerged again in this phase as they helped to create sites for empowerment and resistance. The JMM led large scale movements of protest to regain tribal lands in the 1960s and 1970s. In 1968, a movement among the Santhals in the Santhal Parganas followed, and resulted in, among other developments, the Santhals forcibly harvesting standing crops on lands illegally occupied by moneylenders. The period 1967-74 also saw many struggles under the aegis of the JMM to recover alienated lands from moneylenders and rich peasants in Chotanagpur, amounting to a renewed assertion of strength of a people long exploited.[59]

A history of this period of obvious exploitation can help us understand the nature of the counter discourse of the *adivasis* in Bihar. Faced with economic and political marginalisation, the *adivasi* leadership in Bihar under the Jharkhand party first sought to assert its political views by emphasising a broad convergence of interests with other non-*adivasi* groups. In a region where the *adivasi*/non-*adivasi* distinction had been blurred through decades of migration and where the poorer parts of the Hindu migrant population were as badly off as their *adivasi* brethren, it would have been politically inept to emphasise only an *adivasi* identity. However, the constitutional policy followed by the Jharkhand party through its embracing the language of modernity and the lack of a radical programme in the countryside soon resulted in its decline and its ultimate merger with the Congress. That this happened with disastrous consequences for the *adivasi* people's struggle in the 1960s is evident from the attempts made in the later 1960s and 1970s to evolve new independent political organisations to meet popular demands and, specifically, the emergence of the Jharkhand Mukti Morcha as a radical organisation developing out of the agrarian struggles of the 1960s. This organisation was to attempt a successful cultural revival in the

70s and 80s and to revive the image of the Jharkhand people as the inheritors of their ancestral lands and the forests.

The Sacred Grove and cultural revival

The JMM gave particular emphasis to a cultural revival of *adivasi* rituals related to the land and signalled the revival of the *sahrul puja*. This 'festival of the sacred grove' which was traditionally confined to the villages, now became a grand political event in urban centres, and was accompanied with processions, drum beating and dancing with large crowds lining the streets.[60] In the context of a despoiled landscape, the ritual harked back to the days of an idyllic environmental past. It can be seen as a selective use of memory, where the memory of a pristine environmental past is linked to the solution of contemporary political and economic problems. Given the ecological degradation in Chotanagpur today and the poor state of its village sacred groves, often left with only one remaining tree, this ritual has taken on enormous symbolic significance.[61] It evokes a particular image of the landscape 'older, truer and more directly related to the people' and was used to revive memories of better times and to criticise the inequities of the current regime. The ritual is also a flamboyant assertion of 'tribal' identity and strength and can be compared to the Ramanavami or Moharrum processions[62] in demonstrating militancy.[63] The puja thus became a highly visible, elaborate and ritualised culture of public celebration involving both the performers and the crowds in a collective act articulating the special relationship of the Chotanagpuri peoples with nature and asserting their rights as true custodians of their lands and forests.[64] It is possible to argue, in this context, as Paul Gilroy does with reference to Bakhtin's theory of narrative when he describes the performance of black expressive cultures, that these performances were an attempt to transform the relationship 'between the performers and the crowd in dialogic rituals so that spectators acquired the active role of participants in collective processes which are sometimes cathartic and which may symbolise or even create a community'.[65]

This reinvention of 'tribal' traditions happened, as James Clifford notes, in a complex oppositional context where indigenous populations were threatened by forces of progress and modernity. In-migration had changed the character of Chotanagpur society and by the late nineteenth century it was a society that could not really be categorised as predominantly *adivasi*. However, the Jharkhand movement as a pragmatic means of defence sought to project an *adivasi* identity and they attempted to do so by harking back to an idyllic environmental past and a landscape where their ancestors had left traces and whose spirits still inhabited the land. In their collection of essays, Hobsbawm and Ranger have pointed to ways in which nationalist and ethnic mythologies were historically contingent creations and should be seen as a process. It should be noted that, in recognising ethnicity as a process of construction, we are not seeking to deny its contemporary relevance or the hold it has on the imagination of the peoples of Jharkhand. The use of the term process in this context is to enable us to understand ethnicity as a form of social identity which acquires meaning through a process of conscious assertion and imagining. This is not to say that the communities in Jharkhand did not take up aspects of this identity in order to empower them in the context of a particularly unequal system of power relations.

An understanding of ethnic myths and symbols is thus invaluable for our study. Elements of *adivasi* self-government were also revived or re-invented in the 1970s. The Biasi (assembly) in the Santhal Parganas began to function as a court without fees or pleaders and dealt out simple justice. Traditions of collective farming, preservation of jungles, sacred groves, pastures and common land began to be asserted more forcefully, while common grain pools were also encouraged. There was also a certain amount of distancing from Christian influences which were seen as destructive to tribal traditions.The attitude of the JMM towards non-*adivasis*, however, continues to be ambivalent. While the concept of *diku* is central to the notions of *adivasi* identity and solidarity, the Jharkhand parties today cannot sustain an appeal based on ethnicity alone.[66] Many of the low-caste migrants who arrived in the region in the nineteenth century feel that they have as much a right to be in Chotanagpur as the *adivasis*. Any political programme for Jharkhand therefore has to include these groups. The Jharkhand parties today are, therefore, only partly constituted by ethnic meanings and groupings.

The attitude towards *dikus* is important to our understanding of how the *adivasis* conceptualise themselves and others. Moreover, the concept of *diku* itself is important, for it amounts to a form of boundary-maintenance.[67] The notion of *diku* has not been a stagnant one in Jharkhand. In the nineteenth and early twentieth century, as we have seen, there were already clear notions of difference between *dikus* and tribal groups. The *diku* was clearly seen as a recent immigrant and as an outsider and exploiter who seized tribal lands. In more recent times, while the *diku* is still seen as the outsider who occupies all the government jobs, the boundary differences between *dikus* and non-*dikus* are blurring. An *adivasi* identity is emphasised but these boundary mechanisms are breached with ease. Instead, a more regional identity is taking over based on secondary cultural markers, that is, on a shared history of exploitation, a territorial boundary and a shared culture and life style of the *adivasi* groups. The latter is important because, although there is a growing understanding that the future of non- tribals is assured in the envisaged state of Jharkhand, the parties continue to lay importance on tribal culture and ecology. The appeal to voters is made on grounds of a common economic and cultural predicament.Ethnicity continues to have force in Jharkhand to the extent that it is a politically powerful argument against the way in which state and national political arenas are structured in favour of the dominant outsider groups. Thus, although the parties emphasise a regional identity, ethnic arguments continue to be aired at the popular level.[68] Today, for example, the flag of the Jharkhand party is green in colour, deliberately to emphasise the common cultural and ecological heritage of all Jharkhandi *adivasis*, while the election symbol is a sismandi (a particular kind of fowl sacrificed to a bonga).[69] *Diku* culture, it is argued attaches little value to either of these symbols. But there is little doubt that a homogenous '*adivasi* identity' can no longer be asserted given the history of the region and the impact of low caste Hindu migration though ethnicity continues to rears its head in different guises.[70]

It is not surprising that the main focus of the ire of the Jharkhand parties has been the careless attitude of the Bihar state government with regard to environmental issues in Chotanagpur. One of the most widespread movements in the area in recent times has been motivated against attempts by the Forest Development Corporation to replace *sal* by *sagwan* (teak), since the latter is more valuable as wood in the market. This has grave consequences for the lifestyle of the local people. As we have seen, *sal* products have been useful to them in various ways.[71] There is also a renewed attempt to preserve the sacred groves of the *adivasis* and a growing protest against dam building as at Koel Karo. The effort to prevent the flooding of tribal lands and groves under this project has generated widespread support. The main outcry seems to be directed against the destruction of the sacred groves where the gods reside. In recent years many ethnic movements have legitimised their claims by reference to a global environmentalism (or environmental religion). This involves arguing that the local people are the best stewards of the landscape and have the best claims to control it. What can be clearly seen in the Jharkhand movement, therefore, is a deep social and psychological commitment to an identity of shared pasts, current predicament and common future.

The cultural struggle being waged in Jharkhand today is essentially a movement directed towards transforming the balance of power in the region. In Gramscian terms, it may be represented as a struggle for hegemony in the cultural and the political arena. It is now recognised that the continuous assertion that 'tribal' areas in India were underdeveloped, and its people 'childlike' and 'primitive' (both in colonial and post-colonial state discourse) is an essential element of the discourse of domination. These terms are not used here in a romantic sense but have associations of cultural backwardness, sexual freedom for women and promiscuity as key markers. Another derogatory term used in reference to the tribes is *jangali*, standing for uncouth or uncivilised, but literally meaning 'forest dweller'.[72] These terms form part of an official discourse that aids compliance towards economic and political domination and legitimises the acceptance of one mode of life and the exclusion or extermination of others. The cultural struggle, therefore, has to contest the binary oppositions on which such legitimation is founded.[73] In this context it is useful to see the contemporary *adivasi* cultural revival as political struggle in the same way as Franz Fanon has done in Algeria.

The discourse of the Jharkhand parties now tackles these issues head on, in an attempt to wrest concessions from an authoritarian Indian state. The emphasis on indigenous culture by Jharkhand leaders like Ram Dayal Munda and its current political clout in the region have ensured that the new political equations being formed today include the Jharkhand parties.[74] In the process counter-claims about the inherent originality or purity of *adivasi* culture are made and the history of acculturation with the dominant Hindu culture of the plains is pushed aside. It is in this moment of struggle, as Bhabha notes, that the 'meanings and symbols of culture are appropriated, translated, rehistoricized and read anew'.[75]

The response of the Bihar state government towards the radicalisation of the Jharkhand parties has been to increase repression. In the late 1960s and 1970s, the Birsa Seva Dal, Bihar Prant Hul Jharkhand party and the Jharkhand Mukti Morcha were each set up to contest the tyranny of developmentalism and forced modernisation. Their activities were both in the domain of parliamentary politics and outside it. These organisations participated in elections, while their activists were involved in the forcible cropping of *diku* lands, in sabotaging local transport lines and in organising new forest *satyagrahas*.[76] In 1978, resistance to the planting of teak was sparked off when the forest department undertook to plant teak in 2000 hectares of the *sal* forest. Local knowledge that nothing grew under teak, particularly not the grass roots and tubers on which the local wildlife and people subsisted, sparked off the protest. It was also alleged, that since elephants did not eat teak leaves, they would be forced to seek food in areas where crops grew, thus increasing their depredations. The agitators argued that fruit-bearing trees were being cut down to establish teak nurseries, thus depriving the *adivasis* of a source of food.[77] The problems facing such groups were amply illustrated by the Gua incident in September 1980, when on 4th September police in south-west Singbhum district moved at the behest of local politicians, mine owners and timber contractors to arrest 4,100 tribals and non-tribals for unlawfully cutting trees.[78] The Morcha was driven underground, only to emerge again in new guises and locations. However, such methods have not had continued success. The situation is changing. More recently, the weakness of the ruling party in Bihar has enabled the Jharkhand parties to revive their organisation and has resulted in the forging of new links with the ruling Janata party. This move by the JMM and the astute political manoeuvring of its leadership has given the movement a new lease of life and renewed its strength at the negotiating table. Despite allegations of corruption and bribery on the part of the current JMM leadership, it appears that the marginal voices of the Jharkhand peoples are finally being heard. It is possible to argue that ethnicity will increasingly dominate Indian politics. As the discourse of the nation state in India becomes increasingly totalising, ethnic politics seeks to express itself more forcefully. Though, as has been argued, one should regard ethnicity as a process, there is little doubt that ethnic argument in Jharkhand has come into its own and has great political relevance. A failure to arrive at some constitutional arrangement will therefore have problematic consequences.

This paper has attempted to show the ways in which the recurring image of an older landscape has served as a powerful metaphor in Chotanagpur's resurgence. It directly fuelled protest movements in the late nineteenth century and was part of a new emotive language as a modern political idiom came to be developed by the Jharkhand parties in the twentieth century. The cultural revival signalled by JMM in the 1970s found expression in the festival of the sacred grove which was to become one of the most important markers of Chotanagpuri identity and the site where larger and more powerful hegemonies were constituted, contested and transformed. The paper is also an illustration of the more general problem of historical research and demonstrates the ways in which a post-modern analysis of cultural categories such as ethnicity can be related to the interpretation of the old realities of power politics.

NOTES

[1] These writers have argued quite convincingly that ethnicity, i.e., belonging and being perceived by others as belonging to an ethnic group is an invention. Ethnic groups are part of the historical process and though they may pretend to be eternal and essential they are usually of recent origin and eminently pliable and unstable. They thus constantly change and redefine themselves. See Werner Sollors ed., *The invention of ethnicity*, New York, 1989.

[2] Leroy Vail ed., *The creation of tribalism in southern Africa*, London, 1989, p. 10.

[3] I am aware of the problems of the term as the historian Saul Dubow has noted, 'On account of its capacity to redefine, absorb, and dissolve problematic concepts like race and class, ethnicity has been referred to by one writer as a sort of lightning rod. Like many portmanteau words ethnic or ethnicity can serve as a euphemistic substitute for other appellations. It has become for example variously synonymous with words like population group, tribe, nation, volk, race.' See Dubow, 'Ethnic euphemisms and racial echoes', *Journal of Southern African Studies*, vol 20, no 3, September, 1994, p. 356.

[4] The notable exceptions are E. Hirsch and M. O'Hanlon ed., *Anthropology of landscape*, Oxford, 1995; and Barbara Bender ed., *Landscape perspectives*, Providence, 1993. See also Paul Rodaway, *Sensuous geographies:body: sense and place*, London 1994.

[5] While there is work on landscape as political resource and the mythologising of landscapes, the standard trope of western aesthetics and western historiography has been to see it as a national imaginary, as a repository of a nation's nostalgic past. There is little work on the ways in which ideas about the landscape have fuelled ethnic and cultural resistance against the hegemonising discourse of nation-states or seeing its potential in an oppositional context.

[6] See the work of Richard Grove, *Green imperialism*, Cambridge, 1995; Mahesh Rangarajan, *Fencing the forest*, Delhi, 1996; Shivaramakrishnan, 'British imperium and forested zones of anomaly in Bengal, 1767-1833', *Indian economic and social history review*, vol 33, no 3, 1996; Ajay Skaria, *Hybrid histories: Forests, frontiers and oral*

traditions in western India, Delhi, (forthcoming).

⁷ The concept of a 'tribe' is problematic. The term *adivasi* or original inhabitant rather than 'tribal' is seen as preferable by some historians, notably by David Hardiman, for it is free of the evolutionist implications of the latter term. Hardiman argues that the term *adivasi* relates to a particular historical development: that of the subjugation during the nineteenth century of a wide variety of communities which before colonial rule has remained free or relatively free from the control of outsiders.The experience generated a spirit of resistance which incorporated a consciousness of the 'adivasi' against the 'outsider'. As he notes the term was used by political activists in the area of Chotanagpur in the 1930s with an aim of forging a new sense of identity among different 'tribal' peoples – a tactic that has enjoyed considerable success. See David Hardiman, *The coming of the devi; adivasi assertion in western India,* Delhi, 1987, p. 15. See also Crispin Bates, 'Lost innocents and loss of innocence', in R.H. Barnes, Andrew Gray and B. Kingsbury ed., *Indigenous peoples of Asia*, Michigan, 1995.

⁸ This history had a strong oral tradition and was invoked in cultural terms through songs and proverbs.

⁹ See J. and J. Comaroff, *Of revelation and revolution, Christianity, colonialism and consciousness in Southern Africa*, Chicago, 1991, p. 288.

¹⁰ See the ethnographic writings of S.C. Roy, *The Mundas and their country* and *Oraon religion and customs*, Calcutta, 1912 and 1917.

¹¹ As Arjun Appadurai notes, 'native people confined to and by place have never existed'. Quoted in James Clifford, *Routes, travel and translation in the twentieth century*, Harvard, 1997, p. 25.

¹² Ibid.

¹³ For example, the Doms who were a semi-tribal community were criminalised in Bihar under the Criminal Tribes Act by the British. See A. Yang, 'The case of the Magadhiya doms' in A. Yang ed., *Crime and criminality in British India*, Tucson, 1985.

¹⁴ Classical anthropological theories in the colonial period placed tribes at the bottom of the evolutionary line. As Ernest Gellner notes 'Systematic study of 'primitive' tribes began first in the hope of utilising them as a kind of time machine as a peep into our own (west) historic past, as providing closer evidence about the early links in the great series.' See Ernest Gellner *Thought and change* , Chicago University Press, 1964, p. 18 f. The logical connections between British evolutionism and the establishment of the British empire are obvious. See Johannes Fabian, *Time and the Other; how anthropology makes its object*, New York, 1983, p. 35.

¹⁵ *Ranchi survey and settlement report*, Calcutta, 1912.

¹⁶ E.G. Man, *Sonthalia and the Santhals*, Delhi, 1983.

¹⁷ *Ranchi district gazetteer*, p. 121. See also 'Forests of Chotanagpur', *Indian Forester*, 10, 1884, pp. 890-91

¹⁸ For debates on shifting cultivation in the Indian context see Mahesh Rangarajan, *Fencing the forest*. For a comparative study in Africa see H. Moore and M. Vaughan, *Cutting down trees*, London, 1994.

¹⁹ *Jhuri* refers to partner and can mean either man or woman. As many women left as coolies by the 1940s when this song was recorded, the process was common to both sexes. See W.G. Archer, *The blue grove: the poetry of the Oraons,* London, 1940.

²⁰ Bradley Birt, *The story of an Indian upland*, London, 1910.

[21] As Elwin himself notes in his book *The Baiga*, 1939, he advocated some sort of national park in a wild and largely inaccessible part of the country under the direct control of a Tribes Commissioner. See Verrier Elwin, *The tribal world of Verrier Elwin*, Oxford, 1904, pp. 290-291. See also W.V. Grigson, *The Maria gonds of Bastar*, London, 1949.

[22] This is not to say that it had no positive influence elsewhere as it surely did in north eastern India among the Naga tribes.

[23] See D. Cosgrove, cited in Stephen Daniels, *Fields of vision, landscape imagery and identity in England and US*, Cambridge, 1993, p. 5.

[24] For recent writings on history, landscape and identity see David Lowenthal, 'British national identity and the English landscape' *Rural history*, 2, 1991.

[25] Peter Gow, 'Land, people, paper' in Hirsch and O'Hanlon, *Anthropology of landscape*.

[26] V. Ball, *Tribal and peasant life in nineteenth century*, India, Calcutta, 1880 and H.H. Haines, *The forest flora of Chotanagpur*, Calcutta, 1910.

[27] H.H. Haines, op.cit., p.41.

[28] Leach notes that among the Mende in carving out forest space, men claim that they make farms for women, which 'seems to denote the virgin soil as male and the cultivable space as female' See Leach, 'Women's crops in women's spaces', in E. Croll and D. Parkin ed., *Bush base: forest farm*, London, 1992, pp. 57-76.

[29] In some of the *dhumkarias* a similar slit in the central post supporting the roof could be seen. The central post in one *dhumkharia* visited by Roy had a wooden post with a slit resembling the female organ. Such a post was known as the mistress of the bachelors. See S.C. Roy, *Oraons of Chotanagpur*, Ranchi, 1984 (originally published 1915).

[30] See Daniel Rycroft, 'Born from the soil, indigenous mural aesthetic of the Kherwals in Jharkhand', *Journal of South Asian studies*, August, 1996.

[31] It could be argued, as R. Freeman has done for Malabar, South India, that perceptions of the forest differ according to one's 'class' position. In Malabar he argues that for the elites of settled agricultural regimes 'the forest becomes a symbolic repository for the demonic, antinomian, anti-social of all those lower castes and tribals with whom the higher castes were dependently but ambivalently tied'. See R.Freeman, *Forests and folk, perceptions of nature in swidden regimes of highland Malabar*, Pondicherry, 1994, p. 27.

[32] See M. Leach, op.cit. See also, for similar views, Roy Ellen, 'Rhetoric, practice and incentive in Nuaulu', in Kay Milton ed., *Environmentalism: the view from anthropology*, London, 1993, p. 140.

[33] See Nurit Bird David, 'The original affluent society', *Current Anthropology*, vol 33, no.1.

[34] For an interesting study that attempts such an environmental history see M. Leach and J. Fairhead, *Misreading the African landscape*, Cambridge, 1996.

[35] In his recent reply to Obeyesekere, Sahlins has noted that the post-modern attack on the notion of a bounded and coherent culture has occurred at the very moment when groups such as Maoris, Tibetans, Australian aborigines around the world 'all speak of their culture using that word or some other local equivalent, as a value worthy of respect, commitment and defence' He argues forcefully that no good history can be written without regard for 'ideas, actions and ontologies that are not and never were our own'. M. Sahlins, *How a native thinks*, Chicago, 1995, p. 13.

[36] Simon Schama, *Landscape and memory*, London, 1995, p. 61.

[37] Ibid., p. 7.

[38] See Marshall Sahlins on the notion of remembered landscapes and landscapes of ser-

Vinita Damodaran

vitude in *Anahulu, the anthropology of history in the kingdom of Hawaii*, Chicago, 1992.

[39] S.C. Roy, *Oraon religion and customs* (1928; reprint Calcutta, 1972) p. 251.

[40] Extract from the confidential diary of the S.P. Ranchi 1919, Political Special, File 1919.

[41] Report of the Commissioner of Chotanagpur, 12th May 1921, Political Special, File 1921.

[42] Mark Poffenburger, 'The resurgence of community forest management in the jungle mahals of West Bengal', in D. Arnold and R. Guha ed., *Nature, culture and Imperialism*, Delhi, 1995, p. 348.

[43] Archer collection MSS eur 236/1.

[44] Peter Gow, 'Land people and paper'.

[45] In Gujarat, the Devi movement of the tribals developed similar links with the national movement, see David Hardiman, *The coming of the Devi*, pp. 168-176.

[46] See K.S. Singh, 'Tribal autonomy movements in Chotanagpur', in K.S. Singh ed., *Tribal movements in India*, vol. 2, Delhi, 1983.

[47] J. and J. Comaroff, op. cit., p. 6.

[48] See *Chotanagpur Unnati Samaj ki varshik Mahasabha ka report aur Chotanagpur Adivasi Sabha ka uthpathi* 1937, pp. 1-14.

[49] Political Special 25.4.1946.

[50] See *Indian Nation* 8.3.46 and *Sentinel* 10.3.46.

[51] Interview with Cornelius Ekka Ranchi, April 1992. Ekka remembers going to Jaipal Singh's meeting in 1945, where there was a crowd of nearly 50,000 people.

[52] See K.S. Singh, 'Tribal autonomy movements in Chotanagpur', p.7.

[53] Interview with Jharkhand party leader N.E. Horo, Ranchi, April, 1992.

[54] Verrier Elwin noted that at the time 'there was endless talk of tribal development', Elwin, *The tribal world*, p. 299.

[55] For details of these views see S. Corbridge, 'State, tribe and region: policy and politics in India's Jharkhand 1900-1980', unpublished Ph.D thesis, University of Cambridge.

[56] G.S. Ghurye, *The aborigines, so-called and their future*, Delhi, 1943, p. 293.

[57] This was not just confined to Bihar as Christopher von-Haimendorf has noted a massive invasion of tribal land by outsiders all over India occurred specifically after 1947. See Christopher von Furer-Haimendorf, *Tribes of India: struggle for survival*, Delhi, 1989, p.39.

[58] See Myron Weiner, *Sons of the soil: Migration and ethnic conflict in India*, Princeton, 1978, pp. 165-175.

[59] For details see R. N. Maharaj and K.G. Iyer, 'Agrarian movement in Dhanbad', in Nirmal Sen Gupta ed., *Fourth World Dynamics: Jharkhand*, Delhi, 1982, pp.165-200.

[60] I was told on a visit to Ranchi town in 1992 that the scale of the Sahrul puja I had witnessed there was a recent phenomenon and an 'invention of tradition' in Hobsbawm's terms. As James Clifford notes, 'throughout the world indigenous populations have had to reckon with the forces of 'progress' and 'national' unification. The results have been both destructive and inventive. Many traditions, languages, cosmologies and values are lost, some literally murdered but much has been invented and revived in complex oppositional contexts.' See James Clifford, *Predicament of culture*, Harvard, 1988, p. 16.

[61] On the state of India's sacred groves today see, Paul Spencer Sochaczewski, 'God's own pharmacies', *BBC Wildlife*, January 1996, pp. 68-71.

[62] The celebration of the Hindu religious festival of Ramnavami and the Muslim festival

of Muharram were often used to assert the strength of these communities in urban areas and sometimes resulted in inter-communal rioting.

[63] A reassertion of traditional cultural practices is an intrinsic element of the economy and political struggles of third world peoples. For many minorities, culture is not a mere superstructure, all too often in an ironic twist of Sartrean phenomenology the physical survival of the minority depends on its cultural variable. As Arif Dirlif argues, cultural struggle is an essential counterpart to political and economic struggle. See *Cultural Critique*, 6.

[64] Some scholars studying 'festive culture' have interpreted rituals as manifestations of an evolving folk culture creating meaning and helping people to cope with an alien world, as instruments for the promotion of group solidarity and as public assertions of group power and demands. See Kathleen Neils Conzen, 'Ethnicity as festive culture: nineteenth century German America on parade', in Sollors, *The invention of ethnicity*, p. 46.

[65] Paul Gilroy, *They aint no black in the union jack*, London, 1987, p. 214.

[66] Whereas in 1872, 51.38 % of Chotanagpuris were classified by the British as aboriginals and semi-aboriginals, by 1971 only 30.14% of the region's population belonged to scheduled tribes in Bihar. See *Census of India*, Bihar, 1971.

[67] By mapping the named category of *diku* we can, perhaps, uncover the vocabulary of group differences in the region. What groups call themselves and what they call others is related, of course, both to a language of esteem and a language of insult. Names, as Manning Nash has noted, condense the relevant cultural information into handy social and psychological packages for easy self- and other- identification. See Manning Nash, *The cauldron of ethnicity in the modern world*, Chicago, 1989, p. 9.

[68] As Nash notes, ' Ethnicity is a resource in political economic and cultural struggle..... When economic ends are sought (opportunity, wealth and income redistribution or claims to ownership of a national patrimony) the ethnic group may approximate a political class and exhibit a form of class struggle powered by an ethnic ideology, not a false consciousness but often a true appreciation of the existing state of economic affairs.' Nash, *The cauldron of ethnicity*, p. 127.

[69] P.C. Hembram 'Return to the sacred grove', in K.S. Singh ed., *Tribal movements in India*, p. 89.

[70] Homi Bhabha has emphasised the 'hybrid moment of political change where ideas and forms are re articulated; where there is a negotiation between gender and class where each formation encounters the displaced'. He argues that the agents of political change are discontinuous, divided subjects caught in conflicting interests and identities. See H. Bhabha, *Location of culture*, London, 1994.

[71] See A.K. Roy, 'Sal means Jharkhand, Saguwan means Bihar', *Sunday* (weekly), Calcutta, April 8th, 1979, pp. 46-47.

[72] See Christoph von Furer-Haimendorf, *Tribes of India*, p. 79.

[73] As Caren Kaplan indicates that 'becoming minor' is not a question of essence (as the stereotypes of minorities in the dominant discourse would want us to believe) but a question of a subject position that can only be identified in 'political terms', that is, in terms of the effects of economic exploitation, political disfranchisement, social manipulation and ideological domination on the cultural formation of minority subjects and discourses. It is one of the central tasks of a minority discourse is to define that subject position and explore the strengths and weaknesses, and the affirmations and negations

that inhere in it. See Abdul R., Jan Mohammed & David Lloyd, 'Introduction', *Cultural Critique*, Fall 1987.

[74] See Ram Dayal Munda, *Report on Jharkhand*.

[75] Bhabha, *Location of culture*, p.37.

[76] See K.S. Singh, 'Tribal autonomy movements in Chotanagpur', pp. 14-21.

[77] Ibid., p. 20.

[78] 'Gua massacre of tribals', *Economic and Political Weekly*, XV, 38, 1980.

IV. Cultural Collisions and Competing Knowledges

Different Histories of Buchu:
Euro-American Appropriation of San and Khoekhoe Knowledge of Buchu Plants

Christopher H. Low

Buchu is a herbal medicine used primarily for urinary and kidney problems. It is also thought to be calming on the stomach, a good general tonic, and a useful hangover treatment. As current advertisements are keen to indicate, the use of buchu is derived from the practices of the Khoisan indigenous peoples of southern Africa, a category that includes historic Cape Hottentots and contemporary Khoekhoe and San or Bushmen.[1] In recent contexts, buchu is principally identified as *Agathosma betulina* although other species have been, and continue to be, referred to as buchu. Buchu is characterised by a strong and distinctive smell resultant from the secretion of volatile oils from leaf glands. The smell varies depending on the exact species. Smells range from orange-like to pepperminty and camphoric.

Buchu became known amongst Europeans as much as three hundred and fifty years ago. Over the subsequent period it has persistently featured in the ethnography of the Khoisan and has enjoyed a remarkable passage from seventeenth century herb garden, to English medical drug, to a tonic manufactured in the USA, advertised around the world and sold by the proverbial car boot load. What I wish to examine is the appropriation of buchu by colonists, travellers, merchants and emergent biomedicine and its subsequent rejection and relegation to the status of a herbal medicine. Surprisingly, despite buchu's historic prominence, its precise identity has shifted over history and remains somewhat elusive. I identify and examine the multiple European and indigenous historical names that appear for buchu in different periods as indicators of tensions in the process of colonial scientific data extraction.

The history of buchu is an interesting one for what it brings to the history of economic botany, scientific travellers and the colonial construction of science. In particular it has something to add to recent analysis of 'noise' at the coal face of colonial bio-prospecting[2] and anthropological consideration of distinctive indigenous ways of thinking. Buchu history potentially leads to a number of important questions concerning the indigenous and 'Western' boundary of contact over a medicinal plant. A principal issue revolves around how accurately ethnographers of buchu have represented Khoisan ideas and practice over time. This leads in turn to further questions concerning possible similarities and differences in indigenous and biomedical understanding of medical properties and treatments. The contrast between the Euro-American history of buchu and indigenous use and understanding both highlights the partiality of the process of

Environment and History **13** (2007): 333–361.

Christopher H. Low

data extraction and draws attention to what seem appropriate ways of thinking about Khoisan relationships with buchu.

For reasons of expediency, I have not set out to produce a detailed indigenous history. What I do present, however, is a Khoisan historical outline with an eye to arguments for the continuity of practices and ideas.[3] The paper begins with a presentation of the earliest records of buchu and a reflection upon how these relate to commercial claims of historical Bushmen or Khoekhoe use. A preliminary analysis of this early Khoisan material sets up a later more detailed account of Khoisan understandings. The next section moves chronologically through the introduction of buchu to Europe from the 1700s to the introduction of buchu into the mainstream European medical market place in the early nineteenth century. On the way, I examine settler knowledge of buchu around the same period. A third section engages with questions of 'authentic' buchu as identified by science and commerce and examines how the shape of first European contact with Khoisan has influenced European understanding of buchu. A final section picks up on earlier mentioned Khoisan use of *sâi* (*sā*) plants which seem intimately related to buchu and embed buchu knowledge and use in a wider Khoisan cultural context. This section draws heavily on ten months fieldwork carried out in Namibia in 2000 and 2001 and two months amongst the northern Cape San in mid 2006. In particular I examine the Khoisan role of smell in use of buchu and sâi and how this Khoisan understanding differs from European notions of perfume and scientific effectivity.

At the heart of this history of partiality lies a persistent failure by Europeans to recognise that they are dealing with a Khoisan category of use, not a Western one. The Khoisan category encompasses principally perfume and healing balm but also an associated notion of a spiritual and physical sedative and stimulant. The essential understanding of buchu seems to lie in its ability to pacify, awaken, attract and heal. The key to this understanding lies in the ability of smell to bring about physical transformations. Smell, breath, wind and the personal power of people all overlap in Khoisan thought as they do in epistemologies of other indigenous peoples. If a sick or strong smelling, or otherwise 'potent' person stands near someone else, the other person might become sick because of the movement of the essence of the potent person, as smell or wind, into themselves. Buchu smell moves into people and brings about specific effects in accordance with the context of use and what its smell represents to historically and culturally specific Khoisan peoples.

BUCHU – THE EARLY HISTORY

Numerous commercial enterprises seek to confer indigenous credibility on buchu use by emphasising its ancient San ancestry. Smith supports this assumption in his observation that Hottentot use of buchu was 'almost certainly taken over from the Bushmen'. Whilst this may be right, there is little clear evidence to support this claim.[4] The earliest references to buchu appear in the latter half of the seventeenth century and are part of a caucus of knowledge gathered from sailors and travellers familiar with the Cape. European distinction between Hottentots and San was not a regular feature of Khoisan ethnography until Lichtenstein's observations in the early nineteenth century,[5] and even in much later contexts notions of difference have remained controversial. It was, however, clearly with cattle owning Hottentots that European indigenous relations were first established, and from this contact that early reports of buchu emerged. It was not until around the mid-eighteenth century that travellers penetrated significantly inland and began to furnish detailed reports of Bushmen.[6] On this basis it seems reasonable to claim that it was amongst Hottentots that buchu use was first encountered.

To work through the historical references of Hottentot, or Khoekhoe, buchu use from the seventeenth century to the present is beyond the scope of this article. However, what is known about Khoekhoe use does not change significantly from details presented in some of the earliest information and later accounts repeatedly draw on the early sources. There is also significant evidence to suggest continuity and stability of Khoisan healing ideas over time.[7] In view of this I examine some of the most influential early references to buchu as a way of providing a background to the details that have repeatedly featured across the colonial ethnography and shaped European understanding of buchu.

Possibly the earliest reference to buchu appears in a compilation of material published in Europe in 1668 by a Dutchman, Olfert Dapper. Dapper's authoritative account of Africa reflected a new ethnographic and proto-scientific opening up of the Continent including botanical exploration at the Cape. To briefly highlight the wider context of his account, a 1603 reference to Gouarus de Keyser stands as the first record of a plant collector at the Cape. The earliest mention of a Cape plant concerns the seaweed *Ecklonia maxima* and the first Cape plant conveyed to Europe was *Protea neriifolia*, in 1605. By the later seventeenth century Cape plants were already featuring in European botanical collections, including that of Paul Hermann (1646–1695), a German medic and later director of the botanic gardens at Leiden.[8]

Dapper's sources included information from the first Dutch voyage to the East Indies that passed by way of the Cape in 1595. It is possible that Dapper's buchu details arose from this voyage, but Smith suggests Dapper's information on buchu was 'almost certainly' obtained in 1661 from a visit by Hessequa Hottentots to van Riebeeks fort. Details of the customs of the visiting Hottentots were recorded by Dapper's correspondent in the van Riebeeck administration.[9] Regardless, however, of the exact origin of Dapper's source, buchu only begins to regularly appear in European accounts in the second half of the seventeenth century. Concerning buchu, Dapper simply noted that Hottentot women strewed 'boggoa' over their heads.[10] Shortly after Dapper, Ten Rhyne (1686), a Dutch

Christopher H. Low

physician of the East India Company, recounted details gleaned from a stopover at the Cape on the way to the East Indies, including that Hottentots smeared themselves with oil and animal fat and their heads with the ash of 'bouchou'. This is one of the earliest references to buchu being smeared on the body alongside sheep fat.[11]

In 1695 Johannes Grevenbroek, another Dutch East India Company employee and later free burgher at the Cape, provided a more informative account of buchu, and one very probably drawn from more extensive first hand observation. Grevenbroek mentions 'bochu' in two contexts. In the first he relates how a 'convalescent' is smeared by a 'priest' with sheep fat and sprinkled with 'bochu' powder; 'next the entrails of the animal are plaited and hung round his neck like an amulet; then his blanket is anointed with fat and sprinkled with bochu, and at last he is let go'.[12] Elsewhere Grevenbroek mentions 'bochu' having being rubbed onto the skin together with fat, 'as a protection against the danger of sun, cold or disease'.[13]

In 1706 a mathematician, Peter Kolb, was sent to the Cape by his Prussian patron to research astronomical and meteorological phenomena. Kolb's subsequent account has been highly influential to the writing of Khoekhoe history. It was, he claimed, based on years of contact with different Hottentot groups. Unfortunately, following the ethnographic conventions of his day, he left no account of his travel details and hence the provenance of his contacts. Kolb's contribution to colonial familiarity with buchu is twofold. Firstly, he popularised an early scientific name for buchu, *Spiraea Africana odorata, folis pilosis*, which was the name attributed to buchu around 1692 by Oldenland, one time master gardener at the Dutch East India Company Gardens at the Cape settlement.[14] Secondly, he provides us with more than just passing reference to Hottentot buchu use. In one instance he related an operation in which a testicle was removed and the empty testicular sack was stuffed with a mixture of sheep's fat and powdered 'salutary herbs, particularly the Buchu'.[15] In another, Kolb recorded that when a Hottentot was seized with pain the physician took the caul of a sacrificed sheep and, 'having powdered it with Buchu, twists it in the manner of a rope, and hangs it about the patients neck' saying, 'You will be better now ...'[16]. Kolb additionally observed that newborn babies were smeared with sheep's fat or melted butter which was allowed to soak into the child's pores. Once dried the baby was powdered in buchu from head to foot to form a crust.[17]

Elsewhere again, Kolb related that leaves of buchu or dacha were applied to a poisoned arrow wound, following the rubbing on and swallowing of a mixture of snake venom and spittle. The leaves, he suggested, aided in a cure within months.[18] For headaches he reported that Hottentots often shaved furrows through their hair and applied powdered buchu 'which, they believe, does not a little contribute to remove the pain'.[19] A final reference points to the use of powder and infusions of buchu alongside wild sage, wild figs, fig leaves, garlic and fennel, as treatment for 'most other inward ailments'.[20]

In contrast to these Hottentot sources the earliest reference to Bushmen use of buchu seems to come a century later in an account by Wikar, a hunter who lived north of the Orange River in the 1770s. In 1778 Wikar related that buchu made from camel thorn wood was used in a 'marriage' ceremony he encountered at a bushman kraal.[21]

A more extensive early account of buchu can be found in the journals of a soldier traveller, Robert Gordon, concerning his southern African expeditions into the interior between 1777 and 1786. Gordon noted buchu was rubbed on ostrich feathers that were set in stone animal traps by Bushmen. Elsewhere he observed 'Bushmen hunting stones', on which were placed 'what resembled the heads of people, smeared with buchu'. Around the Doring River area he further observed, 'an old woman practising magic'. She was snorting out 'a devil or evil spirit' from her son, and whilst doing so rubbing his stomach with buchu. She rubbed buchu 'into the noses of several women who were sitting there as well.'[22]

The next clear references to Bushmen and buchu do not appear until the later nineteenth century. From Bleek's and Lloyd's Bushmen research we learn that healers of the Cape /Xam had buchu rubbed onto them to 'make their arteries lie down' when they were suffering from torments associated with the healing process.[23] /Xam also used buchu to pacify the 'rain-bull', a creature of their imaginative twilight world. The rain-bull could be captured if men who went to seize it were smeared in buchu, 'If the bull had smelt buchu, it would have been calm, and gone quietly without struggling'.[24]

INTERPRETATION

Recent accounts of buchu emphasise the credibility of the drug by pointing out that Khoekhoe have used buchu for centuries to treat all kinds of ailments, including, rheumatism, strains, bladder and stomach complaints. They additionally sometimes note the historic use of buchu as a perfume. In their authoritative account of medicinal plants of South Africa, Van Wyk and Gericke observe that:

> A.betulina (and several related plants) were used to anoint the body (after mixing
> the powdered, dried leaves with sheep fat), probably both for cosmetic reasons
> and as antibiotic protection.[25]

On the basis of the indications of buchu use given above these various claims seem to be reasonable. Yet at the same time there is a deep distortion at work which is understandable at a popular level but feeds insidiously into history, leading to greater partiality and misrepresentation further down the line. There is the normative assumption that healing and perfume are words and ideas that can be used interchangeably not only through history but across cultural divides. As indicated in the above examples, whilst buchu clearly does have a role in healing people, this is not quite the same healing known to biomedicine. Despite some formal resemblance between Khoisan and Western uses of perfume, poultices, oral herbal remedies and ointments – antibiotic or otherwise, there remain fundamental differences or tensions in the understanding of cause, effect and desired outcome. If the understanding of illness, how it is caused, spread and healed is different there is historical slippage in justifying parity of current European medical use with past Khoisan use. The relationship between healing and perfume, which holds much of the context of Khoisan use of buchu in healing, has no direct equivalent in current European understanding of buchu – a plant used for healing *and* as a perfume. Although similar ideas have existed in European history, such as in aromatherapy which draws on an overlap of smell and healing, Khoisan understanding remains distinctive.

Collectively early ethnographic references to buchu reveal the sort of contexts in which colonists and travellers were introduced to the plant(s). Given the relatively shallow nature of ethnography and travel accounts that characterise colonialism well into the nineteenth century, it is perhaps not surprising that details of buchu use are not extensive. There is, however, enough information to at least begin to reflect upon historical Khoisan as opposed to colonial notions of buchu. In the absence of an indigenous historical literary account, a locally informed reading back that draws on what interdisciplinary evidence there is, including linguistic, archaeological, historical and anthropological, remains the only option. One factor that helps historical and recent understanding of buchu use is an overlapping history of references, starting with Kolb, to 'Saab', sā or sâi plants which point to buchu not being so much a distinct species, but an indigenous relationship with certain sorts of plants in certain sorts of ritual and healing contexts. I return to this issue in the latter section of the paper.

In terms of what we can glean from the Hottentot sources so far referred to, it would be easy to think of buchu as having been sprinkled on bodies largely to counteract the smell of fat and to make people smell attractive. That the picture might be more complex is suggested by the internal healing power of buchu and fat evinced in Kolb's testicle packing account, in buchu use for poisoned arrow wounds and for headaches and inward ailments. The key to understanding these multiple contexts of buchu use, which I explore later, lies in the transformation of personal attributes. The use of buchu with or without fat confers powers or attraction between a person and the wider world and, depending on the context, brings healing or other potency.

Working with these ideas we can also begin to tentatively unravel the various accounts of Bushman buchu use which articulate the same fundamental understanding of buchu in relation to transformational states and the interconnection of potency and illness. How buchu used in an ostrich feather trap might relate to perfume use, begins to make sense if we think in terms of connections between hunters and animals. Buchu on the feathers may have served to transform or disguise the nature of the trap or to attract or reassure potential animal victims. On the skull, buchu might have been a ritual means of opening a channel between the dead and the living, catalysing the dead into vital participation, although this latter point in particular must remain purely speculative. The phenomenon of rubbing buchu under the noses of participants in a healing session is a feature of recently observed Bushmen dances. In this context, and probably similarly in Gordon's encounter, the buchu is said to 'open people' up. Rubbing buchu on those gathered at healing dances makes them receptive to healing forces and serves as a mechanism of ritual binding.

The /Xam uses of buchu point towards an understanding that the smell pacifies animals. This idea is linked to organisms lying down in repose, which in turn is linked to arteries or organs that are over-excited needing to lie down.[26]

The historical points to recognise are, firstly, that buchu use seems as much a 'Hottentot' phenomenon as a San one over the colonial period and, so far, there is no evidence to justify claims to a purely San origin. Without archaeological evidence, there is no easy way of moving buchu history beyond the colonial ethnography. Furthermore, early Khoisan use of buchu did concern vaguely referenced illnesses, but neither they, nor the way buchu worked, can be assumed to resemble current notions of illness and cure. At the very least it seems a historical disservice to justify buchu use in the current herbal medical market with such blandishments as that included in a current website, which states that buchu was 'first used by the San to make tea'.[27]

BUCHU ENTERS EUROPE

As colonial familiarity with buchu grew, attempts were made to cultivate the plant back in Europe. These early attempts were, however, not indicative of medical interest in the plant as much as a wider imperial botanical curiosity. As the eighteenth century progressed there was increasing interest in *Diosma*, a genus that included buchu species. Only by the 1820s, however, did buchu begin to enter a popular medical market.

The first apparent mention of buchu plants cultivated in Europe occurred in 1706, although the plants were referred to as *Diosma*, meaning 'divine smell', and not buchu. In Amsterdam Casper Commelin described three species of *Diosma* which were allegedly grown from Cape seeds in the University's medical garden. Smith however observes that these were the exact same species that were illustrated at the Cape in the *Codex Witsenius* in about 1692. This, therefore, raises the question whether Commelin's description was copied from the *Codex* or truly related to European buchu cultivation.[28]

In 1753 Linnaeus described a number of *Diosma* specimens that he had been sent a year previously by Tulbagh, Governor of the Cape colony. Linnaeus did not, however, mention an indigenous name or use for the plants.[29]

In Miller's *The Gardener's and Botanists Dictionary*, 1752, sixth edition, he included three varieties of *Diosma*: *Diosma foliis linearibus*, *D. foliis fabulatis acutis* and *D. follis fetaceis acutis*. Miller observed that the first of these had been long known under the title *Spiraea Africana odorata foliis pilosis*, or sweet scented African Spiraea. *Diosma* did not feature in Miller's earlier 1743 edition although *Diosma hirsuta* was cultivated at the Cambridge Botanic Gardens in 1731.[30] Miller indicated that *Diosma* came to England via 'curious gardens in Holland'.[31] For Miller not to have included *Diosma* in 1743 suggests that the plants were very new arrivals to the botanical gardens of Britain and that the Cambridge *Diosma* was probably one of the earliest to have been cultivated in Britain, if not Europe. In the 1759, seventh edition, of his *Dictionary*, Miller extends the above species count with a fourth, noting that other species had been enumerated but only the four were found at that time in English gardens.[32]

British familiarity with *Diosma* grew over the later eighteenth century. In 1766 there were five species of *Diosma* cultivated in Dr. Coyte's Botanic Garden at Ipswich.[33] At the Royal Botanical Gardens, Kew, eight species of *Diosma* were grown between 1759 and 1790.[34]

Between 1772 and 1774 Thunberg, a student of Linnaeus, undertook a series of expeditions that represent the first time a university trained botanist specifically explored the interior of southern Africa with the intention of collecting and identifying plants.[35] In Thunberg's *Flora Capensis* (1818), he includes *Diosma betulina*, noting its strong smell and its use by the Hottentots in the same manner as *Diosma crenata* and *Diosma pulchella*, the latter of which he identifies as 'Hottentottis Bukku'. Thunberg observed that the Hottentots anointed their naked bodies with an intolerably smelly mixture of sheep fat and powdered *bukku*.[36] In the scientific traveller Sparrman's 1786 account of his recent Cape travels, he similarly reported that various species of *Diosma* were used by the Hottentots as 'bucku' although he did not emphasise the role of *D. pulchella*.[37] By the posthumously published revised 1797 edition of Miller's dictionary, no less than 19 species of *Diosma* were included, including the newly mentioned '*Diosma crenata/Hartogia betulina*' and '*Diosma pulchella*'. The dictionary quotes the same information as that included by Thunberg on *D. pulchella* as a powder called 'Bucku', used by Hottentots to anoint themselves. The dictionary includes extensive details on the means of propagating the plants but nothing related to medical usage.

SETTLER KNOWLEDGE OF BUCHU

By the early nineteenth century Cape settlers were clearly highly familiar with buchu. Sparrman noted that buchu and dacka 'were known both by the colonists and the Hottentots to be as efficaceous as they are common'.[38] In Burchell's account of his southern African travels, commenced in 1810, he drew attention to Hottentot use of 'Boekoe-azyn (Bookoo vinegar)' as a wash to clean and heal wounds. He commented that Hottentots and Boers had long used buchu as medicine in vinegar form or alternatively infused in brandy. The early ethhnography suggests that such settler medical knowledge of buchu came out of a wider context of familiarity. Buchu was a part of pragmatic fat application to protect the body from the elements. It was also a 'salutary herb' and one of a number of herbal remedies used by Hottentots who were often thought skilled in such matters. At the same time though, buchu was also an element in superstitious rituals observed amongst Hottentot savages. Despite these multiple identities of buchu, by the time of Burchell's travels buchu had clearly been long since adopted and adapted by settlers and buchu in vinegar or brandy become a folk medicine used by both Africans and settlers. In its colonial folk guise buchu was a remedy thought valuable as a general tonic and antiseptic as much as one for specific conditions. Demonstrating its general applicability, Burchell used buchu vinegar for treating a wrist, the hand of which had been blown off by a shot-gun.[39]

It is perhaps indicative of the inherent flexibility of folk knowledge that settlers used buchu as a broad-spectrum tonic and antiseptic whilst probably aware of its more superstitious applications amongst Hottentots. Similarly, a flexibility in the folk knowledge of what was and what was not buchu, contrasts with the reductionism of some ethnographers, scientific travellers, botanists, doctors and medical merchants keen to identify a 'real' buchu, with an active principal, under a precise scientific name.

Kolb had pinned Oldenburgs *Spiraea Africana odorata, folis pilosis* label to buchu, which by Miller's 1752 *Dictionary* was recognised as *Diosma foliis lineairibus*. In Thunberg's listing and Millers 1797 *Dictionary* buchu is equated most clearly with *Diosma pulchella*. Sparrman, on the other hand, pointed more to the fact that a number of *Diosma* equated to buchu. Burchell similarly noted that buchu vinegar could be made with the leaves of numerous kinds of buchu, such as *D. serratifolia*.[40]. Furthermore he, like the traveller Andersson, reported that plants from the Croton family were also referred to as buchu.[41]

The lack of species specificity suggested by Sparrman and Burchell seems reflective of ideas of buchu amongst colonists of the period. Smith identifies a wide range of buchu plants known to nineteenth century Afrikaaner settlers. Given the prolonged use of buchu by this time, it is highly likely that many of these names reflect earlier settler familiarity. Amongst many examples Smith includes *boegoe-Ankerkaroo*, *boegoebossie* and *boegoekapok*. Smith distinguishes a difference between names indicative of pleasing or medicinal qualities, such as *anysboegoe* and *goeieboegoe* and those considered unpleasant, including *bitterboegoe* and *knoffelboegoe*. This application of the word *boegoe* as suffix or prefix to very different plants might be thought to indicate an Afrikaans tendency to use 'buchu' in a loose generic sense for plants that smell similar or are notable for their smell. Although buchu plants are predominantly smelly this is not, however, always the case. In 1854 Moffat recorded that Doornbergen Bushmen and Bastars used non-aromatic plants, *haasboegoe* and *wolfboegoe*, mixed with fat, as a 'cooling substance' for smearing on their cheeks.[42] This link between plants used by Bushmen and Bastars in indigenous 'cooling' contexts with plants labelled 'boegoe' by colonists, suggests that settler principals of naming may be a good reflection of indigenous use and not a purely Afrikaaner naming process based on ideas of similarly smelly plants. This link between buchu plants and plants used in certain indigenous contexts, as opposed to buchu being plants of a specific species, is born out in the ethnography of sā plants.

From this early period we can, therefore, begin to detect a tension between a scientific desire to identify the exact plant of Hottentot Buchu alongside an awareness that other plants were also used. In the commercial markets of the nineteenth and twentieth centuries chemical claims for the most efficacious buchu bind to commercial claims of authentic Khoisan buchu, whilst, at the same time, merchants draw on a broad range of buchu like plants to feed the market.

BUCHU AND THE EUROPEAN MEDICAL MARKET

Although knowledge of buchu in a medical capacity seems to have been commonplace in the Cape before the nineteenth century it is not clear that this knowledge was known in Britain outside the bounds of travel books, despite indications that the Dutch had been cultivating medicinal buchu. Burchell had presciently observed that the long known virtues of *Diosma* would surely become a part of the *materia medica* of Europe.[43] He was right, but the influence came not directly from dissemination of ethnography but from lines of scientific communication that emerged in the context of colonial expansion. Virtually all the *Diosma* that would later be identified in Britain as medically useful buchu were cultivated in British botanic gardens by the late eighteenth century. The Cambridge garden cultivated *D. crenata* in 1774 and *D. pulchella* in 1787 and Kew Gardens, *D. ovata* in 1790.[44] But it was from 'new information' and a shipment of 'new plants' from the Cape that British medical interest in buchu first arose.

Buchu first appears in a medical context in Britain in 1821, when 'a scientific gentleman residing at the Cape of Good Hope', sent the publishers of *The Gazette of Health* 'a considerable quantity' of the leaves of '*Diosna* [sic] *Crenata* ' The *Gazette* editor, Dr R. Reece, commented that, as the provider of buchu 'wishes the leaves to have a fair trial in the country, a quantity of them is deposited at the Medical Hall, 170, Piccadilly with directions for supplying the faculty with them at a very low price'. The leaves were sent to England with the information that in infusion they were an effectual remedy for gleet, mucous discharges from the bladder, flour albus, rheumatism and gravel.[45] A likely contender for the Cape dispatcher of the leaves is Joseph Mackrill (1762–1820), an English born medical doctor resident in Cape Town at that time.[46]

In subsequent articles Reece recounted details from a further correspondent who both corrected the spelling of *Diosma* and reported that 'natives' at the Cape commonly used an ointment of powdered buchu leaves and grease. Publication of these details indicate that Reece had not been previously aware of Thunberg's or others' familiarity with buchu. Reece went on to elucidate that the Dutch were the first to use buchu as a medicine in Europe. It was, however, from one doctor to another, from colony to metropole, that buchu as medicine was first introduced to the British medical market and subsequently the global market.

Following tests of the leaves that far exceeded Reece's expectations, he published further findings in later additions of the *Gazette* and included detailed accounts of the uses of buchu in his popular *Medical Guide*.[47] His Guide proclaimed that buchu was useful for a host of genito-urinary problems, bowel and prostate afflictions and as an external application for contused wounds and rheumatics. In his 1833 edition, he identified *Diosma crenata* as the buchu plant most esteemed for its medical properties.[48]

This introduction of buchu to the British medical market was allied to increased buchu interest back in South Africa and contributed to a considerable new demand for buchu. By 1826 Lord Somerset, Governor of the Cape, observed that the medical demand for 'Boekhoo' was potentially of great benefit to the interests of the Cape Colony, but to meet the possible demand the plant must be protected.[49] Further demonstrating the contemporaneous popularity of buchu an anonymous note in the Cape *Commercial Advertiser*, 1827, reveals that Moravian missionaries at Genadendal sent *Diosma crenata* leaves to Madras and Calcutta, recommending them on the basis that local Hottentots used them as a remedy for intestinal colic pains.[50] Further testament of the growing prominence and economic profile of buchu is its inclusion in Dr Pappe's *Florae Capensis Medicae Prodromus* which was an expanded commentary of the plants displayed by a South African drug company at the Great Exhibition in London in 1851. Pappe's account of *D. crenata* identifies the active principal of buchu as the aromatic oil 'Diosmin' which has 'a particular smell, and a slightly astringent, bitter, taste'. This focus on Diosmin followed earlier pharmaceutical interest in buchu as indicated by research published in the *Archiv der Pharmazie* 1826. In England the *Penny Cyclopaedia* (1837) attributed pharmaceutical analysis of the leaves to 'Cadet de Gassecourt' and specified that 'Brandes considers the extractive to be peculiar and terms it Diosmin'.[51] As Pappe's notes on buchu suggest, the range of ailments supposedly amenable to buchu treatment rose incrementally with the profile of the drug. In addition to the maladies previously mentioned, Pappe noted buchu had been prescribed for cholera morbus.[52]

By the mid-nineteenth century buchu based medicines started to become extremely popular in the USA, particularly as the active ingredient in 'Helmbold's Fluid Extract of Buchu', first advertised in 1850. H.T. Helmbold, a retail druggist based in Philadelphia and New York, was a thrusting business man with big ideas. Although he was declared bankrupt by 1872, his 'Extract' continued to be manufactured by his brother for a number of years. In 1869, 1870 and 1871 he purportedly spent $500,000, in each year, on buchu marketing and at one stage his income from buchu was estimated around a million dollars per annum. His 'Extract' advertisements appeared across North America, in Europe and Asia, on sites ranging from inaccessible looking Rocky Mountain faces to the pyramids of Egypt. The label on his product claimed use of the fluid, as 'an invaluable remedy' used by 'the United States army, and in all the state hospitals and public sanitary institutions, as well as in private practice'.[53] In 1878 Helmbold revealed his winning formula to one of his managers. For twenty five gross – each bottle held about three and a half fluid ounces: Medley (short buchu, 2 parts, uvarjasi [?] 1 part) 63 lbs.12 oz.; Cubebs..21 lbs.; Liquorice root-cut..7 lbs.; alcohol 18 gal., 9 fl.oz.; caramel 10pts; molasses 5 pts.; oil (peppermint) 8 fl. Drams; water 112.5 gal. The peppermint flavour was used to give the 'highly concentrated compound' a buchu like flavour.[54] Helmbold shared a vigorous buchu market with other druggists including the manufacturers of 'Buchu-Paiba', E.S. Wells Company of Jersey City, New Jersey.[55]

Buchu maintained prominence on the US and British medical markets from the heady days of Helbold's worldwide touting to well into the twentieth century. Buchu was recognised in the U.S. and British Pharmacopeias and derivative compounds and their uses were listed in a host of pharmaceutical publications and herbal guides including the *American Journal of Pharmacy* (1888), *King's American Dispensatory* (1898), *The British Pharmaceutical Codex* (1911) and *The Working Man's Model Family Botanic Guide* (1924). Continuing popularity of buchu into the twentieth century is further evidenced by its inclusion in the *Herbal Manual: the Medicinal, Toilet, Culinary and other Uses of 130 of the Most Commonly Used Herbs* (1936) by Harold Ward, a British medical herbalist.

Over the nineteenth and twentieth centuries the story of buchu becomes increasingly complex as buchu disappears into a mass of synonyms and vernacular references which are indicative of both competing voices in the construction of science and the continued popularity of buchu at a folk and household level. From the early nineteenth century onwards, a key, although still not consistent change in naming, followed the influence of botanist Carl Ludwig von Willdenow (1765-1812), who promoted replacement of the genus *Diosma* by *Barosma*, meaning 'heavy smell' in Greek.[56] To add to the confusion, in 1950 Pillans published justification for further revising the genus *Barosma* to *Agathosma*.[57] Despite this change *Barosma* continues to be popular to the present day.

The commercial story of buchu over the nineteenth and twentieth centuries predominantly concerns short buchu (*Barosma betulina* or round leafed buchu), long buchu (*B. serratifolia* or narrow-leaf buchu) and oval buchu (*B. crenulata* or oval-leaf buchu – synonyms: *Diosma crenata*, Linnaeus; *Diosma odorata*, De Candolle). In the light of extensive chemical testing around the turn of the nineteenth century,[58] *B. serratifolia* and *B. crenulata* were both expunged from the British Pharmacopoeia, the former by 1898 and the latter by 1911.[59] *B. serratifolia* was deemed particularly inferior to both *B. betulina* and *B. crenulata* because it was found not to contain diosphenol which was considered an important constituent.[60] Over a similar period numerous substitute plants were tried on the market in response to shortages of 'real' buchu and competitive merchants wishing to improve, consolidate or expand their product range. By 1900 available alternatives had included *Diosma vulgaris, Diosma succulenta, Barosma pulchella, Barosma Eckloniana* and *Adrenda fragrans*.[61]

In 1910 two successive short crops of buchu in South Africa, which remained the only source, had led to unprecedented price rises of short buchu to 6s.5d per lb. and UK merchants were finding it hard to compete with the American demand. The problem of supplying *B. betulina* in Britain was exacerbated by US customs making it difficult or impossible to import long and oval buchu into the States. Only good quality short buchu was readily admitted.[62] This led to a particular shortage of short buchu for the British market.

The continuing overall demand for buchu was such that during World War I experiments were conducted at the National Botanical Gardens, Kirstenbosch, to assess the viability of commercial cultivation.[63] In Cape Town in 1920 prices ranged from 9s. to 11s. per lb, a strong price reflecting still growing demand in the US and additionally in Australia. In 1921 an anonymous contributor to the *Pharmaceutical Journal* observed that many colonists were keen to cultivate medicinal plants, and particularly the profitable buchu. The Kirstenbosch gardens assisted in the economic cultivation of buchu by supplying surplus seeds to enquirers whenever possible.[64]

By 1953 buchu had been excluded from the *British Pharmacopoeia* on grounds of it having been superseded by better drugs. The 1954 *British Pharmaceutical Codex* noted that *B. betulina* was still imported into Britain alongside other species which were offered as buchu, the most important of which was *B. bathii*.[65] Buchu continued to feature in the *British Pharmaceutical Codex* until 1963.[66] Since the 1960s buchu has maintained a presence only in pharmacopoeias that include herbal remedies, such as the *Extra Pharmacopoeia* (1967).[67] Buchu features in both the *British Herbal Pharmacopoeia* (1990)[68] and its companion the *British Herbal Compendium* (1992).[69] In terms of the recent regulatory status of buchu, in the 1990s buchu featured on the UK General Sales List, was accepted for specific indications in France and permitted as flavouring by the Council of Europe and the US administration.[70]

'REAL' BUCHU

Two aspects to the story of buchu assimilation help us better understand the processes and transformations of this indigenous knowledge through history. The first of these concerns scientific and commercial desires to identify and promote the real or best buchu. That this story has inconsistency in naming and species selection points to the mechanism of the construction of science. The second concerns how the structure of colonial ingress into Khoisan culture, through certain southern Cape Hottentot groups, has distorted images and understanding of Khoisan buchu relationships.

Thunberg emphasised *Diosmas betulina*, *crenata* and *pulchella* as buchu plants with *pulchella* seemingly the most esteemed. *D. pulchella*, which although its smell is distinctive strongly resembles *betulina*, later drops off the radar in the commercialisation of buchu which instead includes *serratifolia*. The first medical buchu imported to Britain was *Diosma crenata* which is also very similar to *betulina*. Over the nineteenth and twentieth century *betulina*, *crenata and serratifolia* vied for position as the 'true' or most efficacious buchu. Harvey and Sonder suggested in their influential 1885 *Flora Capensis* that 'Buku' was a referent to a number of Cape species, 'but *Barosma* crenulata is considered to possess the medical virtues of the tribe to a stronger degree than others'.[71] But despite Harvey and additionally Pappe having identified *B. crenulata* as the primary buchu plant it was *B. betulina* with its stronger Diosmin content that was in demand for medical preparations in the US. As the twentieth century progressed *B. betulina* gradually began to take precedence over *crenulata and serratifolia* as the primary buchu plant.

Where accounts of buchu become historically distorted becomes evident in claims concerning buchu such as the following made on a current website: 'Best known is the real buchu, *Agathosma betulina*, which is widely used in medicine.'[72] What, however, does 'real' mean – that *A. betulina* was most important to the Khoisan and that the link between indigenous buchu and a precise species had been made and is consistent? The evidence clearly reveals the fallacy of these assumptions. The predominance of *A. betulina* is contingent upon European identification of efficacy. This is immediately evident when we recognise that the earliest professional botanical evidence, from Thunberg, prominenced *D. pulchella* over *D. betulina*. Focussing on *A. betulina* draws attention away from the different phases and inconsistencies of European understanding. It also suggests that real Khoisan usage should be fixed at some time and at some place, around the south western Cape, where Europeans first encountered Hottentots.

The scientific name variations, including changing of the genus, species synonyms, and preferences in the construction of scientific labelling, such as *B. crenata* versus *B. crenulata* are principally revealing of competing scientific and commercial voices. If one further considers different spellings of 'buchu', one can flesh out a broader account of how indigenous knowledge moves from its shifting indigenous contexts into the wider world.

Published variations for buchu include:

boggoa (Dapper, 1668)[73]
puchu (Schreyer, 1669)[74]
bouchou (Ten Rhyne, 1686)[75]
bochu (Grevenbroek, 1695)[76]
buchu (Kolb 1719)[77]

pucku (Bövingh, 1708)[78]

bugga (Buttner, 1725)[79]

buku (Thunberg 1773)[80], bukku (Thunberg, 1818)

boegoe (Wikar 1779)[81]

bucku (Sparrman, 1786)[82]

boeghoe (le Vaillant, 1791)[83]

buckee (Percival, 1796)[84]

bookoo, buku, boekoe (Burchell, 1824)[85],

boekhoo (Records of the Cape Colony, 1826)[86]

buχu (Hahn, 1881)[87],

booko (Wood, 1918)[88]

p/nkaou: two Nama varieties, d/nhora or d/khonsa or Haas buchu and p/kabourie (Laidler, 1928)[89]

borgoe, bergboegoe, Fontein-boegoe, Olifantsboegoe (Watt and Breyer-Brandwijk, 1932)[90]

boochoo (http://www.herbsorganic.co.za, 2005)[91]

bugu (http://www.beneforce.com, 2005)[92]

boechoe, boekoe, buccho, bucchuu, bucco (unprovenanced seventeenth and eighteenth century sources listed by Smith)[93]

International variations include:

boegoe (Afrikaans)

ibuchu (Xhosa)[94]

bucco (Spanish, French, Dutch)[95]

bukko (Italian and Danish)[96]

bucku, bukko (German)[97]

õli-bukopõõsas (Estonian)[98]

The variety apparent in this list is striking but seems more understandable when one considers the factors at work in naming. Anthropologists have distinguished flexibility of ideas as an intrinsic feature of historical and pre-historical Khoisan culture. When one considers variation in dialects encountered by ethnographers in different regions, alongside this flexibility, such variation becomes less surprising. Yet in addition to this one must also take into account both the transformative role of specific translators and the lack of consistency in early European national and international spelling, to say nothing of spelling Khoisan dialects.

The second factor of importance to understanding the assimilation of buchu concerns how notions of real buchu or the role of buchu have become predominantly associated with the geographical and cultural contexts in which buchu was initially discovered. Historical partiality has encouraged notions of Cape Khoisan knowledge as the original Khoisan knowledge which distilled out in somehow less virtuous forms to other Khoisan. Burchell may have encouraged this situation with his suggestion that 'in the countries lying beyond the geo-graphical boundary of that genus [*Diosma*], other plants of various genera are, of necessity, made use of'.[99] 'Of necessity' suggests hardship in the absence of better quality or more suitable alternatives. Smith seems to have fallen prey to this distortion in his observation that: 'Beyond the distributional boundaries of the Diosmeae, the common name has been applied to a number of other aromatic species which were used as a body perfume by Bushmen, Hottentot and Bastard tribes'.[100]

The transition from plants first identified at the Cape under a broad name of buchu to essentially *Agathosma betulina*, with provisos that other plants might be used, in a sense suggests that all Khoisan might seek out and use *A. betulina* unless it is unavailable, in which case they use another similar plant. This is interesting because it conflates first European arrival at the Cape with discovery of a Khoisan population who subsequently, through their primacy of 'discovery', are promoted in history as somehow original Khoisan. This has led to an often implicit ethnographic message that buchu is principally a Cape phenomenon and in its 'correct' Cape manifestation relates to *A. betulina* and any use of non *betulina* plants beyond the Cape is somehow shadowy and less pure knowledge. The implied usage of less effective plants must, however, be recognised as a result of European bioscience having privileged the scientific effective properties of *A. betulina*.

Whilst it may well be appropriate to associate the word 'buchu' with southern Cape populations, as we have seen plants from beyond the early contact zone, the southern Cape, historically carried an Afrikaaner 'buchu' label, suggesting that they too may have been used in a 'buchu' manner by local indigenous groups. In 1777 Gordon observed that the 'Geissiqua people [...] sat pounding red buchu from camel-thorn bark'.[101] This again suggests that beyond the southern Cape non 'buchu' plants had a role as buchu. Certainly in recent contexts the term buchu is used by Khoisan at least as far north as Namibia's Walvis Bay and is not tied to plants of the southern Cape. My fieldwork indicates that Khoisan from the northern Cape upwards use plants known as sâi (or sä) in a category of practice that compounds ritual, healing and perfume and overlaps with the way buchu is and has been used elsewhere. These plants have a long history of use. Buchu seems therefore to be a southern Cape variant of the broader buchu/sâi cultural phenomenon. Earlier ethnographers equally seem to have recognised overlaps between buchu and sâi.

Following Smith (1966), van Wyck and Gericke suggest the name San may be derived from Bushmen applying perfuming 'San, Son or Sab' plants to the body. Whilst the claim may be hard to sustain they make the observation that these Khoi words originally referred to the aromatic *Pteronia onobromoides* (family Asteraceae) that was used as a perfume and, furthermore, that the mean-ing later shifted to any shrub.[102] Shearing observes that *Pteronia adenocarpa* is known as 'Boegoekaroo',[103] and other commentators again include *Pteronia erythrocaetha*[104] as buchu. Again, the evidence is unclear but at the very least suggests, coupled with wider links between *Pteronia* and buchu, that variously termed sâi or sā plants relate to plants that overlap with buchu in name and perfuming role.

SÂI AND BUCHU, A SHARED HISTORY

As with buchu a history of sâi can be used to highlight the tensions and the process of colonial data extraction. At the same time the history adds to understanding of the context of buchu.

From the early eighteenth century Kolb noted that *Saab* was used 'to acquire greater swiftness of foot'.[105] Being 'swift of foot' was a classically received exoticising epithet common in seventeenth and eighteenth century travel accounts. Nonetheless, the suggestion is that there is more to *Saab* than simply smelling good.

Francis Galton's 1850 diary of his travels across central South West Africa presents a later reference to *sāāb*. His entry demonstrates a nineteenth century interest in the useful and a different notion of sāāb as a general reference to powders.

> Chou wāb, […] or better khou waap is the name of the herb the Namaqua powder and use as scent, it may be quexonum? the gun-resin? spoken of in the admiralty manual..
>
> They call all their powders sāāb [continuous hyphen over both 'ā' 's]; one sort is made from the inner peel of the coffee Chou [?] tree one which they mix with the Khou waap seems to be a lichen? They are mixed in the specimens I have.[106]

Galton's notes demonstrate the centrality of the commercial eye to colonial travel. The manual he referred to may well have been *A Manual of Scientific Enquiry; Prepared for the Use of her Majesty's Navy and Adapted for Travellers in General*, published in 1849, the year before Galton sailed for Africa.[107] If it were, he would have been alerted to the importance of 'Bucku of the South-African Hottentots'. The manual encouraged readers, 'To determine the different kinds collected by the natives'. [108]

In 1881 the philologist Theophilus Hahn included buchu in a list of regional word comparisons:

	Khoikhoi /Kham-Bushman [/Xam]	
buchu	sāb	tsā[109]

The use of the term *sā* remains evident in the early twentieth century. In 1918 Hoernlé referred to *sāp* as a generic term for powder, elaborating that Hottentots make the powder by grinding sweet smelling bark, or roots, or leaves, which they rub freely on clothes or skin.[110]

In 1928 Fourie, medical officer to the administration of South West Africa, noted that amongst Bushmen:

> a deceased person is believed to move about in the form of a ghost at night. Buchu (tsa) is accordingly sprinkled over the grave to make the spirit of the departed happy so that it may not return at night to molest others... .[111]

Hahn's and Fourie's references support both an interchangeability in ideas of buchu, *sa* [*sãb*] and *tsa* [*tsã*], and the suggestion that the powder is intimately related to more than just perfume or medicine. In 1929 Dorothea Bleek published a comparative Bushman dictionary in which she listed:

> *tʃã*[112] buchu, scented herbs used for toilet or ceremonial purposes, *Lepidum ruderale, L., Ocimum feticulosum*, Burch. s. sã, tsã (Na. Sãb) CII.(DB) Ex.: *tʃãtʃa, peliostomem leucorhizum*, E. Meyer, *tʃãba, Bouchea pinnatafida*, Schauer. [113]

Bleek's inclusion adds to our list of buchu plants whilst also demonstrating in her use of the phrase 'used for toilet or ceremonial purposes' a potential epistemological cleavage of Khoisan ideas. It assumes a separation of perfume use from matters that seem ceremonial. Might it not be ceremonial to put perfume on? More to the point, might not the two actions or uses be operating the same ideas of the function of buchu?

These various references to buchu, *tsa* and *sã* suggest that travellers and researchers have long been confronted with people who not only use the terms, buchu, *tsa* and *sa* apparently interchangeably but that the names are applicable to many different plants. The variety of botanical Latin names attributed to the plants speaks of the variety of plants that fit the indigenous category, but also of the inconsistency of early and later botanical references and probably of the inability of some Europeans to identify the plants shown to them. A reference by the traveller and artist Thomas Baines in his 1864 *Explorations in South West Africa* begs the question as to how many ethnographers really knew much about the plants they reported:

> Unfortunately, we are neither of us deeply skilled in botany, and besides this, the "Flora Capensis" of Harvey and Sonder, in Chapman's possession, has only reached its first volume. Lindley's "School Botany" gives us considerable help, but this refers only to strictly British vegetation.[114]

Recent researchers readily equate sâi with buchu. Haacke and Eiseb, authors of a Khoekhoegowab-English glossary include, 'sâi.i(/b/s)*n*. buchu (powder)'.[115] Sian Sullivan, an anthropologist who has carried out extensive fieldwork amongst northern Namibian Damara, suggests that sâi is referred to as buchu in historical sources.[116] As we have seen, however, both terms are historical and contemporary and essentially co-exist. What seems more relevant is their geographical provenance. Barring Kolb's reference, sâi or *sāb* seems predominantly a northern Cape or Namibian term. Perhaps this represents disappearance of the term within the southern Cape.

Summary of alternative sā derivatives

Saab	Kolbe 1734 [1719]
sāāb[117]	Galton 1850
sāb	(Khoikhoi); *tsā* (/Kham Bushmen), Hahn, 1881
tsa	Fourie 1928
tʃ̃ã	Bleek 1929
sâi	Haacke 1999 cf. Sullivan 1999

In the 1990s Sullivan recorded nearly 40 species of sâi plants that were used by Damara women. The aroma of these plants seemed key to their selection, as it is for most buchu.[118] This considerable variety again points strongly to the two names 'buchu' and 'sâi' relating more to overlapping characteristics or roles of plants than to specific species.

A further clue to the meaning of sâi [*sā*] might also lie in language. In Khoekhoegowab sā means 'to gather, glean, collect, pick up; peck up (of: bird)'.[119] The meaning of sâi may, therefore, lie somewhere around gathering a particular sort of plant in a particular context. Although both the type of click and the tone of apparently similar words is crucial to meaning in Khoisan languages, there is sufficient evidence of variation in clicks and pronunciation to warrant consideration of such apparent root affinities (Köhler (1963), Haacke (1986, 1997), Traill (1986)).[120]

Linguistic relationships might also provide some insight into the ideational context of buchu and sâi use. Haacke's and Eiseb's Khoekhoegowab dictionary gathers words together in clusters of meaning. Perhaps revealingly, sâi sits between a large range of, on the one hand, words for rest, linked to lying down, and on the other, words linked to cooking and related to boiling as in the rising up of steam.

A crucial life-giving juxtaposition between the reclined, asleep or dead and the rising, standing and alive is played out in notions of Ju/'hoan (!Kung) healing. Like other Bushmen the Ju/'hoansi envisage that a healer dances to awaken the *n/um* or healing potency that lies dormant in their stomach. A healer told Lee that, 'You feel your blood become very hot just like blood boiling on a fire and then you start to heal'. Lee notes that *n!um*, to boil, is related to boiling of water and ripening of plants. He proposes that there is a symbolic association between still cool water that boils and activating medicine and plants which are dormant 'becoming nutritionally potent when ripe'. He sees this notion extended in joking metaphor to nubile maidens who have reached menarche and are considered '"ripe" for intercourse and impregnation'.[121]

A similar link between *sāi* as something that boils and awakens love is indicated in Hahn's, *Tsuni-//Goam* in relation to later nineteenth century Khoikhoi. Hahn observed that the name of *Heitsi-eibeb's* son *!Urisib* is derived from *!ū* which he links to the colour white, the ostrich egg, and *!Uris*, the white one, also called *Suris*, the sun. *Suris* he continued: 'gives the root *su,* to broil, to be hot; Soris or Suris, therefore, means the broiling one, the heating one, the inflaming one.' Derived from *sū* comes: '*sāi,* to boil', *sūs,* a cooking pot and '*Sureb* or *soreb* (masc.), *sores* (fem.), the lover, the sweetheart, the one who is inflamed- viz., with love, or who inflames with love'[122]

The Khoisan role of buchu and sâi that these linguistic relationships seem to support was something I first encountered as associations of ideas. These associations seem to exist regardless of the validity of these possible linguistic relationships. As the Khoikhoi and Bushmen material suggests, they appear to be associations which are played out in different Khoisan dialects.

BUCHU AS PERFUME

The evidence suggests that buchu operates between day and night, between life and death, between being asleep and being awake. It restores the sick and calms the excited. It anchors those in dangerous liminal states. European accounts of buchu commonly begin and end with the normative and shallow observation that the Khoisan use it as a perfume. Whilst this is, in one sense, certainly accurate, it does nothing to unpack the role of perfume in the wider context of Khoisan plant use nor its relationship to Khoisan notions of potency.

The idea of perfume in both Khoisan culture and beyond is tied to attraction and in some sense ripening. It is tied to the emitting of pheromones stimulating reproductive activity and the fragrance of flowers which encourage fertilisation. Sullivan encountered sâi primarily in the context of Damara women and has interpreted the role of sâi from a highly gendered perspective. Sullivan sees sâi as 'emphatically a constituent of the separate but contiguous realities [...] which might be rendered as female'.[123] Sullivan elaborates that use of sâi powder as perfume on the body, garments and bed-clothes imparts female beauty and attractiveness. All aspects of procurement, preparation and cosmetic use are controlled by women and 'underpinned by forceful symbolic significance'.[124] This emphasis of feminine Damara relationships with sâi seems related to Hahn's much earlier claims cited above, that the word '*sāi,* to boil', is related to *sores* (fem.), the lover, [...] or who inflames with love'.

Vedder noted in the early twentieth century that during their first menstruation Khoi girls were instructed in the preparation of buchu perfume. They were given their first tortoise shell buchu holder to fill with perfume, which from then on they would keep attached to their clothing. Schmidt observes that 'this powder box was the symbol of her femininity, and buchu the symbol of her feminine potencies, of fertility and giving life.[125]

The theme of giving life ties in tightly with wider Khoisan roles and understandings of buchu. Sullivan relates a myth in which a deceased person is revived when their heart is placed in the ash of a fire and buchu is sprinkled over it. Additionally she observes that Vedder attributed buchu with calming and taming properties. Sullivan concludes that, ultimately, sâi 'is a substance with power; this power is conferred through its association with the fertility of women'.[126]

Sâi does have a gender specific role which is particularly prominent in female realms of perfume use, menarche and fertility, but the broader context in which this sâi use sits lies in strong smelling plants being envisaged as potent. As the revealing world has demonstrated to the Khoisan, smell as the holder of essence can transfer qualities between organisms and across barriers of life and death. Khoisan understandings of smell sit in a web of relationships linking wind, breath, smell and even arrows. Each person and certain potent animals are envisaged as possessing their own specific winds. A person's wind represents their identity and their power to act in a particular manner in the world, bringing about particular results. The notion of wind crosses over with notions of spirit helpers that inhabit Khoisan. Many Namibian Khoisan refer to these as / gais (b) or gais. If one has the wind of the mamba or the mamba gais, there is something of the mamba inside you. If you see a mamba it will not harm you because it will recognise one of its own. I was told if such a person sees a mamba they must say, 'I am one of you, do not harm me' and the mamba will let them pass. Similarly Khoisan healers have long been noted for addressing storms and rain in a personal manner, saying 'do not harm me, I am one of you'. Such people, known as /nanu aob or 'rain men' amongst Damara, hold the essence of the lightning or rain.

The key to understanding buchu use in recent and historic contexts seems to lie in a set of ideas that might be termed 'a family of resemblances'. Wind is one aspect of this family, another is breath.[127] In Khoekhoegowab the word for breath and soul is /om. Breath is a gift from the divine that both gives life and demonstrates life. It is a vital force that holds the identity and hence potency of a person. Many Khoisan believe the potency, and in some sense breath, is held in the heart. Often the lungs and heart are envisaged as interrelated wind holding organs. That bad thoughts can be expelled harmfully from the heart in the wind of words, or that to lose one's heart is to lose life, are both common Khoisan beliefs founded in deeper notions of wind. The wind of people and organisms and the wind that blows, holds people together in an existentially self evident mesh of revealing relationships.

Arrows are equally tied to ideas of potency that can move on the wind between organisms. Arrows are a medium for acting at a distance. The unseeable activity of poison on the arrow tips works within an organism. The arrow connects the hunter with their prey. The notion of shooting and receiving arrows of power

Christopher H. Low

is played out in numerous contexts including struggles with divinity or unsee-able malevolent attack that induces fear in wandering bushmen, in movement of sickness and healing potency between and within people, and even in love arrows fired by Bushmen towards their intended to be accepted or rejected ac-cordingly. Buchu is, like arrows, a way of moving potency.

CONCLUSION

There are cross over points between European understandings of buchu and Khoisan. Smell is clearly an essential distinguishing characteristic of buchu plants to Khoisan and also to Europeans, as the meaning of the Latin names attributed to buchu indicate. Science knows the world around it by refining knowledge that our senses equip us to detect. But in a commercial context 'Western' scientific knowledge of buchu transformed the commonality of the experience and knowledge of buchu. For Khoisan, medical uses of buchu were and in some cases still are, occasions of personal transformation. 'Treatment' is founded in a way of knowing still firmly rooted in notions of smell and potency. In contrast to this, scientific and commercial users of buchu applied a scientific or pseudo-scientific rationale to the apparent efficacy identified in some Khoisan practices and extended the range of buchu uses enormously. In the nineteenth and early twentieth century, preparations of buchu included tinctures, elixirs and fluid extracts. These remedies were indicated for some diseases that Khoisan might have recognised but understood in an entirely different manner.

Khoisan use of buchu demonstrably binds illness to realms of potency abroad in an interlocking world of organisms, phenomena and spirits. The Khoisan dif-ference from scientific understanding represents a divergence. One path follows smell as meaningful potent phenomenon tied to identity, the other as chemical property with physiological potency.

The European uptake of buchu is an account of science appropriating and transforming indigenous knowledge. A significant distortion of indigenous knowledge is evident in the need of commercial science and buchu drug pedlars to identify a true and most efficacious buchu which they then misrepresented as that which had been the most desirous and thought most efficacious amongst Khoisan. It is significant that, as buchu became sidelined from orthodox phar-maceutical use, the indigenous historical usage of buchu has become all the more emphasised, and particularly so in a highly marketable Bushman context. Buchu has thus been more recently sold less on scientific efficacy than its image as an implicitly healthy, indigenous alternative medicine.

In line with other accounts of scientific appropriation, the buchu story de-tracts from notions of clean scientific data gathering. Contextualising the word 'buchu' amongst its alternatives points to the partiality of the historical legacy. That buchu and not sâi seems the most popular term for the phenomenon, at least in recent British and American contexts, probably speaks of the consider-able influence of Cape ethnography on Khoisan history. The variation in Latin names for buchu over time is testament to an impressive netting of information but the need to find a true and scientifically potent buchu distorted recognition of buchu as a type of human/plant relationship, culturally mediated around a very broad range of strong smelling southern African plants. Closer consideration of what perfume means in European contexts – a powerful scent, still often derived from animal parts, that through pheromonal action operates in a biologically and socially potent manner – brings us closer to Khoisan understanding of buchu. However, to really begin to appreciate Khoisan buchu use, one has to turn to the details of Khoisan life as a culturally rich people with continuing close ties to a rural African environment. Trying to pin down whether buchu or sâi refers to particular plant species, a role for particular plants or even the gathering process of powerful plants opens up the cracks in representation of Khoisan.

Understanding the context of buchu and sâi use requires an appreciation of the links between notions of wind, smell, breath, arrows and healing. That this web of relationships may look untidy and perhaps unconvincing in certain elements, highlights how ethnography, predominantly bound to Euro-American categories of enquiry, has remained blinkered to certain indigenous perspectives.

NOTES

I am very grateful to the ESRC for the funding of the doctoral and postdoctoral work upon which this paper is based.

[1] 'Khoisan' is a European constructed compound of old Nama *khoi* (modern *khoe*), meaning people in most Khoe languages and San (or Sān, Saan) the word Khoekhoe ('Khoekhoen' meaning people of people) use for Bushmen; see A. Barnard, *Hunters and Herders of Southern Africa* (Cambridge: Cambridge University Press, 1992), 7. I use 'Khoisan' as opposed to 'Khoesān' as it remains more readily recognised. Aware of the pejorative nature of the word 'Hottentot', I use 'Hottentot' for historical accuracy, as a simple replacement with Khoekhoe would misrepresent the complexity of the material. In my experience only the most politicised Bushmen insist on being called 'San'. The vast majority I have encountered prefer to be known as 'Bushmen'. This may or may not represent repossession of the word. I use San and Bushmen interchangeably.

[2] Londa Schiebinger, *Plants and Empire: Colonial Bioprospecting in the Atlantic World* (Cambridge (Mass.): Harvard University Press, 2004).

[3] The paper builds on accounts of buchu in my DPhil historical analysis of Khoisan medicine: Chris Low, 'Khoisan Healing: Understandings, Ideas and Practices' (DPhil diss., University of Oxford, 2004).

[4] C.A. Smith, 'Common Names of South African Plants', in *Botanical Survey Memoir*, 35, ed. E. Percy Phillips, (Pretoria: Dept. Agricultural Technical Services, 1966), 134.

[5] See Elphick for further discussion of this point: Richard Elphick, *Khoikhoi and the*

Founding of White South Africa (Johannesburg: Ravan Press, 1985), 4.

[6] There are arguments that the difference between Khoekhoe and Bushmen has historically been economic (see John Wright, 'Sonqua, Bosjesmans, Bushmen, aba Thwa: Comments and Queries on Pre-Modern Identifications', *South African Historical Journal* 35 (Nov. 1996): 16–29. However, evidence seems to suggest that there is validity in envisaging a distinctive Bushmen culture founded in a hunter-gatherer culture that persists in certain dimensions despite significant social change. The Bushmen I am looking for historically are people connected to hunter-gatherer culture as described by the Marshall family, Richard Lee, Megan Biesele, Mathias Guenther, amongst others.

[7] Chris Low, 'Khoisan Healing'.

[8] Mary Gunn and L.E. Codd, *Botanical Exploration of South Africa* (Cape Town: A.A. Balkema, 1981), 5, 12, 14, 27.

[9] Smith, 'Common Names', 138.

[10] Olfert Dapper. 'Kaffrarie, of Lant der Hottentots', in *The Early Cape Hottentots: described in the writings of Olfert Dapper (1668), Willem Ten Rhyne (1686) and Johannes Guielmus de Grevenbroek (1695)*, trans. I.Schapers and B. Farrington, ed. I Schapera (Cape Town: The Van Riebeek Society, 1933), 40.

[11] Schapera, *The Early Cape*: 115 and Smith, *Common Names*: 136.

[12] Schapera, *The Early Cape*: 245.

[13] Schapera, *The Early Cape*: 263.

[14] P. Kolb, *The Present State of the Cape of Good Hope*, trans. Mr. Medley, 2 vols. (London, 1731), I: 96; Oldenland cited by Smith, 'Common Names', 138.

[15] Kolb, *Present State*, 113.

[16] Kolb, *Present State*, 133; similarly: 99.

[17] Ibid., 141

[18] Ibid., 305

[19] Ibid., 305

[20] Ibid., 309.

[21] H.J. Wikar, *The Journal of Hendrik Jacob Wikar* (1779), with an English translation by A.W. van der Horst, ed. E.E. Mossop (Cape Town: The Van Riebeek Society, 1935), 63.

[22] R.J.Gordon, *Cape Travels, 1777 to 1786*, ed. Peter R. Raper and Maurice Boucher (Houghton: The Brenthurst Press, 1988), 87, 122, 216.

[23] Roger Hewitt, *Structure, Meaning and Ritual in the Narratives of the Southern San*, Quellen zur Khoisan-Forschung (Hamburg, Helmut Buske Verlag, 1986), 294.

[24] Dorothea Bleek, 'Customs and Beliefs of the /Xam Bushmen', *Bantu Studies*, 7:4 (Dec. 1933): 382.

[25] Ben-Erik van Wyk, Bosch Van Oudtshoorn and Nigel Gericke, *Medicinal Plants of South Africa* (Pretoria: Briza Publications, 1997), 34.

[26] Khoisan envisage that illness can be caused by moving organs. Often these are said to 'stand up'; see Chris Low 'Khoisan Healing'.

[27] http://www.evitamins.com/product.asp?pid=475 (accessed March 15, 2007)

[28] Smith, 'Common Names': 136.

[29] Ibid.

[30] J. Donn, *Hortus Cantabrigiensis or a Catalogue of Plants Indigenous and Exotic*, 5th edn. (Cambridge, 1809), 50.

[31] Philip Miller, *The Gardener's Dictionary*, 6th edn. (London, 1752), listing *'Diosma'*.

[32] Philip Miller, *The Gardener's Dictionary*, 7th edn. (London, 1759), listing *'Diosma'*.

[33] Coyte, *Coyte's Botanic Garden* – Facsimile edition of Hortus Botanicus Gippovicensis (Ipswich, 1988): 26.

[34] 'Kew Record Books' 1793–1847 MS.

[35] V. S. Forbes, ed., *Travels at the Cape of Good Hope 1772–1775* (Cape Town: Van Riebeek Society, 1986), xix.

[36] Carol Pet Thunberg, *Flora Capensis*, 2 vols. (1818), II: 139–43.

[37] Andrew Sparrman, *A Voyage to the Cape of Good Hope …from the year 1772 to 1776*, 2nd edn. (London, 1786): 184.

[38] Sparrman's Voyage, cited in Oxford English Dictionary, online: http://dictionary.oed.com: 'dagga'(accessed March 15, 2007).

[39] William J. Burchell, *Travels in the Interior of Southern Africa*: reprinted from the original edition of 1822–4, 2 vols. (London: The Batchworth Press, 1953), I, 330–31.

[40] Ibid., 331

[41] Burchell, *Travels*, 275; John Andersson, *Lake Ngami; or, Explorations and Discoveries, during Four Years' Wanderings in the Wilds of South Western Africa*, 2nd edn (London: Hurst and Balckett, 1865), 259.

[42] Smith, 'Common Names', 141.

[43] Burchell, *Travels*, 331.

[44] Donn, *Hortus Cantabrigiensis*, 50; 'Kew Record Books' 1793–1847 MS.

[45] Richard Reece, ed., *The Gazette of Health*, 6:62 (to Feb 1, 1821): 799.

[46] Gunn and Codd, *Botanical Exploration*, 239.

[47] R. Reece, *The Gazette of Health*: 6:63 (to March, 1821): 812–13; 64, April 1821; First Addition to the Appendix of the Gazette of Health (May 31, 1822); *The Medical Guide for the Use of the Clergy, Heads of Families, and Seminaries and Junior Practitioners in Medicine* (London, Longman, 1833).

[48] Reece, *Medical Guide*, 151.

[49] Records of the Cape Colony, Jan–Feb 1826, Vol XXV, printed 1905: 224.

[50] Cited by Smith, 'Common Names', 137.

[51] Constantini, 'Ueber die *Diosma* crenata', *Archiv der Pharmazie*, 19:3 (1826): 255. *Penny Cyclopaedia* (1837), IX, 5/1. The reference is probably to the French pharmacist Charles Louis Cadet de Gassicourt (1769–1799).

[52] L. Pappe, Florae *Capensis Medicae Prodromus; or An Enumeration of South African Plants used as Remedies by the Colonists of the Cape of Good Hope*, 2nd edn. (Cape Town, W. Brittain, 1857), 7.

[53] 'Who was Henry Helmbold', *Druggists Circular* (Nov. 1912), listed on http://www.bottlebooks.com/helmboldstory/Helmbold.htm (accessed 27 Sept 2005); 'Supreme Court of New York. A.L. Helmbold v. The H.T. Helmbold Manufacturing Co.', *The American Law Register (1852–1891)*, 26:3, New Series Volume 17 (Mar. 1878): 170–71.

[54] 'An Ex-Manager', 'Who was Henry Helmbold'.

[55] William H. Helfand, 'Historical Images of the Drugs Market', *Pharmacy in History* 28:1 (1986): 55.

[56] William H. Harvey and Otto Wilhelm Sonder, *Flora Capensis: Being a Systematic Description of the Plants of the Cape Colony, Caffraria & Port Natal* (London: Reeve

& Co., 1885): 373, 393

[57] N.S.Pillans, *Journal of South African Botany* 16 (1950): 55–183.

[58] See e.g. Y. Shimoyama, 'Chemistry of Buchu leaves', *American Journal of Pharmacy* 60, (12 Dec. 1888), http://www.swsbm.com/AJP/AJP_1888_12.pdf

[59] Annotations, 'Advancing Buchu Prices', *The Pharmaceutical Journal and Pharmacist* (20 Aug 1910): 271.

[60] 'Buchu Folia, B.P.', *The British Pharmaceutical Codex* (London, The Pharmaceutical Press, 1911), no page number, http://www.henriettesherbal.com/eclectic/bpc1911/barosma.html (accessed 15 March 2007).

[61] E. M. Holmes, 'Buchu Leaves', *Pharmaceutical Journal* (Oct. 1900): 70.

[62] Annotations, 'Advancing Buchu Prices', *The Pharmaceutical Journal and Pharmacist* (20 Aug. 1910): 272

[63] Anon., *South African Journal of Industries*, 2 (1919): 748.

[64] Editorial Articles, 'The Cultivation of Buchu', *The Pharmaceutical Journal and Pharmacist* (22 Oct. 1921): 308.

[65] *British Pharmaceutical Codex* (London: The Pharmaceutical Press, 1954), 108.

[66] *British Pharmaceutical Codex* (London: The Pharmaceutical Press, 1963), 102–104.

[67] R.G. Todd, ed., *Extra Pharmacopoeia*, 25th edn. (London: The Pharmaceutical Press, 1967), 1509.

[68] *British Herbal Pharmacopoeia*, vol. 1, (Dorset: British Herbal Medicine Association, 1990).

[69] Peter Bradley, ed., *British Herbal Compendium*, vol. 1 (Dorset: British Herbal Medicine Association, 1992), 43–4.

[70] Ibid.

[71] Harvey and Sonder, *Flora Capensis*, 373.

[72] http://www.plantzafrica.com/plantab/acmadeniaheterophylla.htm (accessed 14 Feb. 2007).

[73] Schapera, ed., *The Early Cape*, 40.

[74] Cited by G.S. Nienaber, *Hottentots* (Pretoria, J.L. Van Schaik Beperk, 1963): 222.

[75] William Ten Rhyne, 'A Short Account of the Cape of Good Hope and of the Hottentots who inhabit that Region', Schapera, ed., *Early Cape Hottentots*, 115.

[76] Schapera, ed., *The Early Cape*, 245.

[77] *The Present State*: 96, 113, 133, 141, 305, 309.

[78] Cited by Nienaber, *Hottentots*, 222.

[79] Ibid.

[80] Ibid; Thunberg, *Flora Capensis*, 143.

[81] Cited by Nienaber, *Hottentots*, 222.

[82] Sparrman, *A Voyage to the* Cape, 145, 184.

[83] Le Vaillant *Travels into the Interior Parts of Africa by the Cape of Good Hope in the Years 1780, 81, 82, 83, 84 and 85* (1791), 68.

[84] Cited by Nienaber, *Hottentots*: 222.

[85] Burchell, *Travels*, 275, 331.

[86] Records of the Cape Colony, Jan–Feb 1826, Vol XXV, printed 1905: 224.

[87] Theophilus Hahn, *Tsuni - //Goam: The Supreme Being of the Khoi-Khoi* , (London:

Trübner & Co., 1881), 23.

[88]Wood, George Bacon and Franklin Bache *The Dispensatory of the United States of America USA*, revised by J.P. Remington et al (Philadelphia and London: J.B. Lippincott Co., 1918): 39, http://www.swsbm.com/Dispensatory/USD-1918-complete.pdf (accessed 3 Feb. 2007).

[89] P.W. Laidler, 'The Magic Medicine of the Hottentots', *South African Journal of Science*, XXV (Dec. 1928): 441.

[90] Mitchell Watt and Gerdina Breyer-Brandwijk, *The Medicinal and Poisonous Plants of Southern Africa: being an account of their medicinal uses, chemical composition, pharmacological effects and toxicology in Man and animal* (Edinburgh: E&S Livingstone 1932), 90.

[91] http://www.herbsorganic.co.za/pages (accessed 18 March 2005).

[92] http://wwwbeneforce.com/buchu (accessed 29 Sept. 2005).

[93] Smith, *Common Names*, 135.

[94] Ben-Erik van Wyk and Nigel Gericke, *People's Plants: a Guide to Useful Plants of Southern Africa* (Pretoria: Britza, 2000): 139.

[95] Wood, *Dispensatory of the United States*: 39; Henriette's Herbal Homepage, www.ibiblio.org/herbmed/php/get?id=418 (accessed 20 Sept. 2005); http://liberherbarum.com/pn0872.HTM (accessed 20 Sept. 2005).

[96] http://www.liberherbarum.com/pn0872.HTM; http://www.botanic-garden.ku.dk/eng/index.htm (accessed 18 March 2005).

[97] Wood, *Dispensatory of the United States*: 39; http://www.heilfastenkur.de (accessed 18 March 2005)

[98] http://www.liberherbarum.com/pn0872.HTM (accessed 20 Sept. 2005).

[99] Burchell, *Travels*, 275.

[100] Smith, 'Common Names', 140.

[101] Gordon, *Cape Travels*, 339.

[102] van Wyk and Gericke, *People's Plants*, 215.

[103] David Shearing, and van Heerden, K. Karoo. *South African Wild Flower Guide 6* (Kirstenbosch, Botanical Society of Southern Africa, 1994), 174, cited by http://www.museums.org.za/bio/plants/asteraceae/pteronia.htm (accessed 21 Sept. 2005).

[104] http://www.museumsnc.co.za/McGregor/othermusems/victoria.htm (accessed 21 Sept. 2005).

[105] Kolb, *The Present State*, 116.

[106] Francis Galton, MS, Journal, 1850 (University College London archives).

[107] *A Manual of Scientific Enquiry; Prepared for the Use of her Majesty's Navy and Adapted for Travellers in General*, ed. Sir John F.W. Herschel (London, John Murray, 1849).

[108] Herschel, *A Manual*, 415.

[109] Hahn, *Tsuni - //Goam*, 7.

[110] A.W. Hoernlé, 'Certain Rites of Transition and the Conception of !Nau among the Hottentots', in *Harvard Studies* (1918): 71; Hoernlé included a " ' " above her *ā* which is not available typographically.

[111] L. Fourie, *The Native Tribes of South West Africa* (Cape Town: Cape Times, 1928), 104.

[112] ∫ denotes a glottal stop

Christopher H. Low

[113]Dorothea F. Bleek, *Comparative Vocabularies of Bushman Languages* (Cambridge: Cambridge University Press, 1929), 224. CII.(DB) refers to the provenance of her information: central Kalahari / Botswana and Dorothea Bleek.

[114] Thomas Baines, *Explorations in South West Africa being an Account of a Journey in the Years 1861 and 1862 from Walvisch Bay, on the Western Coast, to lake Ngami and the Victoria Falls* (London: 1864), 220.

[115] W. Haake and E. Eiseb, *Khoekhoegowab-English, English Khoekhoegowab Glossary/ Mîdi Saogub* (Windhoek: Gamsberg Macmillan, 1999), 35.

[116] S. Sullivan, 'Perfume and Pastoralism: Gender, Ethnographic Myths and Community-Based Conservation in a Former Namibian Homeland', in D. Hodgson *Rethinking Pastoralism in Africa: Gender, Culture and the Myth of the Patriarchal Pastoralist* (Athens: Ohio Press, 2000), 152.

[117] [hyphen over both 'ā' 's]

[118] Sullivan, 'Perfume': 152.

[119] Haacke and Eiseb, *Khoekhoegowab*, 33.

[120] Oswin Köhler, 'Observations on the Central Khoisan Language Group', *Journal of African Languages*, 2 (1963): 227–234; W.H.G. Haacke, 'Preliminary Observations on a Dialect of the Sesfontein Damara', in *Contemporary Studies on Khoisan 1*, Q.K.F. 5.1, ed. R. Vossen and K. Keuthmann (Hamburg, H. Buske, 1986), 375–396; W.H.G. Haacke, E. Eiseb, and L. Namaseb, 'Internal and External Relations of Khoe-Khoe Dialects: a Preliminary Survey', in *Namibian Languages: Reports and Papers*, ed. W.H.G. Haacke and E.E. Elderkin (Windhoek,1997), 125-210; A. Traill, 'Click Replacement in Khoe', in *Contemporary Studies on Khoisan 2*, Q.K.F. 5.2, ed. R. Vossen and K. Keuthmann (Hamburg: H. Buske, 1986), 301–320.

[121] Richard B. Lee, 'Trance Cure of the !Kung Bushmen', *Natural History* (Nov. 1967): 31, 33.

[122] Theophilus Hahn, *Tsuni-//Goam*, 141.

[123] Sullivan, 'Perfume', 152.

[124] Ibid., 153.

[125] Cited by Sullivan, 'Perfume', 153.

[126] Ibid.

[127] Chris Low, 'Khoisan wind: hunting and healing', in 'Wind, life, health: anthropological and historical perspectives', *JRAI* Special Issue (2007): 71–90.

Changes in Landscape or in Interpretation? Reflections Based on the Environmental and Socio-economic History of a Village in NE Botswana

Annika C. Dahlberg and Piers M. Blaikie

INTRODUCTION

This is a schematic environmental history of a single village, Kalakamate in North East District, Botswana, over the main part of the present century. The paper interprets part of a body of primary data collected from 1990 to 1994, and focuses on specific aspects of the interface between social and environmental change, and on the challenges involved in analysing local people's interpretation of a changing environment. (The present paper essentially corresponds to the first six sections of paper III in the doctoral thesis by Dahlberg [1996]. Only a few recent references have been added.) The study mostly relies on interviews with older people, and these are corroborated, tested and interpreted using secondary data sources such as aerial photographs and records of rainfall, human population, and livestock. The current scientific debate about range ecology of semi-arid areas is highly relevant to the questions raised and forms part of our interpretation of the data. Clearly there are serious methodological and epistemological challenges in bringing together such sources. How are opinion and fact to be identified and differentiated? Under what circumstances should 'closure', i.e. a single and most credible interpretation, be sought, and when should a regime of additive pluralism be allowed to remain? When should one data source be privileged over another?

These problems are unavoidable when views grounded in different cultures engage. Often in this study, a resolution of seeming contradictions requires further detailed information which is simply unavailable, or the physical processes operating are not fully understood by anybody – scientist or farmer. However, on some issues the inclusion of information from different sources provides detail enough for a resolution to be approached with some certainty. The environmental history of this part of Africa is particularly interesting in the light of the post-colonial discourse on people and the environment (Dahlberg 1994, Thomas and Middleton 1994). Where colonial administrators and post-colonial consultants saw widespread degradation caused by overstocking and indigenous ignorance, contemporary range ecologists and rural sociologists more often see environmental resilience and successfully applied indigenous knowledge.

Environment and History **5** (1999): 127–174.

Annika C. Dahlberg and Piers M. Blaikie

For many areas of Southern Africa detailed environmental time series data are sparse. Aerial photographs seldom exist for periods before the 1950s, and for both early and late photo-series the area covered was often limited. In addition, photo-coverage is usually repeated only at intervals of a decade or more, making it problematic to differentiate between continuous change and short-term variation. Historical documents, when they exist, can be of great value. However, they seldom provide authoritative or detailed information on the environment. Ecosystem monitoring is becoming more common, but it is usually either very large-scale and/or conducted for a single land-use purpose, and in all cases time-series are still short. Also, the understanding of semi-arid ecosystems has changed fundamentally in recent years, with dominant ecological dynamics claimed by many to be event-driven and characterised by drastic events, random variation and resilience (Behnke and Scoones 1993). This puts new demands upon existing data, and renders our spatial and temporal gaps even more problematic, since the events themselves and subsequent changes may be obscured by the sparsity and irregularity of data points in space and time.

Rehearsals of data inadequacies in sub-Saharan Africa are of course commonplace, and provide few surprises. They are stated here to outline the opportunities and constraints involved in constructing an environmental history, and to explain the methods we adopt in the interpretation of diverse and contradictory information. Interpretations that focus on the local environment have been sought, and influencing factors from the social and physical sphere will be discussed in relation to observed and inferred changes of such environmental factors as availability and condition of land for grazing and farming (including crop productivity and soil fertility), water availability and signs of soil erosion.

A long time horizon assists us in assessing the significance and probability of the many different views and also the limitations of policy and regulation. Also, to the extent that the narratives of local farmers and new findings in range ecology are correct, a long time horizon is essential for most statements on environmental change in the semi-arid regions. One of our more unsettling conclusions is that the history of policy decisions, which were often based on now discredited hypotheses, is full of unexpected outcomes in how it affected this village. The most important variables which influenced environmental outcomes were either single events, both in the social sphere (e.g. the proclamation of the Protectorate, and later of Independence), and in the physical sphere (e.g. droughts), or policies enforced without sufficient attention to possible environmental impacts (e.g. prohibition of fires, erection of fences). The approach adopted has enabled us to explain many of the apparent contradictions, and to combine the diverse sources of data. In several cases, contradictory descriptions of environmental change were resolved when the spatial and temporal generalisations made were analysed. With this approach, many earlier assumptions of land degradation in the area can be questioned and, at least in some cases, refuted.

THEMATIC FRAMEWORK

Any environmental history requires an overall framework which identifies the main components of the environment, the society which perceives it and imbues it with symbols and meaning and which transforms it through time, and the processes acting upon both. Figure 1 identifies these components in a context specific to Botswana. They are mostly based on the narratives of the local people, but labelled by the researchers. We have also indicated the importance of separating different types of change (trends, cycles and events). Most environmental histories which choose to use both narrative and formal secondary sources have to interpret the interpretations of the informants, which in turn requires some exogenously constructed logical 'architecture', or thematic framework. Leach et al. (1997) also outline a similar epistemological approach. However, our differs in emphasis somewhat since it seeks 'closure' and convergence between

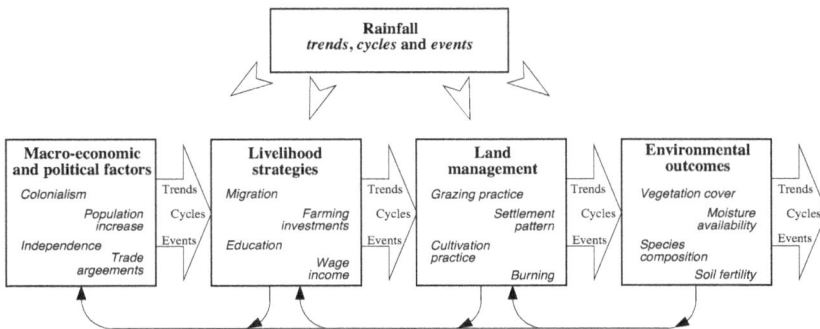

FIGURE 1. Components and processes of environmental change

Environmental outcomes can be characterised by, for example, species composition and frequency, vegetation cover, type and magnitude of soil erosion and moisture distribution. Changes may occur as events, cycles or trends as a result of natural and/or socially induced causes. People affect the environment directly through land management decisions concerning such practices as burning, cultivation, grazing of livestock and settlement pattern. The choice of specific practices depends on the physical and social setting, and variations in management may occur as isolated events, or as cycles or trends, depending on physical factors like rainfall, and changes in people's livelihood strategies. As the structure of a household changes over time, strategies may change. This is, however, often interrupted by events at the individual or family level which may have far-reaching implications for land management. Livelihood strategies also change along with general shifts in the community and wider social sphere. Macro-economic and political factors influence and shape patterns of change at the local level. Events may have a short-term or long-term influence, and cause different impacts depending on the local setting. A trend which may be clear at the national level may be difficult to identify at the local level. (See also Thomas & Sporton 1997.)

scientific and folk explanations, rather than to construct a 'counter-narrative' to dominant, scientific and western ones. This helps to resolve a number of contradictions in the attribution of cause and effect by informants themselves. Wherever possible, we constructed socio-environmental relationships from the informants' narratives. There is a constant danger that such interpretations reduce the independence and authenticity of the local narratives, and instead become a ventriloquisation by the researchers. The focus of this study is on the four main components: macro-economic and political factors, livelihood strategies, land management, and environmental outcomes – and on the linkages between them.

STUDY AREA

The North East District was chosen because historical documentation of this area, having been early settled by the British, was comparatively good. Environmentally the district is of special interest, following '...more than half a century of official lamentation about the condition of the range and predictions of imminent ecological disaster' (Fortmann 1989: 197). A description of the village of Kalakamate (fig. 2), has been presented in more detail elsewhere (Dahlberg 1995).

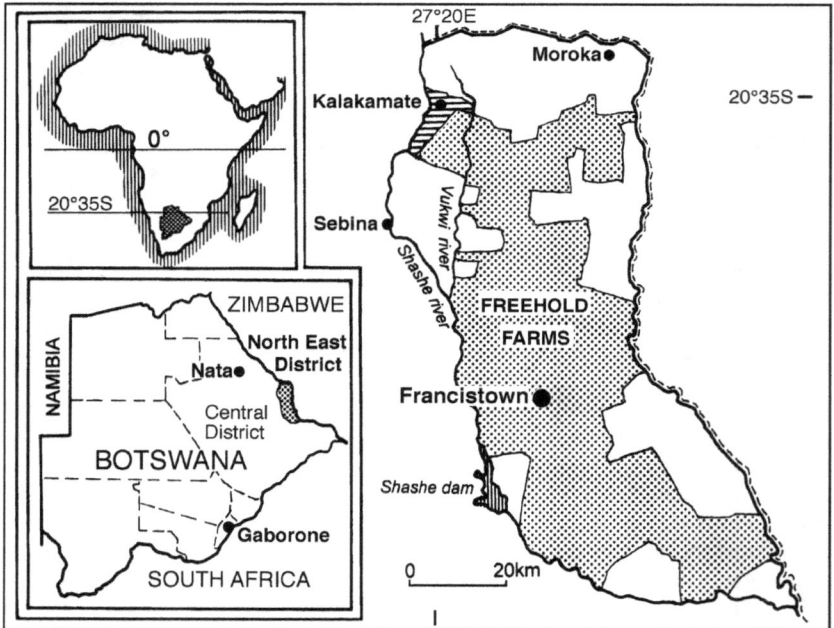

FIGURE 2. Botswana with North East District and the village of Kalakamate

Depending on the period selected and how the borders are defined, the area of Kalakamate has varied between today's approximately 70 km² and perhaps as much as 90 km². Traditionally the eastern and western borders were the seasonal rivers Vukwi and Shashe respectively. However, early this century a large tract of land to the east was acquired by the Tati Company, a private company founded when the Bechuanaland Protectorate was created in 1885. Until the 1950s this land was accessible to village livestock, but then the land was fenced, and in the late 1980s the village borders were changed again when a veterinary cordon fence was erected in the north and north-east. The tributaries of the two major rivers cut across a gently sloping land with scattered granite kopjes and ridges of dolerite intrusions and amphibole rocks (Botswana 1975). Soils have developed mainly from gneiss and granite, but also from basic rock, and may vary markedly over short distances. In most parts of the study area soils are shallow with a high content of gravel, moderately to well drained and with textures varying from loamy sand to sandy clay loam. Regosols, luvisols and leptosols are the common soil types, while narrow strips of alluvium are found along rivers and streams. To the south a large expanse of vertisol, a heavy dark clay, is found (Botswana 1984, Radcliffe 1990, and for local descriptions of soil types see Dahlberg 1996: III).

Rainfall is extremely varied, both between and within years, and the last 15 years have been characterised by recurring agricultural droughts. The area has long been settled by Kalanga-speaking people, agro-pastoralists whose history and language differ from that of the Tswana in the south (Werbner 1975). Since the turn of the century different Setswana-speaking groups, predominantly the Khurutse, have also settled in the village. All land is affected by human use, but areas not used for cultivation, now or recently, are covered by dense woodland and bush-savanna. *Colophospermum mopane* dominates among the woody species, with other common genera being *Acacia*, *Combretum*, *Commiphora*, *Terminalia* and *Grewia*. Common grasses belong to the genera *Aristida*, *Eragrostis*, *Digitaria*, *Tragus* and *Urochloa*.

METHODS OF DATA COLLECTION

Data were collected from archival documents, rainfall measurements, records of livestock and population, aerial photographs from 1964 and 1988 (Department of Surveys and Lands, Gaborone), field observations and interviews with 42 villagers, i.e. 6% of the village population. (Aerial photographs are available also for 1971 and 1981, and preliminary inspection of these indicates that the main results would not be influenced by their inclusion.)

Old people were targeted for the interviews (see table 1), and approximately 35% of the villagers aged 50 or more were interviewed. This naturally means that younger people's use and perception of the landscape, likely to be at odds

Annika C. Dahlberg and Piers M. Blaikie

Characteristics	Women	Men	Total
Residence:			
Born in Kalakamate – stayed there always	13	4	17
Born in Kalakamate – been away for period >10 yrs	1	10	11
To Kalakamate as child – stayed there always	2	–	2
Moved to Kalakamate at marriage	9	3	12
Occupation:			
Only (ever) a farmer	20	2	22
Farmer now – other occupation before	2	10	12
Farmer and civil service job	–	2	2
Farmer and manual job	3	2	5
Too old and sick to farm	–	1	1
Number interviewed	25	17	42
Median age	61	70	63

TABLE 1. Respondents (Kalakamate, 1992 and 1994)

with those of the elders, were not recorded. However, by concentrating on the old men and women, one can reach further back in time to identify the practices, events and circumstances likely to have had an impact on the transformation of the landscape over the past fifty or sixty years. A roughly equal number of men and women, as well as an even spatial distribution in terms of homesteads and fields, was aimed for. The interviews were conducted in the growing seasons of two hydrological years, 1991/92 (a dry year) and 1993/94 (a year with average rainfall), and the different characteristics of these years were taken into account during the interviews and when analysing responses.

The interviews were open-ended and often long, sometimes lasting many hours, in which case they would extend over a few days. Most were conducted at the fields, or when walking through the surrounding grazing land. A middle-aged man from the village, a former teacher, assisted as an interpreter in the two local languages Kalanga and Setswana. Efforts were made to avoid mis-understandings due to prejudice, suspicion, differences of culture and gender, as well as pure translation errors, but of course these will still have occurred. It should also be noted that the semi-structured nature of the interviews meant that different topics were highlighted with different respondents, and that on several issues responses were collected only from a part of the sample group. (For a list of respondents and questions see Dahlberg 1996: III.) What we present is our interpretation based on this procedure.

Aerial photographs can provide important and detailed information about environmental features and about past and present land use. With rapidly improving techniques for the interpretation and presentation of data, their value is even more enhanced. Black and white photographs for the study area were obtained from 1964 (*c.* 1:36,000), 1971, 1981 and 1988 (1:50,000). An advanced stereoscope (Interpretoskop B, Zeiss, Jena) giving a three-dimensional view of the landscape was used for interpretation. This way a high spatial resolution is achieved, allowing objects of 1-2 m to be detected. Different features and land-cover types were drawn directly on transparent overlays and checked in the field (1992 and 1994). For analysis and presentation, GIS software ARC/ INFO and ArcView were used to geocorrect the overlays (using the official topographic map over Botswana, scale 1:50,000, UTM grid coordinate system). To this end the overlays were scanned, and a number of ground control points identified which could be recognised in the photo as well as in the topographic map. These were then used to produce the transformation equation with which the overlays could be digitally converted to match the topographic map. For a complete match, height differences in the terrain must be taken into account, but this level of accuracy was not necessary for the present purposes.

So-called 'hard' data also entail an interpretation, and one should acknowledge the subjectivity of all actors involved in the assessment of a particular landscape and its chains of causes and effects. Scientists choose how to use and classify aerial photographs (for a discussion of problems involved see e.g. Larsson and Strömquist 1991, Dahlberg 1996), and how to order records of such parameters as population, livestock and rainfall. Colonial administrators chose when and where to travel, and what to record. Different groups have their distinctive cultural and professional repertoires, and when interpreting the data, the agendas of the actors have to be acknowledged.

THE INTERFACE BETWEEN ENVIRONMENT AND SOCIO-ECONOMIC CHANGE: LIVELIHOODS

The concept of livelihood strategy is useful for linking changes in environmental management to social change, and therefore also in studies of the causes of environmental change (Blaikie 1985, Blaikie et al. 1994). Here we focus on the access enjoyed by different farming households, and individuals within them, to a range of income opportunities. Of these, a selection is made to provide a portfolio which forms a 'livelihood' (numerous examples are presented e.g. in Scoones and Thompson 1994, and Steenhuijsen Piters 1995). Some of these income opportunities will involve the use of local natural resources (e.g. the cultivation of crops), while other will not (e.g. the migration of young males to sites of non-agricultural employment). Some will involve a cash transaction (e.g. wage labour or the sale of agricultural produce) and some will not (e.g. the collection of firewood for household use). In the present context it is important to understand the different livelihood strategies and their components since they all have significant, though sometimes indirect, impacts upon environmental management.

Annika C. Dahlberg and Piers M. Blaikie

The household livelihood strategy is also a useful concept for tracing the influence of more general socio-economic changes and events at the regional, national or international scale upon local changes in natural resource use. These changes are mediated through a range of local decision-making strategies which have environmental impacts. However, generalisations about processes aggregated to the national or even international level, although useful for communicating diverse experiences of a shared reality, are partly fictitious and should be treated with caution.

Population growth and livelihoods

One of the trends least mentioned by villagers, but which may affect their livelihoods, was the growth of human and livestock populations. These two factors are those which have dominated range ecology and neo-malthusian thinking, over the past sixty years or so. Increases in the populations of humans and livestock in the tropics and sub-tropics have often been assumed to result in increased pressure on natural resources, and thus to heighten the risk of land degradation. Although this may indeed be a credible explanation in some instances, the relationship is often far from clear, as recent studies of the Machakos District in Kenya have shown, where population pressure contributed to an intensification of land use and the incentive to invest in longer term conservation practices (Tiffen et al. 1994). A broader range of adaptations to population growth which include intensification, a rise in the value of land relative to labour, and new conservation technologies, have been empirically studied in a number of case studies of farming systems elsewhere in Africa (Turner et al. 1993). More specifically, it is important to differentiate between an overall population increase in a country or district, and an increase of population density in specific parts of these areas. Due to such factors as alienation of land (e.g. private farming or state control), migration, and urbanisation, the latter may differ substantially from the former.

The national population of Botswana has increased from a count of 120,778 in 1904 (Vanderpost 1992), to 1,326,796 in 1991, giving an average annual growth rate of 2.79% for the whole period, and 3.49% for the period 1981 to 1991 (CSO 1982, CSO 1992). The part of the country that later became North East District has experienced drastic changes in population densities (Tapela 1982). It is believed that agro-pastoralists and hunter-gatherers occupied the land for at least a thousand years. In the 19th century the land was heavily depopulated and population densities were very unstable due to war (ibid.). At times of calm people would move back, and be joined by groups from other areas, only to flee or move again in the face of renewed raids and land shortage. At the turn of the century the area of the present North East District was divided between company land, the Tati Concession, and the so called Tati Native Reserve, severely limiting local people's use of land for agriculture and grazing (Werbner 1970).

Changes in Landscape or in Interpretation?

Figure 3 (a–d) shows recorded changes in resident population at the scale of nation, district, chiefdom and village. These records should be treated with caution due to changing enumeration areas and lack of detailed data on migration (Tumkaya 1987). According to the records, the population of Kalakamate grew from 457 to 678 between 1964 and 1991, i.e. an average annual growth rate of 1.47% (BNA CENSUS 39/4, CSO 1992). This is much lower than the national and district average for the same period. It is also lower than the average for the Chiefdom of Ramokate, of which Kalakamate is a part, although these figures do not cover exactly the same period. Population density has increased in Kalakamate, due to a combined effect of population growth and a decrease of the area available for grazing and cultivation. However, density estimates for the northern part of the district are in the range of 25–35 persons/km^2 (NEDC 1986, Hoof and Maas 1991), whereas an estimate for Kalakamate is approximately 10 persons/km^2. A trend of population increase has occurred in all areas, but the slope of the trendline varies for the same, or roughly comparable, time periods. While an exponential trendline fits the national data, this is not as evident for smaller areas (fig. 3).

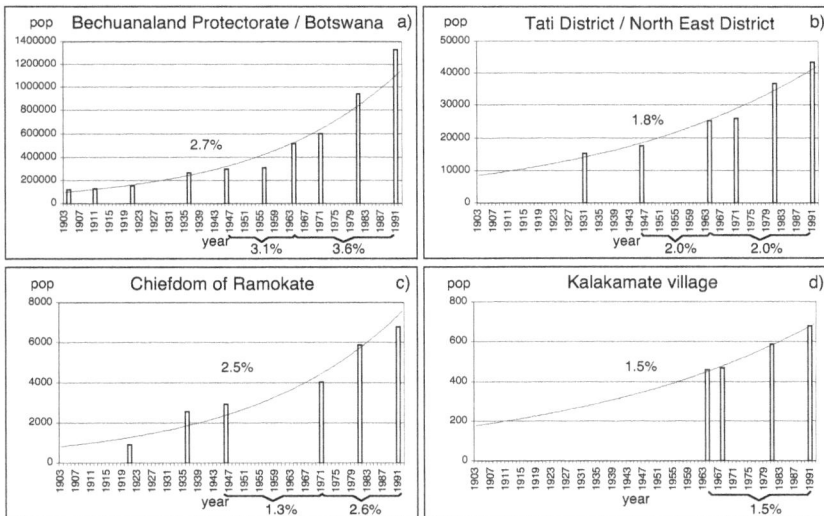

FIGURE 3 a–d. Change in resident human population in four areas.

To facilitate comparisons between the graphs, we have – in spite of the sparsity and unreliability of data – computed curves as a least squares exponential fit over the whole period considered, and the numbers next to the curves give the corresponding growth rate in percent per annum. We also give equivalent average annual growth rates as numbers below designated intervals on the time axis. In diagram B, numbers for 1931 and 1946 refer to Tati District (North East District after 1966) including Francistown (which then had a small population), while numbers for later years exclude Francistown. For sources see note 1.

Annika C. Dahlberg and Piers M. Blaikie

Even if population has shown an overall increase, variations between areas must be considered in any explanation of perceived environmental change. Locally, changes in livelihoods, including seasonal and temporary migration, have had a profound impact, arguably greater than an aggregate population increase. Male migration to the mines in South Africa was especially high during the first half of the century (Schapera 1947), but fluctuated widely depending on the economic and political situation (Tumkaya 1987). This was characteristic also for Kalakamate, where 70% of the interviewed men had worked outside the area for periods varying between 7 and 46 years, and as many as 40% had been away for more than 20 years. In Kalakamate, as elsewhere, the effect of male migration on local land use will vary, depending on factors such as length of absence, frequency of visits home, age and sex of remaining family members and local sources of income (cf. Hesselberg and Wikan 1982).

The majority of the interviewed men were able to return to the village every year during the ploughing season, for periods varying between one and three months. There was no indication that migration had changed such cultivation practices as type of crops, or methods of ploughing and planting, and remittances were seldom used for such agricultural improvements as fertiliser, fencing material or improved seeds. However, money from wages did add to food security, especially necessary in times when the household could not plough as much land as needed or at the most favourable time, e.g. because of labour shortage. Wages were also used to pay for help with clearing, ploughing and harvesting, and especially to build up the household herd of livestock. The latter served as an insurance, and also improved a household's chances of better yields. Concerning increased opportunities to acquire cattle, the aggregate statistics of cattle numbers discussed below suggest that the ability to increase herds on a stable basis was limited to a few households, and did not result in a trend of rising cattle numbers in the village. It should be noted that households also utilised other strategies to overcome the labour shortage resulting from the husband's absence. The wife may stay with her parents-in-law and help in their field, or receive help from them, her own parents or other relatives to plough her own field.

Since Independence in 1966 there has been a marked drop in migration to South Africa (CSO 1991), and in 1991 only 1% of the population in Kalakamate was listed as working abroad (CSO 1994), while instead both men and women look for work in the towns of Botswana. This is a further change from earlier times, when only men migrated, while the women stayed at home looking after fields and livestock. Throughout Botswana livelihood strategies have changed, reflecting a diversification of income sources away from farming and livestock, and a generational change in perception of the role of agriculture. It is common today to hear villagers say that in dry years they do not bother to plant, that at the most they will plant only a small area, and that the young people lack an interest in farming. Instead, wages and drought relief are used to buy the necessary staples. These changes have two important environmental impacts. Land management has changed, not so much in actual practices (e.g. what crops people grow) as in patterns over space and time (e.g. where and when to cultivate), and this may have direct and indirect influence on the environment. Furthermore, changes in livelihood strategies influence the way people perceive and value the environment, and thus environmental changes occurring today cannot be judged according to criteria of good and bad used several decades ago (and vice versa).

Livestock numbers and livelihoods

Turning now to changes in stocking populations and densities, the differences between national and regional trends and local variation are very pronounced. At a national level the cattle herd of the communal rangelands has increased at least eight-fold since the turn of the century (Abel 1992). Dramatic fluctuations, caused by drought, diseases and market factors are superimposed on this trend of increasing cattle numbers (Roe 1980). Figure 4 (a–d) shows recorded cattle numbers for years with available records for different areas in north-eastern Botswana. Approximate geographical location and size of these areas are shown in figure 5. These data should be treated with caution. As sources differ, so do methods and timing of counting, the area included in a certain designated region may vary substantially over time, and the records make little mention of such events as eviction of cattle, temporary removal to cattle posts and freehold lands, or the opening up of new grazing areas (see notes to fig. 4). Furthermore, many archival sources are vague about exact dates, and it is sometimes impossible to be sure to which year certain cattle numbers refer.

As a first example of this uncertainty, Francistown Veterinary Region (later called Francistown Agricultural Region) covers both communal land and private ranches, in both North East and Central Districts, and also includes an extensive salt pan. The area and definition of this region has varied widely over time. Furthermore, commercial ranching has gradually become more important over the last four to five decades due to the eradication of tsetse fly, the drilling of wells, and new policies favourable to cattle rearing. The dramatic increase for the whole area therefore says little about changes in stocking rates on communal land. The other areas shown in figures 4 and 5 are all communal land, while Kalakamate village is part of Ramokate's chiefdom, which in turn is part of the former Tati Native Reserve, i.e. the area in northern North East District not taken up by freehold land. It should be noted that cattle numbers presented for these areas may actually lead to an overestimate of the grazing pressure on communal land, since for at least part of the year many of these animals would be grazing on land belonging to the Tati Company, either illegally or by rental agreements (Schapera 1971). In these areas any trend of increasing cattle numbers is hard to discern. It may be noted that whereas in 1936 the Tati Native Reserve held 34% of the cattle of Francistown Veterinary Region, this proportion has fallen

Annika C. Dahlberg and Piers M. Blaikie

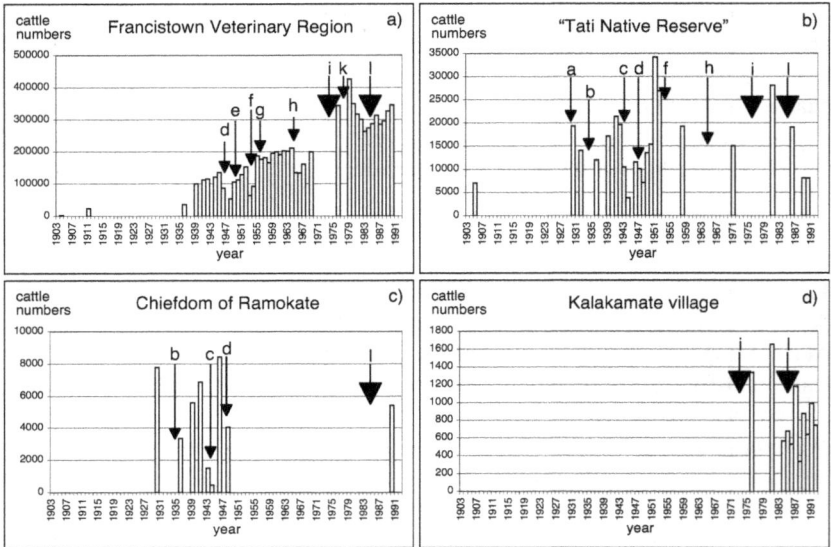

FIGURE 4 a–d. Change in cattle numbers in four areas

Numbers for the area previously called Tati Native Reserve 1981, 1986 and 1989, and for Kalakamate 1981 are estimates based on available numbers given for individual communal areas. For sources see note 2. Arrows and lower-case letters represent events which had a strong effect on cattle numbers:

a) 1930: Cattle numbers for Tati Native Reserve include cattle kept at cattle posts in Matsiloje, an area SE of Francistown – but these did not exceed 3000 (Kinlund 1995).

b) 1933–35: Severe droughts, high cattle mortality (Roe 1980).

c) Nov–Dec 1943: Enforced destocking of cattle in the Tati Native Reserve. 6,732 head of cattle were removed from the whole reserve, of which 1,065 were removed from Ramokate's area (BNA S.238/9).

d) 1946–47: Severe drought with heavy loss of cattle. Also some voluntary removal of cattle from the reserve (BNA S.238/7/2).

e) 1949: The colonial government started up ranching ventures around Nata (Blair Rains and McKay 1968).

f) 1951/52 and 1953/54: Severe droughts.

g) 1955 to 1956: The area of the Bokalaka (i.e. between Tutume and the border to Southern Rhodesia [Zimbabwe] and the Tati District) was included in the Francistown Veterinary Region (Fortmann et al. 1983). From 1956 and onward the pace of borehole development in the central and western areas of the Francistown Veterinary Region increased (Blair Rains and McKay 1968, Roe 1980).

h) 1963–65: Severe droughts.

i) 1970s: Almost all years gave abundant rainfall. Also, Nata Crown Lands were included in the Francistown Veterinary Region (Fortmann et al. 1983).

k) 1975–1978: Areas around Nata were allocated for commercial ranches within the TGLP scheme (i.e. the Tribal Grazing Land Policy) (Sandford 1980, White 1993).

l) 1982–1989: Almost all years with below average rainfall, severe agricultural droughts.

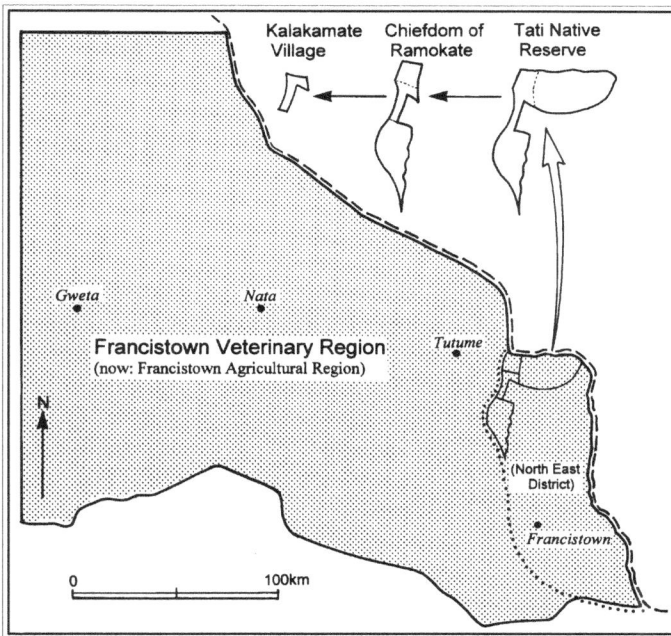

FIGURE 5. Areas referred to in figure 4

The grey area shows the whole Francistown Veterinary Region, the broken straight line shows the national border, while the dotted line shows the district boundary.

drastically and (except for fluctuations in the 1940s) fairly evenly to 9% in 1981, 6% in 1986 and 2% in 1989.

An environmental impact ascribed to increasing cattle numbers in one area cannot simply be extrapolated to others. Neither can changes in livelihood opportunities through time be inferred in a general manner. It is well established that ownership of cattle varies very much among households in Botswana. Most households own no or very few cattle, while some own large herds, and it is those with few who are likely to suffer worst in a drought or epidemic (Roe 1980, Sandford 1980, White 1993). The local history, as told by villagers in Kalakamate, relates how fortunes have fluctuated over time. All but one of those who owned cattle at the time of the interviews, or who previously had owned cattle, stated that over the last decades their herds had declined, often by 70%–100%. Many animals had died in the recent droughts in the 1980s and in 1992, but other unrecouped losses, experienced in droughts in the 1940s and 1960s, were also recounted. Villagers perceive lack of rainfall as the main

Annika C. Dahlberg and Piers M. Blaikie

cause of death of cattle, but they were also concerned about shortage of grazing land. However, this is not seen as caused by any increase in cattle numbers, but by the alienation of grazing land through the fencing of freehold land, by the veterinary cordon fence, and to some extent by reduced accessibility to grazing lands because of the recent centralisation of the settlement, as discussed below.

Changing land use and livelihoods

In many parts of eastern Botswana, the common practice since at least the mid-19th century until shortly after Independence was large and nucleated settlements. Land use was divided, with cultivated areas situated at a distance from the settlement, and pastures located even further away (Silitshena 1979). Since Independence there has been a move back to the lands, with people staying permanently in small scattered settlements close to their fields and cattle-posts (Silitshena 1982). The ways in which these spatial patterns change differ throughout the country and cannot be assumed, or extrapolated from findings in other areas (see e.g. Hesselberg 1985). As discussed further on, the exact nature of these patterns can have a strong impact on the local environment, and on how environmental change is perceived.

Land use in Kalakamate can be characterised as a mixed farming regime, where livestock holding is combined with cultivation, and where people depend to a large extent on local natural resources for building material, fuel and supplementary foods. Most households work the land for crops, although few (if any) are solely dependent on agriculture. In agreement with other studies (Roe 1980), respondents state that farming practices have changed very little over the last sixty years. Tractors are still uncommon, the application of kraal-manure and fertiliser remains virtually nil, soil conservation measures are basically restricted to contour ploughing, only a couple of improved crop varieties are sown, and it is only in the last decade that wire fencing of fields has started to become common.

While reported farming practice may have changed little, its spatial pattern within the village has changed considerably, and in some ways in the opposite direction from what has happened in areas further south, mainly settled by Setswana-speaking groups. The traditional settlement pattern among the Kalanga of the north-east was one of extended families with homesteads scattered throughout the area (Werbner 1970, 1975). Each cluster of huts was surrounded by the families' fields, and livestock was herded on the nearby grazing land.

As part of the process of improving livelihoods after Independence, the state initiated and strongly encouraged the nucleation of settlements, to facilitate access to such amenities as water, schools and clinics. This, in combination with other outside interventions, resulted in a change of the spatial pattern of natural resource utilisation. The change in settlement and cultivation pattern in Kalakamate between 1964 and 1988, as interpreted from aerial photographs, is shown in figure 6 (a–b). Classifications (also for 1988) are based on what could be identified in the photographs to make the comparison as unbiased as possible. Thus for 1988 the class 'field abandoned some 5–20 years ago' will include some, but not all, of the fields cultivated in 1964 and later abandoned.

Individual compounds are usually clearly visible in the aerial photographs, although the number of households per compound cannot be determined. Compounds were counted throughout the village (table 2). Two areas have experienced an increase in the number of settlements: the area around the kgotla (the chief's office and the village meeting place), where the school, the clinic and other services are now located (area indicated by arrows A in fig. 6); and the area by the crossroads just south of this (B). From here a tarred road runs south, and along this (area C), as also further south (area D, along the old main road), the number of compounds have decreased. The area to the east (E) has lost all permanent habitations. Of the two compounds found in area E in the 1988 photographs, one provides temporary housing during the growing season, and the other had been completely abandoned by 1992. Hesselberg (1985), studying two other villages in north-eastern Botswana, found a similar pattern of movement to the main village, especially by the poor and landless and by the most wealthy who can employ extra labour.

Areas designated as 'fields not cultivated last few years' are easily identified in the photographs since here woody vegetation is very sparse and fences are still clearly visible. However, one cannot determine whether these fields are abandoned, or simply left fallow for one or a few years because of lack of rainfall or other constraints. Fallow here means resting, i.e. periods without cultivation. Areas classified as 'field abandoned some 5–20 years ago' are former fields left fallow for more than approximately five years. These areas have become so overgrown with dense thornbush that people very rarely invest the labour needed to clear them. Such land is usually left for periods of 20

Location	Number of compounds	
	1964	*1988*
near kgotla (A)	26	76
at crossroads (B)	19	40
area just south of B (C)	13	11
area even further south (D)	25	15
area to the east (E)	28	2
Total	111	144

TABLE 2. Location and number of compounds in 1964 and 1988

Letters in brackets refer to the main text and figure 6.

Annika C. Dahlberg and Piers M. Blaikie

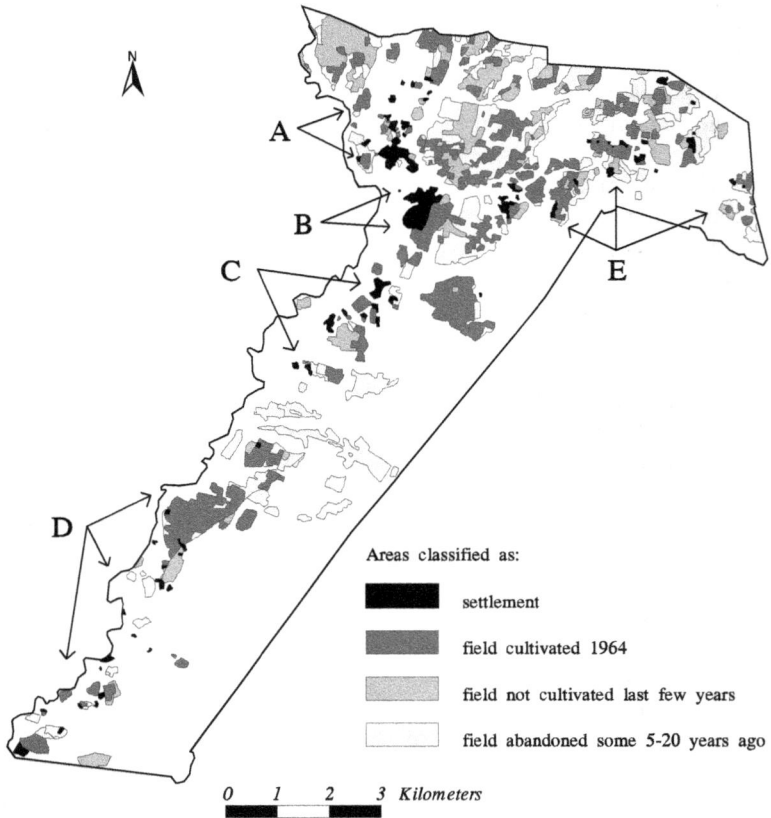

FIGURE 6 a. Settlements and cultivated land in Kalakamate 1964.

Figs. 6a and 6b are based on interpretation of aerial photographs, and are geocorrected using the official topographic map over Botswana (scale 1:50,000, UTM grid coordinate system). In both figures the remaining unmarked white area is classified as grazing land, i.e. areas with varying densities of woody vegetation and grass cover.

to 40 years or longer, until the thickets of thornbush have been replaced by a more open shrub- and woodland. These areas can often be identified on aerial photographs many decades after abandonment. However, as can be seen in the maps, some areas ploughed in 1964 can no longer be identified as fields. This is mainly because in some areas, usually with soils derived from basic rock, the secondary vegetation quickly merges with the surrounding woodland. The

Changes in Landscape or in Interpretation?

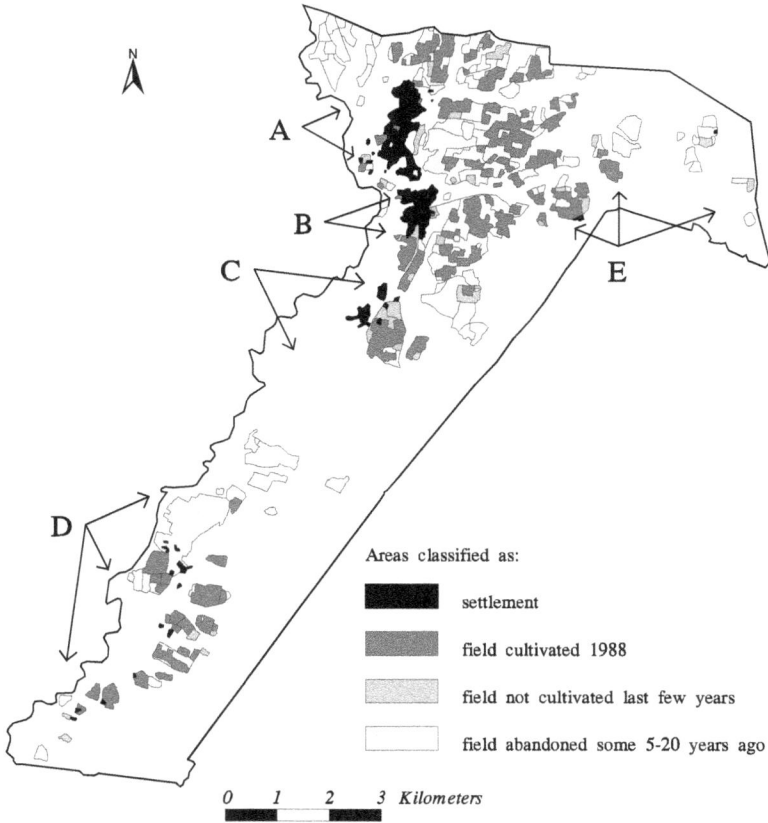

FIGURE 6 b. Settlements and cultivated land in Kalakamate 1988

remaining unmarked white areas in figure 6 were classified as grazing land, and analysis of characteristics and changes in these areas are presented in section 7 of Dahlberg (1996: III).

In general, proximity to the borehole, the school and other services, was the main reason for building a new homestead in or close to the main settlement, and with the move of settlement followed a move of areas of cultivation. To have one's field close to the homestead reduces time and labour input in working the field, and makes it much easier to guard the crops against birds, baboons and livestock. Many respondents told about how they had cleared new land after having moved their compound, i.e. it was a labour investment most found worthwhile. This has resulted in some areas having been almost completely abandoned, with settlement and cultivation now concentrated to two quite distinct zones (see fig. 6). Additional reasons why land in the eastern part of the village has been all but completely abandoned were presumably the erection of the private fence, and later of the veterinary cordon fence in the north and east. These effectively cut off the possibility of watering cattle in the Vukwi river, the only major water source in this part of the village. Cattle are used for ploughing, and also at other times of the year people prefer to keep them fairly close to the homestead, which is problematic without close access to water.

Surveys and reports discussing environmental change at national and district level emphasise the growing shortage of land for cultivation and grazing (Fortmann et al. 1983, Arntzen 1986, Asselman 1986, DTRP 1987). In many areas, especially close to towns and large villages, land pressure (and shortage) is undoubtedly a factor with important environmental implications. In Kalakamate, and possibly in other less densely populated areas, constraints connected with land shortage are different. Villagers in Kalakamate also complain of land shortage, but what is implied here is a wish for land which has been allowed to lie fallow for a suitable time, and which is situated at convenient walking distance from the homestead. The recent nucleation of settlement and fields has meant that land fulfilling these criteria is becoming scarce, a situation different from one where land with potential for cultivation actually does not exist. Differences in actual and perceived land shortage, in different areas or at different spatial scales, will have different implications for the local environment and for future land use.

A few major events at the local and national level deserve special attention. As described by Roe (1980), the colonial government set aside funds for water development, with the primary target of securing water for cattle. A colonial report from 1943 states that in the part of the native reserve to which Kalakamate belonged, the chief, aided by Government funds and local labour, had constructed one dam for the watering of cattle, and sunk three new wells through bedrock (Thornton 1943). Three years later, and referring specifically to Kalakamate (BNA S.238/7/1), it is described how both villagers and cattle use water from the rivers and streams during the wet season. When these dried up, usually at the end of June, there were plentiful wells, or pits, dug in the banks of the rivers and streams, which provided water. All in all the officer counted 17 wells in the village, dug by the people themselves, '[In this area] water is plentiful and stock are not driven long distances over the grazing ground.' (ibid.: 3). Villagers gave a similar description of the water situation in the past, and stated that water, at least for human consumption, was to be found in most years. However, water availability became even more secure after Independence, when a well was drilled and a dam constructed near the main settlement. This easy and secure access to water throughout the year, for both people and cattle, has influenced livelihood strategies, as well as land-use patterns and the environment.

Grazing practice, or rather the spatial pattern of grazing, has of course been affected by the changes in settlement pattern and water availability described above. Traditionally people grazed their cattle in and close to the village, but it was also common, at least among those with large herds, to keep some animals at remote cattle posts. However, most respondents stated that today, even if they could find trustworthy people to look after their animals, and afford to pay them, their herds are now too small to send any animals away. The location of homesteads and access to water have always been important for determining where cattle are to be grazed. The nucleation of settlements, and the location of the borehole and a dam nearby, has thus contributed to a concentration of cattle. The erection of private and veterinary fences has emphasised this even further. Previously cattle could move freely in all directions outside the village boundary, but now grazing opportunities to the north and east have been curtailed by outside intervention.

According to villagers, access to pasture has also been affected by regulations concerning the burning of grazing lands. In 1977 the Herbages Preservation Act (also termed Prevention of Fires Act) was laid down by the Government, as one of several laws intended to protect the range, including wildlife populations (Vegten 1979). The law states that nobody is allowed to burn grass in communal areas (Arntzen 1985). This, together with other regulations intended to preserve natural resources on communal land, is supposed to be enforced by institutions such as the Land Boards and the Agricultural Resource Boards. To a large extent this is, as yet, not an active part of their agenda, and in most areas there exists little control over management (Gulbrandsen 1984, Arntzen 1985). However, villagers in Kalakamate were well aware of this law, and claimed they adhere to it, in spite of its perceived negative impact on grass cover. It was stated that people hardly ever start fires any more, except when clearing a field, and that sometimes such fires may spread by mistake.

Concluding remarks on livelihood changes

Following the schematic presentation of components and processes of environmental change presented in figure 1, the present section has described changes within the social sphere which could be expected to have an impact on the environment. The importance of separating spatial scales has also been stressed. For example, population increase has occurred at all levels, from national to village, but the rate of increase varies, as does density distribution within any particular landscape. Livestock numbers have increased steadily at national and regional level. However, on communal lands in the former Tati Native Reserve, and more specifically within the study area itself, no trend is evident. Instead, drastic fluctuations caused by outside intervention, rainfall events and shifts in livelihood strategies seem to have been the norm. Agricultural practices have changed very little during this century, while the spatial pattern of cultivation and grazing has changed more.

During the colonial period what is now Botswana was characterised by a low rate of infrastructural investment. The peaceful transition to Independence in 1966, the discovery of diamonds, generous development aid and favourable trade agreements with Europe resulted in rapid changes which have had a marked impact on land use and environment. In Kalakamate, improvements in education, health care and infrastructure, as well as other changes in livelihood opportunities and strategies, follow district and national statistics, albeit at a slower pace than in urban areas (CSO 1994). Spatial distribution of settlements changed, contributing to other changes in land use. Villagers commented on how the diversification of income opportunities has reduced the dominant role of agriculture, livestock and other local natural resources in their lives, and also the knowledge and skills associated with the use of these resources. Modernisation of roads and transport, and a new spatial pattern of settlement, cultivation and grazing, have changed how people move through the landscape, further contributing to altering perceptions of the local environment.

MULTIPLE INTERPRETATIONS OF RAINFALL

After the above discussion of trends and events in the social sphere which have influenced livelihood strategies, land management and the environment, we now turn to a discussion of how rainfall pattern and events are perceived to have influenced these same components. In this semi-arid region, rainfall is characterised as low and extremely erratic, as a dominant driving force of ecosystem dynamics, and as having a major impact on the agricultural and pastoral aspects of livelihood strategies and therefore upon local land management strategies and decisions (Garanganga et al. 1994). However, farmers and scientists observe and measure the amount and distribution of rainfall in different ways, and they also often interpret the environmental impacts of rainfall variation differently. Views and explanations vary also among villagers and scientists themselves, and within both groups these change with time. This study reveals examples of consensus, as well as of contradictions. The latter are often more interesting, and an approach was adopted to explore the possibilities of a convergence of initially contradictory views.

The respondents, most of them farmers, described rainfall as the factor of overriding importance when explaining environmental change, and placed any discussion of their land management practices firmly within the context of rainfall. Similarly, range ecologists and others have long recognised that rainfall – along with environmental parameters such as soil type, and management practices such as burning and grazing pressure – is an important aspect of the ecosystem dynamics of semi-arid environments (Skarpe 1992). Until recently, most modelling of environmental change in semi-arid areas assumed a condition of equilibrium which depended on stable and predictable relationships between variables through time (e.g. when studying vegetation succession, carrying capacity and land degradation). By applying the 'correct' management practices, the dominant paradigm ran, carrying capacity could be sustained and the process of vegetation succession controlled to suit a particular set of management objectives (see discussion in Abel and Blaikie 1989). Traditionally, certain environmental changes such as bush encroachment, a decrease of grass cover or specific changes in species composition, were seen as sufficient indicators that the land was being mismanaged and degrading (e.g. Walter 1985).

Over the past ten to twenty years, many of these assumptions have been re-evaluated, and a new understanding of semi-arid ecosystem dynamics is gaining ground, where outcomes at particular points in time and space are seen as the results of particular events, such as drought (Behnke et al. 1993). This approach has been found to be especially valid in areas characterised by low average annual rainfall and high inter-annual variation (Shepherd and Caughley 1987, Ellis and Swift 1988). Here, climatic variability (such as rainfall fluctuations and perturbations through drought) are seen to be the main factors controlling plant biomass, livestock numbers and human strategies for exploiting the environment (Ellis et al. 1993, Scoones 1995). The concept of long-term equilibrium, broken only by human management practices, has been replaced by one where a non-equilibrium dynamic is identified as an inherent driving force of these ecological systems. Although much of the ecological theory behind this paradigm-shift is not new (see e.g. Holling 1973), its application to semi-arid rangelands is quite recent. This has come about through a re-examination of physical data, partly through the recognition of the unsatisfactory predictive performance of equilibrium type modelling (e.g. Westoby et al. 1989), and partly through a wider recognition of indigenous knowledge systems and practices – and perhaps most importantly, the rather poor record of range management initiatives (see e.g. Sandford 1983, Behnke et al. 1993).

Farmers' recollections

Villagers and scientists agree on the importance of rainfall amount and variability. However, agreement about a local chronology of rainfall events, including long-term patterns of change, is often – but not surprisingly – lacking. Almost all villagers claimed that rainfall has declined since their youth. It was stated that 'before', rainfall varied less between years, and annual amounts were, on average, much higher than at present. Most said that random variation, between and within years, has always been the main characteristic of rainfall, but stressed that underlying this a downward trend had occurred, '...droughts have become more common', '...but the heaviest rains fell in the 1940s', and '...the rain has become more unreliable than before'. This perceived decline was also illustrated

Annika C. Dahlberg and Piers M. Blaikie

and substantiated by descriptions of primary and secondary effects on natural resources and land use, and by evidence of changes in moisture availability (see table 4).

Initially many stated there were no droughts in 'the old days', but would later recount stories told by their parents and grandparents, of droughts so bad that '...people had to eat the skins of animals', and 'there was one drought that was extremely bad and it was called "don't ask me"'. In their construction of a rainfall chronology, hardships experienced during droughts in the '40s, '60s, '70s, '80s and '90s were described. Some years were described as bad by all, e.g. the extreme drought in 1946/47, while other drought events were only remembered by some. The perception of an overall decline in rainfall was not shared by all respondents, there were a few who denied the existence of any trend, 'This is the way it has always been; some years we get enough rainfall, whereas other years there is too little or none'. It is the view of these few informants which is in accordance with official rainfall records.

Official records

The use of official rainfall records poses several problems. With rainfall variation in space in any one year being extremely high, it is important to use data recorded as near the study area as possible. On the other hand, reliability and long time series are equally important. In Botswana only a few stations have recorded rainfall since the beginning of this century, and it is these records that are most reliable and complete (Department of Meteorological Services, Gaborone). For the study area, Francistown is the nearest such station, situated 66 km from Kalakamate. Records from rain gauges located closer to Kalakamate go back only to the late 1950s or later, and contain gaps in the data of months and even years. In Sebina, located 26 km from Kalakamate (and 48 km from Francistown, see fig. 2), rainfall has been recorded since 1958/59, and these records are almost complete. In figure 7 (a–d) and table 3 these data are compared with those from Francistown.

In Botswana the rainy season covers the period October to April, with very little rain falling in the remaining months. Therefore, throughout this study, rainfall is presented and compared as annual rainfall by hydrological year, i.e. July–June. When discussing effects of rainfall variation on crops, pasture and other natural resources the intra-annual distribution of rainfall is extremely important (Vossen 1987). This is discussed further later on. The comparison between Sebina and Francistown in figure 7 is therefore given for annual totals and separately for the periods of early, mid and late rainy season.

In table 3, the median, mean, and coefficient of variation (CV) of annual rainfall for Sebina and Francistown are presented. For Francistown these values are given for the periods 1922/23–1993/94 and 1958/59–1993/94 respectively. These values indicate that the average rainfall has been almost identical, with

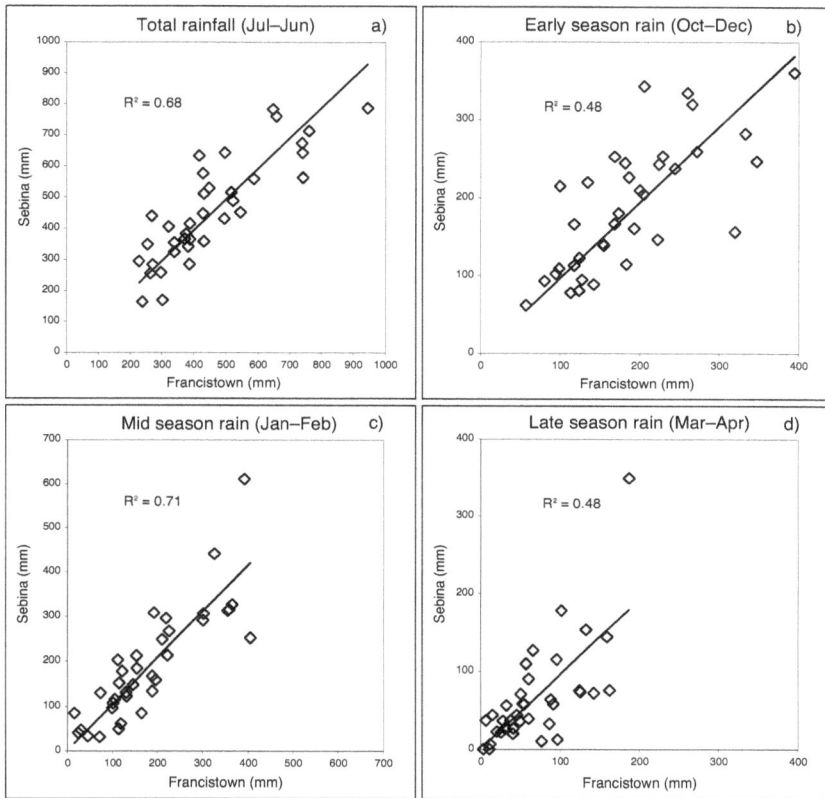

FIGURE 7 a–d. Correlation of rainfall recorded at Sebina and Francistown for the period 1958/59 to 1993/94

	Francistown 1922/23–1993/94	Francistown 1958/59–1993/94	Sebina 1958/59–1993/94
Median (mm)	426.4	423.4	435.3
Mean (mm)	458.4	456.2	458.5
CV (%)	39	38	37

TABLE 3. Median, mean and coefficient of variation (CV, i.e. standard deviation/mean) of annual rainfall at Sebina and Francistown

Annual rainfall is measured by hydrological year July–June.

Annika C. Dahlberg and Piers M. Blaikie

a slightly smaller spread for Sebina. Figure 7 shows that, considering the high inter-annual variability of rainfall, there is a reasonable correlation between the rainfall data from the two locations. (It may be noted that during the last 20 years the agreement has been better. The difference has been less than 10% in 10 years, and 10–20% in 8 years. However, in the recent dry years 1991/92 and 1993/94 the difference was great, with very low values from Sebina. Also, there are considerable local variations in the intra-annual rainfall distribution, as seen in figure 7.) Although it is recognised that local rainfall on a daily, and sometimes even monthly, basis will often deviate from data recorded at a nearby location, the comparisons presented here show that for longer periods the general pattern of variation is comparable between these particular stations. Thus, for the purpose of this study it was considered most appropriate to use rainfall records from Francistown, since they cover a longer time and can be expected to be more reliable.

In figure 8 annual rainfall data from Francistown for the period 1922/23–1993/94 are presented in more detail. The intra-annual distribution is shown by dividing the rainy season into early (Oct–Dec), mid (Jan–Feb) and late (Mar–Apr) rainy season, and adding the rainfall before and after the rainy season at either end of the bars.

Figure 9 shows the five year running mean of annual rainfall for the same period, with the calculated linear trend. The data from Francistown are in accordance with regional studies, where statistical analyses of rainfall records for the last hundred years and more have shown no trend of declining, or increasing, annual rainfall in Southern Africa (Lindesay and Vogel 1990). Instead a cyclic

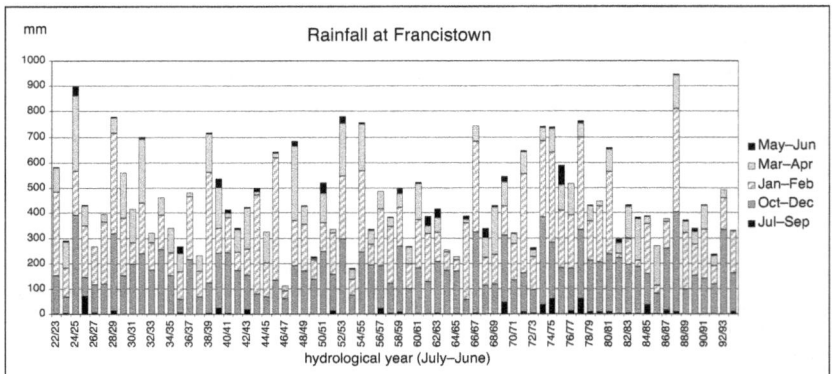

FIGURE 8. Annual rainfall (July–June) and intra-annual distribution of rainfall at Francistown for the period 1922/23–1993/94

The grey-scale codes for periods Jul–Sep and May–Jun are the same.

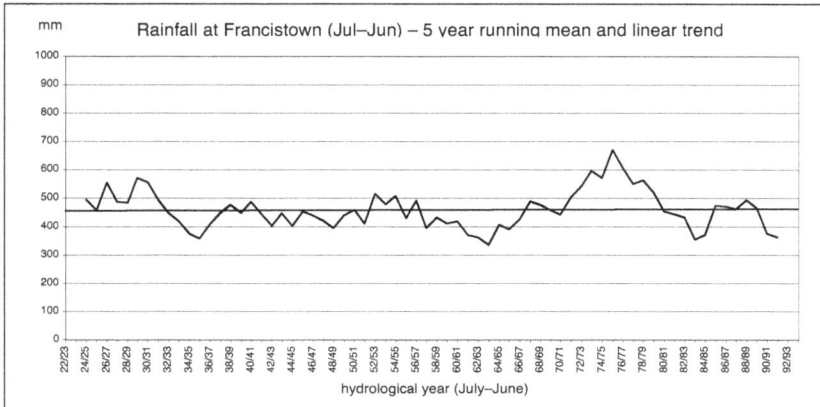

FIGURE 9. Five year running mean (centred) and linear trend of annual rainfall for Francistown 1922/23–1993/94

pattern of rainfall fluctuations, with a quasi-oscillation of 18 years between alternating wet and dry decades, is indicated (Tyson 1986). Existing rainfall records for Botswana, dating back to the early 1900s, show the same pattern, especially for the eastern part of the country (Tyson 1979, Bhalotra 1985). However, it should be noted that within the cycles, a high random variation occurs on a seasonal and annual basis (Tyson 1986).

Dealing with contradictions: a method towards achieving closure

The assessments of rainfall variation over time presented above are contradictory on several counts, both when comparing farmers' views with official records, and when comparing perceptions among the respondents. To resolve these contradictions completely, more data would be needed. The high spatial variability of rainfall calls for precipitation measurements from the study area, and even from different locations within this area. Verification of the change in moisture availability and its effects as described by farmers would necessitate records of local physical and socio-economic parameters. Lacking these, a methodology towards achieving closure (i.e. a single and most credible intrepretation) must build on a discussion of socio-environmental relationships, in the context of what the stated observations of rainfall really measure. (See also Leach et al. 1997, for a review and critique of different approaches to contested and diverse interpretations of environmental change.)

Annika C. Dahlberg and Piers M. Blaikie

While scientists measure rainfall as an exact amount at a specific place and time, to local farmers rainfall, and its variation over time and space, is perceived in relation to what it can afford them, i.e. what the effects of rainfall are for them in terms of their livelihood strategy. The impact of rainfall is evaluated for each particular set of socio-economic circumstances, i.e. in relation to a particular livelihood opportunity, or rather to the specific livelihood strategy that an individual can manage at a particular point in time. In addition, we suspect that rainfall is often used as a metaphor for a general change in livelihood circumstances. Thus, a statement of declining rainfall may not primarily reflect a particular environmental change, but instead a feeling of regret for 'the good old days'. This is especially likely when the respondents are older people, who have lived through dramatic socio-economic and political upheavals. This is the **first** possibility for explaining contradictions between the local and the official – local people think that they are stating empirical facts, but they are not, and scientific and official versions must be the arbiter for discovering the truth. Certainly it is problematic to gauge to what extent local people are wrong, and it brings up a number of post-modern challenges over representation, authorship and the privileging of knowledge claims (Rosenau 1992).

Put simply, we all invent narratives, which are our own versions of a complex reality, and our cultural and professional repertoire is usually well stocked with them. Their function is partly to reduce uncertainty and ambivalence, which often necessitates that reality is simplified, e.g. through a reduction of influencing factors and interdependences. Usually, the authors' unstated agenda is to work out an acceptable version of their own relations to the subject matter, and therefore explicitly or implicitly put themselves into the narrative. The other important aspect of narratives is that they are repeated over and over again, becoming stereotypes and formulas for coping with contradictions, uncertainty, and a state of flux. They are also adapted to the views of others and to external events, and although they form a stable set of assumptions, they are seldom static.

Older people the world over have their own narratives, so much so that the narratives often become the stereotype by which their authors are perceived. Older people in Botswana are no exception, as they try to make sense of their ageing and their changing relationships to family and wider society. The 'good old days' is one typical narrative, and the 'ignorance or fecklessness of youth' is another. Both appear in older people's versions of the environmental history of Kalakamate. However, here we treat informants' accounts as both narrative and as a version of a shared truth about the environment. Thus, when most respondents say that rainfall has declined, this might be expected – that is what they would say. The next step is not to dismiss their accounts with incredulity and attribute them merely to a narrative of ageing and change, but to check them with other narratives, including scientific ones. This would include the adaptation of narratives said to describe longer periods to recent events and short-term 'trends'. For example, in this particular case it must be stressed that the interviews took place after a period of over a decade characterised by droughts (see fig. 8). Furthermore, this period had followed on a decade with several years of above average rainfall. Although efforts were made during the interviews to avoid such biases, the low rainfall experienced over the last 15 years is likely to have influenced perceptions of long-term changes.

While the first possibility refers to what may be called imperfect storage of observational data, a **second** possibility is that the changes perceived are due to the informant (or the researcher) not fully realising the influence of differences in 'measuring conditions' then and now. With different measuring conditions we mean for instance that observations 'then' were governed by the interests, needs and capabilities of a young person, while observations 'now' are those of an old person. Apart from age, several other factors which influence an individual's perception may also change with time, e.g. where in the village the person lives and moves, as well as family conditions and wealth. These are important aspects in the assessment of most, if not all, information obtained in an interview-survey about change.

We now turn to other explanations which are based upon the premise, not that one version is right and the other wrong or inaccurate, but that each group of 'observers' is experiencing different aspects of a real environment and and/ or valuing their importance differently. Many villagers state that average annual rainfall has declined, while official records show no such trend. For explaining such contradictory experiences, a **third** possibility is that rainfall actually has declined in Kalakamate. That is, an environmental (or social) variable is observed in several places (or at several times), and it is wrongly taken for granted that the data represent the same reality. However, it is very unlikely that rainfall in a particular village should deviate essentially from the overall pattern of the region, and thus the following discussion is based on the assumption that average annual rainfall has not decreased in Kalakamate. It should be noted that for several other socio-environmental variables mistaken generalisations of this kind are common.

A **fourth** possibility is that the internal dynamics of the environmental variables studied have changed in a way that affects the farmers perception and experience, but not the apparently contradictory scientific measurements considered. For rainfall, it could be that the intra-annual distribution pattern has changed. It is well documented that distribution of rainfall within the rainy season varies between years (Bhalotra 1985, 1987). Thus, over the year, for different areas and for different purposes of land use, available moisture may differ dramatically between years with similar annual rainfall. Vossen (1987) found that the correlation between annual rainfall and cattle and crop performance improved substantially when the rainy season is divided into three sub-seasons. Early and late season rainfall, which have a strong influence on the biomass of grass and browse available during the dry season, are more important for livestock survival than mid-season rainfall. Crop yields show a strong correlation with

Annika C. Dahlberg and Piers M. Blaikie

total seasonal rainfall, since farmers will shift the planting period and the crops and/or soils chosen depending on how rainfall fluctuates. However, the mid-season rainfall is a period of crucial importance, since in a year with poor early rains farmers rely on being able to plant in January, while in a year when early rainfall is good, crops will reach the stage with the highest water requirements during the mid rainy season (Vossen 1987).

Rainfall records from Francistown for 1922/23–1993/94 were analysed for a possible trend of early, mid and late rainy season rainfall respectively. In figure 10 (a–c) the five year running mean for these periods is shown along with the linear trend. These data show a slightly confusing picture since they indicate that early rainfall has increased, while mid season rainfall has remained virtually the same, and late season rainfall has declined. If anything, the situation for crops has improved, while grass cover may be expected to be depleted earlier in the dry season, but then regenerate quicker the following rainy season. Considering the high year to year variation for all these sub-seasons it is doubtful whether the indicated (but slight) trends are reflected in any long-term changes of the environment; also the spatial variation between Francistown and Sebina is large enough to allow quite different slight trends at Kalakamate.

Annual precipitation, intra-annual distribution and variation over time are some characteristics of rainfall which affect the environment. Many other rainfall characteristics, e.g. intensity of rainfall, length of dry spells and when in the

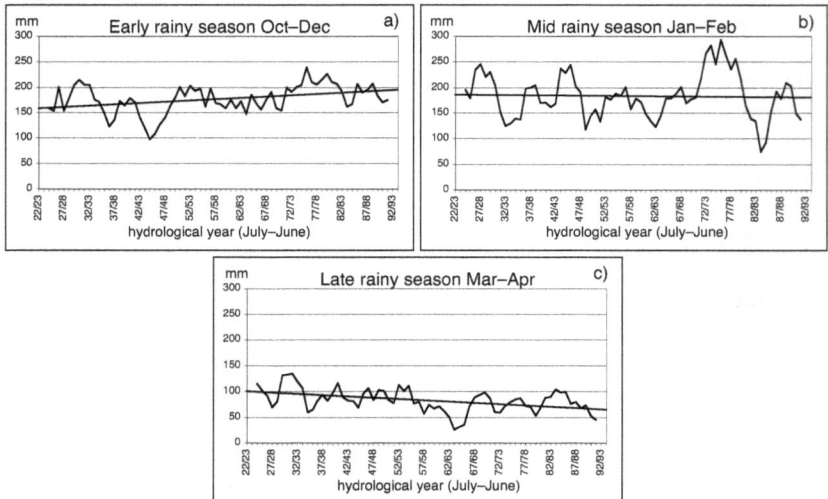

FIGURE 10 a–c. Five year running mean and linear trend for early, mid and late season rainfall for Francistown 1922/23–1993/94

year they occur, are also important. Added to these are the synergistic effects of different rainfall characteristics acting together with other climatic variables, such as temperature (Jackson 1989). An in-depth study of all these factors is not within the scope of this study, and thus we cannot state with certainty that there has not been a change in some aspect of rainfall. However, had a measurable such change occurred, with an environmental effect akin to that of an overall decline in rainfall, we assume that it would have been reported in the studies consulted. Thus, we find it highly unlikely that the farmers' perception of declining rainfall is due to a permanent shift in rainfall pattern or characteristics.

A **fifth** possibility is that, while rainfall characteristics (and several other environmental factors) probably have not changed significantly over the past seventy years, livelihood strategies and patterns of land management have. Thus, rainfall may have declined when seen in relation to its relative sufficiency for an altered set of requirements. This possibility is examined in relation to the impact the nucleation of settlements, the fencing of the private ranch, and the erection of the cordon fence has had on the location of fields. As shown in figure 6, fields are now concentrated around the kgotla (the 'chief's place'), and the other nucleated settlement areas to the south. The key characteristic to be examined here is the possible change in produce output, especially in terms of yields, that may be interpreted as an effect of a decline in rainfall. We will discuss the possible decline in yields as an effect of changes in available moisture and/ or a reduction in soil fertility, and we hypothesise that the change in the general location of fields has had a negative impact on these parameters for individual households. An increased competition for land in a particular area may force households to clear land which has been rested for an insufficient time, to make do with a soil type less suitable for cultivation than those previously ploughed, or it may constitute one reason for continuous cultivation of the same field in spite of a foreseeable or experienced fertility decline.

Looking first at possible changes in moisture availability, this is important to the farmer in terms of its effect on such resources as crops, grass, browse and fruits, as well as on the amount of water in wells and rivers. This fits the scientific understanding which states that it is the amount of moisture available to plants and animals which is of direct environmental, and therefore agricultural, importance (Huntley 1982). This, in turn, depends on many factors, e.g. evaporation, transpiration, deep percolation (Bate et al. 1982), and the balance between infiltration and storage in the soil (Stafford Smith and Pickup 1993).

Infiltration and storage depend on the texture and consistency of the soil, surface sealing, a presence or absence of a pan horizon, distance of this horizon from the surface, salinity and micro- and macrorelief characteristics (Tinley 1982, Casenave and Valentin 1992). Concerning soil storage of water, Biot (1988) found that this is largely a function of soil depth and the capacity of the soil to retain water per unit volume. In addition, other physical variables such as vegetation cover and type of plant community play an important role in determining the distribution of surface, soil, and ground-water. Thus, the variability over space and time of a multitude of environmental variables affect moisture availability. On the whole, farmers are aware of this, and their observations were found to constitute an important element in their descriptions of rainfall variation through time.

Farmers' awareness of different relationships between the biotic and abiotic are often expressed in their land management decisions. For example, the considerations behind a farmer's choice of where to locate a new field include, among many others, locally recognised relationships between different types of soils and crops under conditions of varying rainfall. If other factors, e.g. distance between field and settlement, are disregarded, most farmers said the best location of a field is one with at least two different soil types, which is illustrated in box 1 (numbers within brackets in the following text refer to the number of a particular quotation in a box).

Farmers express preferences about which crops to plant on what soil type (1, 2), as well as strategies concerning which soil type to plough and plant depending on actual and predicted rainfall distribution in any particular year (3, 4, 5). Although a loamy soil is, overall, considered best, specific crops are said to do better in sandier soils, and although clay-rich soils are the most productive under beneficial circumstances, most respondents stated that access to a sandier

---

1 'Motlhabana soil [light red, sandy]* can produce better crops of groundnuts and monkey nuts than the mokata soil [red, sandy loam].'

2 'Red corn grows well on mokwakwa soil [light grey, some clay]. This is the only crop that does well on this soil, but only if rainfall is not too low.'

3 'Motlhabana soil [light red, sandy] is good when there is little rain – but when there is too much rain the crops sink in this soil. Mokwakwa soil [light grey, some clay] gets dry when there is little rain, but it is good when there is much rain. We have these two soils in our field.'

4 'When the rains are late, or rainfall is very low at the beginning of the season, then we only cultivate the part of the field that has mokata soil [red, sandy loam]. The other half of the field is more rich in clay, and this is too hard to plough if it is dry. However, if there is enough rain this soil will produce a good harvest.'

5 'When choosing land for a new field you look at the soil and the vegetation, which will tell you if a good crop can be had from that land. The presence of many mokoba [*Acacia nigrescens*] and modumela [*Kirkia acuminata*] trees tells you that there is enough water available. The best soil is loam with some sand, red in colour [mokata] – on this soil you can get all kinds of crops.'

* Throughout the paper all local soil names are given in Setswana.

BOX 1.

soil was vital, since it will produce at least some yield even in years with low rainfall. Loamy soils have a higher water holding capacity than sandy ones, but the fact that sandy soils have low water retention may result in a better crop response under dry conditions with scattered rain showers, as compared to a soil with a higher clay content (Radcliffe 1990, Walker 1993).

Although a detailed soil survey would have been preferable, some information as to soil differences within the study area can be gleaned from the maps of geology (scale 1:125 000), soils and land suitability (1:250 000) produced for the area (Botswana 1975, 1984, 1989, Litherland 1975, Radcliffe 1990; and see Dahlberg 1995 for Kalakamate specifically). It should be noted that the maps of soils and geology only list the order of dominance of different soils and rocks for different units, recognising that the spatial variation is much larger than what can be shown on the maps. This is especially true in the study area where numerous occurrences of basic rock cause marked variations in soil quality.

The land around the main settlement in the north-west is dominated by acidic rocks (granite and gneiss) and soils are mainly regosols. In the eastern corner and in the southern half of the area dolerite intrusions and basic and ultrabasic rocks (e.g. serpentinite and amphibolite) are quite common, and soil types are more varied, with regosols, luvisols, leptosols, cambisols and vertisols occurring in a patch-like manner. Regosols are shallow coarse sands to loamy sands with much gravel and stone in the topsoil, and due to their extremely low waterholding capacity are prone to crop failure especially in years when rain fails at the peak of the growing season, and even in good rainfall years large yields cannot be expected (Radcliffe 1990). The higher content of fine sand, silt and clay found in the other soil types indicate better water holding capacity, although most of these soils are also characterised as poor.

On the land suitability map (Botswana 1989), where expected output of rainfed cultivation of sorghum has been the main ground of classification, the whole of Kalakamate falls under the categories 'marginally suitable' and 'permanently unsuitable'. The authors acknowledge that in spite of this rating 'traditional rainfed cropping is, in fact, commonly practised [in the area]' (Radcliffe 1990: 54), and stress the fact that the evaluation is related to an assumed average set of conditions, and that the results 'do not indicate the chances of success or failure associated with cropping [each] land unit' (loc.cit.) under conditions of fluctuating moisture availability. However, the land suitability map shows that although almost all land in Kalakamate is classified as 'permanently unsuitable', the area designated as slightly better is located in the southern part of the village, far from the main settlement. It should be noted that a large part of this 'better' land is taken up by heavy clay soils, which in most years are either too hard or too muddy to plough.

The maps of soils and geology also give some indication as to where soils with a slightly better nutrient content may be found. Soil derived from acidic rock can be expected to have lower nutrient levels than those from basic and

Annika C. Dahlberg and Piers M. Blaikie

ultrabasic parent material. Thus, the likelihood of finding soils with a higher nutrient content would increase as one moves to the east and south. However, many factors influence soil fertility, and past land use is among the most important. Respondents were asked what would cause them to abandon cultivated land, and a decline of soil fertility was very seldom given as a direct reason. A field ploughed for a long time was said to become 'old', and signs of this were that yields decreased, top soil was washed away, and most importantly that the field would become infested by troublesome weeds, especially the grass motlhwa (*Cynodon dactylon*).

This grass is a hardy perennial pioneer which colonises bare ground, is tolerant to drought, fire and grazing, and, once established, spreads rapidly (Skerman and Riveros 1989). Such infestation was described as a common reason to clear new land. When asked specifically about soil fertility, some respondents said that the occurrence of motlhwa in a field was a sign of nutrient depletion, while most simply stated that the weed 'will win over the crops' and yields decline. This grass is common in Botswana, and often occurs in dense swards which suppress the growth of crops (Phillips 1991). Farmers described how on recently cleared land, which previous to clearing had been rested for a long time, weeds were not a serious problem, and motlhwa was seldom encountered. However, after cultivating the same land for some time the field would gradually become infested by motlhwa. When this happened, farmers would gradually abandon the parts of the field where motlhwa appeared, increasing the field in another direction, until finally it was time to clear a whole new field. When this was not possible, due to land and/or labour shortage, the farmers would plough and plant in spite of the weeds, or make do with a smaller field. This general description was substantiated by a number of histories of fields for individual households and families, which often could be corroborated in the aerial photographs.

In most cases mentioned in the interviews, a move of settlement led to the clearing of a new field, even if this meant leaving still fertile land and, in some cases, having to cultivate on a less preferred soil type, 'My parents' field was on motlhabana soil [sandy], and this is better for crops than what I have now which is seloko soil [heavy clay]. We chose the present field because it is close to the house'. Only in a few cases did respondents actually state that they had ended up with less fertile soils, but many complained about the increasing shortage of so called 'virgin' land (i.e. land that had been rested long enough). With more households competing for land in the same area it is very likely that the scope for 'a best choice' has become more limited.

Furthermore, the preference for having easy access to water and other services is one reason why people in the future may be less inclined to abandon their fields because of erosion or weed infestation. In addition, the practice of wire-fencing one's field, which with favourable government loans is becoming increasingly common, may have a similar effect. Respondents were very positive towards such fencing, since it more securely keeps livestock away from the growing crops, but it may also lessen the incentive to allow for fallow periods. This generalised picture does not give any definite verification of our hypothesis. However, it does suggest that the possibility of acquiring fields on soils where productivity could be expected to be higher has been reduced as people have become concentrated in a limited area. It also suggests that, at least at present, people may be more ready to accept a decline in yields, rather than locating their field or compound further away from the nucleated settlement areas.

We shall now look into other aspects of changing farming practice and strategies, in a broader examination of the possibility that local interpretations of change are structured in relation to differences between informants, as individuals or as members of different groups. This is our **sixth** possibility of explanation of contradictions. Characteristics such as age, gender and wealth throw some light on the, often contradictory, interpretations of changes over time. Respondents perceive drought periods differently, and therefore remember them differently depending on these and other characteristics. The influence of wealth is the most obvious, and must be included also when responses are compared by age and gender. For example, if a household's herd is large, the loss of a few animals is not so serious as if it only had a few head of cattle to start with. If the animals are in poor condition following a drought, the household may still manage to get the field ploughed if there is money to rent a tractor. Other characteristics being equal, a woman (particularly one without other sources of income) may experience a drought harder than a man, since it is she who usually is responsible for feeding the family, while it is the man who has more easy access to outside income (Hesselberg 1993).

The quotations in box 2 illustrate some of the socio-economic circumstances which influence the individual's perception of rainfall. Amount, distribution and variability of rainfall will have a different impact on people depending on the particular period in the socio-economic history of a community (6), and on the particular period in the life-cycle of a household or individual (7). That is, at any

---

6   'Before, a drought would affect people more, since they could not buy food –
    therefore 1937 and 1947 were the worst drought years.' *(man born in 1912)*

7   'I would not have cared so much about the earlier droughts, since I did not
    have any cattle to lose then.' *(man born in 1927)*

8   'Before, the people used to be free, but now they are starving. Now people
    have to buy food if the harvest fails, so if you do not have money you will
    starve. Before people would keep reserve food in their granaries, then they
    were not forced to buy anything. The head of the family would keep the
    granary and distribute grain in times of drought.' *(woman born in 1908)*

BOX 2.

9 'At times, e.g. this year *[1993/94]*, there is rain, at least here in Kalakamate.'

10 'This year *[1993/94]* we got nearly as much rain as when I was young, but it stopped in the middle.'

11 'This year *[1993/94]* we were not able to plough the whole field because we were sharing a span of cattle with many other people. We do not have any of our own. And before it was our turn the cattle had to be vaccinated after which they have to be kept aside for two weeks, and after that it was too late to plough.'

12 'This year *[1993/94]* is a drought year, I cannot think of any year that has been drier than this.'

BOX 3.

point in time differences in livelihood opportunities will shape the individual's, and/or the community's, interpretation of changes in available moisture (6,7,8).

Risk reduction is another important aspect of livelihood strategies. In Kalakamate, local management of natural resources is adapted to the erratic variation of rainfall, and also to meeting socio-economic shifts. In an unpredictable physical and social environment, strategies of risk avoidance are the norm. Perception of rainfall variation will depend largely on how well different risk avoidance strategies can be put into effect, and this will vary between people and over time. A dry year may not be perceived so negatively if the farmer has been able to implement successful risk-reducing strategies. For example, in the growing season of 1993/94 there was some rainfall in November and December, slightly better conditions in January and February, whereafter not a drop fell until next December. As demonstrated by the quotes in box 3, farmers who had early access to draught power could successfully practice a strategy of early ploughing, and look forward to a fair harvest (9, 10), while those who could not described 1993/94 as a dry or bad year (11, 12).

However, it should be noted that individual experiences of good and bad years may also be caused by the high variability of rainfall over very small distances (Jackson 1989). In the same year, fields in one part of the village can receive, for example, above average rainfall, while those in another part receive much less. The extremely local nature of rain showers also means that a scattered distribution of fields, within an extended family or the village as a whole, would be a strategy of decreasing the risk of wide-spread crop failure.

Alternative interpretations of a perceived decline in rainfall

In tying together the different threads presented above, the point of departure will be the respondents' own words. Those who claimed that rainfall had declined were asked how they could be sure of this, and apart from direct statements, a

Direct statements
Before, there were fewer droughts
Before, the rain was more reliable
Before, the rains were heavier

Primary effects
Before, people got bigger harvests
Before, there was plenty of fruit
Before, not so many cattle died

Secondary effects
Before, people made the effort to plough
Before, people had granaries
Before, the land would recover faster after a drought

Evidence
Before, there was water running all over the ground
Before, you could not take the car off the road, it would sink
Before, there was more water in the rivers
Before, there were more springs with water
Before, the watertable in the borehole was higher

TABLE 4. Changes attributed to declining rainfall

range of evidence was presented, in addition to perceived primary and secondary effects. A selection of these answers is listed in table 4, and these are analysed in the context of the relationships between rainfall, moisture availability, livelihood strategies and land management practices discussed above.

The first statements, about droughts and the reliability of rainfall, are not accurate in a scientific sense, but they may well be a reality for these particular respondents. Rainfall intensity has not been measured in the past, but an analysis of the number and distribution of rainfall days shows no trend. However, if soil characteristics and vegetative cover have changed, then a change in the rate of runoff and infiltration is likely to follow. This would create a different situation in terms of plant growth, erosion and hydrology, even if rainfall intensity is the same. In other words, different causes can create similar effects.

Several people claimed that 'the harvests were larger before', i.e. comparing those of their childhood and youth with the present. Others refuted this, saying that a good rainfall year in recent times (e.g. 1987/88) gave as large a harvest as in a good year when they were young. One factor directly related to yield is the kinds of crops grown. The interviews revealed that very little had changed in this respect. All respondents maintained that they plant the same crops as their parents did, except tobacco and dagga. A few new varieties of maize, sorghum and beans had been introduced over the last decades, but these had not replaced the previously used varieties. A farmer decides which crops and varie-

Annika C. Dahlberg and Piers M. Blaikie

ties to plant in a particular year, or on a particular part of the field, depending on realised and expected rainfall. Many mentioned that access to new varieties had increased the possibility of a good harvest. However, the harvest will vary between farmers in one year, and for the same farmer between years, leading to different perceptions of rainfall in relation to harvest in any particular year.

Of course, a decline in harvest can be caused by many factors, such as a decrease in soil fertility, in available labour and draught power, in manure and viable seeds, or an increase of soil erosion. Initially, few said they had observed any signs of soil erosion, but upon further discussion almost 70 per cent said that in their field some soil was being washed away. Farmers stated that, apart from ploughing along the contours, there was little that could be done to stop erosion, but then, very few saw it as a significant problem. If the field is on sloping land soil will be washed away, 'but it is not a threat to the lifetime of a field', and only one or two farmers made any connection to declining yields. In the study area most land is flat or gently sloping, and neither aerial photographs nor fieldwalks gave any indications of severe erosion in fields. What was, however, seen as a threat to the harvest was the infestation of the weed motlhwa (*Cynodon dactylon*). Depending on the magnitude of infestation, yields of the same crop will vary from field to field in the same year.

Farmers did not recognise fallow periods as part of their cultivation practice, but in reality it is included in their management of the land, since people would shift the location of their fields when weeds became too troublesome, or when grazing became scarce. It would seem, however, that these periods of resting the soil may already occur less frequently, and be of shorter duration. Increased competition for farming land in certain parts of the village will make people less inclined to abandon cleared land. It was described how farmers now deal with weeds by manual weeding, by an extra ploughing in the winter, or by 'planting on top of the weeds, even though this gives a lower harvest'. The erection of proper wire fences around the fields may further reduce farmers' willingness to abandon an old field. However, as discussed below, outside incomes have meant that many families are less inclined to plough and plant every year, which could result in more frequent short fallow periods.

As we have shown, the availability of grazing and farming land has also become cumulatively more restricted. The establishment of a private ranch has sharply reduced grazing and agricultural land to the east and southeast; the nucleation of settlements has resulted in people leaving the outlying areas; and the veterinary fence has curtailed movement to the north and north-east. Not only are people forced to cultivate the same land for a longer time, there is also less choice in available land, 'Things are not equal [to how they were before], and so you cannot simply compare harvests'.

The size of harvests, as well as of livestock herds, is also related to availability of labour. Many of the old people in this study remember a time when cultivation and livestock rearing involved the whole extended family. Today most young children attend school, either locally or as boarders with relatives in a larger village or town, and therefore take less part in such activities as herding, weeding, guarding the field and collecting wild produce. Many of the respondents' children who were beyond school-age have moved away to earn a living in Francistown or in other towns and large villages in Botswana. Lack of labour due to migration was mentioned as one reason for temporarily, or even permanently, abandoning a field. Several respondents complained that their children were not interested in agriculture, and that the whole culture of helping family and neighbours was disappearing. As mentioned above, this is a typical stereotype narrative about the 'young generation', but in Botswana the educational and economic development since Independence has in fact resulted in young people seeking a future in other venues than traditional farming.

The statement 'before, [in the past] people made the effort to plough' is explained by villagers in many different ways. Although a decline in rainfall is principally identified as the reason for this lack of action, there are others too. Irrespective of a real change in rainfall, or of other factors affecting agricultural productivity, for most households an increase in prosperity is seen to lie in activities other than agriculture. In the period 1930 to 1960 remittances from the men working in South Africa were mainly invested in cattle, or kept as a safeguard for drought years. Today money pays for education, travel, housing, household goods and an increased dependence on imported foods. Although it is still important to retain access to agricultural land, and land and livestock act as security measures, wages and government subsidies, including drought relief, are replacing the role previously played by the village and household granary. Other changes perceived as primary effects of declining rainfall, e.g. in the abundance of natural produce such as grass, firewood and fruit, are discussed in Dahlberg (1996, III, section 7).

Some villagers suggested direct visible evidence of a decline in rainfall, claiming that the amount of surface and ground water had declined. However, many respondents denied this, describing how the amount of water in rivers and streams has always varied within and between years, and that this general pattern of variation has not changed. Whether the streamflow has changed or not is very difficult to ascertain. Changes in vegetation and soils could cause a change even if rainfall remains the same, and such effects may explain claims that certain small streams have dried up. The drying up of springs described by some can also have been caused by local environmental changes. It is also conceivable that the reduced dependence on surface water for people and cattle, since the borehole was drilled, has meant that people no longer are as aware of fluctuations in ground and surface water. Some people complained that the pits dug in the streambanks contain less water now, but several respondents offered alternative explanations: 'Now we prefer to go to the tank for water, so no one bothers to keep the pit free from sand that has been washed in. If this sand was removed perhaps there would be as much water there again'.

Annika C. Dahlberg and Piers M. Blaikie

Concluding remarks on rainfall

Before going on to summarise the whole paper we can list the following tentative conclusions about rainfall trends in Kalakamate:

i   Farmers' descriptions of environmental change during their lifetime are, in a majority of cases, dominated by a perception of a decline in average annual rainfall, and this belief is used to explain changes in a multitude of environmental and social parameters.

ii According to official records average annual rainfall has not declined.

iii Changes in other rainfall characteristics, likely to have had an effect similar to that of an overall decline in rainfall, have not been identified.

iv We suggest that farmers' perceptions of declining rainfall is instead explained by, on the one hand, changes of other physical parameters such as moisture availability, soil fertility and vegetative cover; and on the other hand, different and changing livelihood opportunities and strategies. In some cases change has resulted in reduced options for utilising rainfall and for avoiding the risk of moisture shortage. In other cases the change is in the increased range of choices in livelihood opportunities outside agriculture.

v   Accurate descriptions of past climatic change, in relation to its impact on livelihoods and land-use strategies, are desirable for an understanding of the present and predictions for the future, but difficult to collect without lengthy narrative and (sometimes) speculative interpretation.

vi There is some evidence that, because of the nucleation of settlement pattern and the exclusion of other potential agricultural land through privatisation, cultivation now takes place more often on soils which give poorer yields.

Three of the four components in the diagram of relationships of relevance to environmental change (fig. 1) have been discussed in some detail in this paper. Trends and events at the national, district and village level have been related to their effects on livelihood strategies and land management. The fourth component in the diagram (environmental outcomes) has been described and discussed more briefly. For details about observed and perceived changes in such environmental features as the cover, density and species composition of the woody vegetation and grasses respectively, and in the abundance and availability of firewood, fruits and thatching grass, see Dahlberg (1996, III, section 7).

SUMMARY AND CONCLUSIONS

Numerous changes have occurred in the social and environmental sphere in Kalakamate and in Botswana as a whole. However, the direction, magnitude and persistence of these changes vary from case to case. Population has increased, but at the village level the rate has not been high enough to have any pronounced effect on the environment. In contrast to national figures and to the situation in other parts of Botswana, livestock numbers in Kalakamate, as well as in surrounding areas of communal land, have hardly increased. Instead, official records and villagers' narratives both tell of a continuous fluctuation, which can be convincingly attributed to rainfall variation and events and developments generated outside the village. Turning now to climate change, and more specifically rainfall, official records and the perception of villagers are at a first glance contradictory. Official records show no clear trends, but villagers claim to have experienced a decline. We suggest that the seeming contradictions may be overcome by pointing out the differences in how rainfall is measured – by official sources in terms of mm at a rain gauge over a period of time, and by farmers in terms of how they can utilise the available moisture. The latter may well have changed because the location and character of farming activities have changed, primarily due to the nucleation of settlements. In a similar manner, grasses, trees, fruits and other natural resources are used and perceived differently owing to changes in livelihood strategies and changes in the individual's experience of these resources in daily life.

Our approach, to attempt closure and resolve contradictions in diverse narratives, leads to a remarkable convergence between local and scientific accounts and interpretations of environmental change, especially when related to recent developments in semi-arid range ecology. The following conclusions have been shown to be consistent with both sources:

a) Variations in rainfall, including events such as drought, are the main driving force of ecosystem dynamic.

b) Changes in the physical environment are event-driven, e.g. by drought, but also by other random events or planned interventions. In range ecology terms this is called an environment at dis-equilibrium.

c) Stocking densities have a relatively low importance in explaining environmental change.

d) Few, if any, of the observed changes are irreversible and of a kind that would fit any of the current definitions of land degradation.

e) A flexible strategy of land management, including a package of options to be used in an opportunistic way, is both logical and sound.

The cover and density of woody vegetation has increased, especially in parts of the village, and more specifically the spatial distribution of different vegetation types has changed. This is primarily caused by changes in the spatial pattern of settlement and cultivation, a shift initiated at the national level. Likewise, the erection of fences, which cut off access to water and grazing land, and also the reduced frequency of burning, were caused by outside intervention – single events caused long-term local changes of land use and environment. These and other examples demonstrate how people adjust to events and trends

Annika C. Dahlberg and Piers M. Blaikie

in the socio-economic sphere, and how resulting shifts in livelihood opportunities cause changes in land use, with both immediate and gradual effects on the environment at the local level.

Increased access to income opportunities outside the village has reduced the dependence on agricultural and pastoral income earning activities in general, and on certain natural resources specifically. This may have influenced the abundance and distribution of species, but it has also altered people's perception of the environment. Depending on individual circumstances, the strategies, perceptions and interpretations of change will differ from person to person and from place to place. Simply speaking, narrators, including the scientists, are perceiving aspects of the same reality in different ways, or are generalising from different experiences and/or interests. We have tried to achieve closure by explaining how these differences in perception and explanation came about, not so much by proving one account wrong and the other right, as by establishing why they differ. The inclusion of local narratives is informative in its own right, and also presents outside actors (scientists, consultants, policy-makers) with complementary and often improved means for perceiving, measuring and evaluating the environment.

Both local and scientific sources provide narratives, and both have to be 'de-narratised' when comparing them. Local narratives are not told for the same reasons as those told by scientists. Villagers are experiencing their surroundings and trying to make sense of their lives, while scientists collect data with the specific purpose of trying to reduce uncertainty and increase predictability. In the case of semi-arid areas of Africa, scientific data and their interpretation have recently been subjected to a scientific critique which emphasises uncertainty and complexity. The villagers of Kalakamate seem to have come to the same conclusion.

ACKNOWLEDGEMENTS

Without the help and interest of the people of Kalakamate this study would – obviously – not have been possible. Without their patience our understanding of components and processes of change would have been even less complete. The fieldwork was greatly enhanced by the assistance of numerous people at the District Agricultural Office in Masunga, the Regional Range Ecology Office in Francistown, and the Department of Environmental Science, University of Botswana. Colleagues at Stockholm University, the University of East Anglia, and the University of Oslo have graciously read earlier drafts of the paper and suggested many improvements. Our reviewer, David Thomas, showed great insight and gave valuable comments, as did Erling Dahlberg, as always. Peter Kinlund has generously shared collected archival data as well as advice, and Mats Leine's knowledge of cartography was invaluable when creating maps from the aerial photographs. We also thank Karin Weilow and Hans Drake who drew some of the maps and figures, and Lennart Esklund who provided guidance through the digital world.

The Office of the President, Republic of Botswana, kindly authorised the research, and the fieldwork was financed by the Swedish Agency for Research Cooperation with Developing Countries (SAREC), and by grants from the Scandinavian Institute of African Studies, the Swedish Society for Anthropology and Geography (SSAG), and by the C F Liljevalch Jr. research foundation.

NOTES

[1] Sources for figure 3 a–d:

Bechuanaland/Botswana: 1904–71 (Tumkaya 1987), 1981 (CSO 1982), 1991 (CSO 1992).

Tati/North East District: 1931 (BNA S.238/4), 1946 (BNA S.87/3), 1964 (Tumkaya 1987), 1971 (CSO 1973), 1981 (CSO 1982), 1991 (CSO 1992).

Chiefdom of Ramokate: 1921 (BNA S.17/7), 1936 (BNA S.238/7/1), 1946 (BNA S.87/3), 1971 (CSO 1973), 1981 (CSO 1982, 1983), 1991 (CSO 1992).

Kalakamate village: 1964 (BNA CENSUS 39/4), 1968 (BNA BNB 8151), 1981 (CSO 1982, 1983), 1991 (CSO 1992).

[2] Sources for figure 4 a–d:

Francistown Veterinary Region: 1904, 1911 (in Fortmann et al. 1983), 1936 (BNA S.86/15), 1939 (Fortmann et al. 1983), 1941–42, 1944–46, 1948–55 (in Fortmann et al. 1983), 1956–57 (PRO DO 112/31), 1960–61 (PRO DO 102/37), 1962–67, 1969, 1976 (in Fortmann et al. 1983), 1979–90 (Botswana 1980–91).

Tati Native Reserve: 1905 (BNA R.C.11/5), 1930 (BNA S.130/6), 1932 (in Fortmann et al. 1983), 1936 (BNA S.238/6), 1939 (BNA DCF 6/15), 1941–42 (BNA S.238/8), 1943–44 (BNA S.238/9), 1946 (BNA S.238/7/1), 1947–48 (BNA S.238/7/2), 1949–52 (in Fortmann et al. 1983), 1958 (PRO DO 35/4494), 1971 (Egner 1971, App C), 1981 (RAO R12D), 1986, 1989 (Maas and Jansen 1990), 1990 (Veterinary Office 1992).

Chiefdom of Ramokate: 1930 (BNA S.130/6), 1936 (BNA S.238/6), 1939 (BNA DCF 6/15), 1941 (BNA S.238/8), 1943–44 (BNA S.238/9), 1946 (BNA S.238/7/1) 1948 (BNA S.238/7/2), 1990 (Veterinary Office 1992).

Kalakamate village: 1976 (Greenhow 1976), 1981 (RAO R12D), 1984–92 (Veterinary Office 1992).

REFERENCES

Abel, N.O.J. 1992. What's in a number? The carrying capacity controversy on the communal rangelands of Southern Africa. Unpublished PhD thesis. Norwich: School of Development Studies, University of East Anglia.

Abel, N.O.J. and Blaikie, P.M. 1989. Land degradation, stocking rates and conservation policies in the communal rangelands of Botswana and Zimbabwe. *Land Degradation and Rehabilitation* 1: 101–123.

Arntzen, J.W. 1985. Agricultural development and land use in eastern communal Botswana: The case of Kgatleng District. Working Paper 50. Gaborone: National Institute of Development Research and Documentation (NIR).

Arntzen, J.W. 1986. Land shortage: The need for comprehensive land use planning. *Botswana Notes and Records* 18: 133–137.

Asselman, G. 1986. National Conservation Strategy consultation report for North East District. Draft, unpublished report. Botswana, Francistown: District Office Lands.

Bate, G.C., Furniss, P.R. and Pendle, P.G. 1982. Water relations of Southern African savannas. In B.J. Huntley and B.H. Walker (eds) *Ecology of tropical savannas*, pp. 336–358. Berlin: Springer Verlag.

Behnke, Jr. R.H. and Scoones, I. 1993. Rethinking range ecology: Implications for rangeland management in Africa. In R.H. Behnke Jr., I. Scoones and C. Kerven (eds) *Range ecology at disequilibrium. New models of natural variability and pastoral adaptation in African savannas*, pp. 1–30. London: Overseas Development Institute.

Behnke Jr., R.H., Scoones, I. and Kerven, C. (eds) 1993. *Range ecology at disequilibrium. New models of natural variability and pastoral adaptation in African savannas.* London: Overseas Development Institute.

Bhalotra, Y.P.R. 1985. *Drought in Botswana.* Department of Meteorological Services, Ministry of Works and Communications. Gaborone: Government of Botswana.

Bhalotra, Y.P.R. 1987. Climate of Botswana. Part II: Elements of climate. 1. Rainfall. Department of Meteorological Services, Ministry of Works and Communications. Gaborone: Government of Botswana.

Biot, Y. 1988. Forecasting productivity: Losses caused by sheet and rill erosion. A case study from the communal areas of Botswana. PhD thesis. Norwich: School of Development Studies, University of East Anglia.

Blaikie, P.M. 1985. *The political economy of soil erosion in developing countries.* Harlow: Longman.

Blaikie, P.[M.], Cannon, T., Davis, I. and Wisner, B. 1994. *At risk: Natural hazards, people's vulnerability, and disasters.* London: Routledge.

Blair Rains, A. and McKay, A.D. 1968. The Northern State Lands, Botswana. Land Resource Study 5, Land Resources Division, UK: Ministry of Overseas Development.

BNA BNB 8151. North East District Council Development Plan 1968–72. Gaborone: Botswana National Archives.

BNA CENSUS 39/4. Completed questionnaire forms. Population census for the Bechuanaland Protectorate 1964. District Office, Francistown. Gaborone: Botswana National Archives.

BNA DCF 6/15. Boundaries between area under control of different sub-chiefs in the Tati Reserve. Gaborone: Botswana National Archives.

BNA R.C.11/5. Tati District: Selection of Reserve in. Gaborone: Botswana National Archives.

BNA S.17/7. Census in Bechuanaland Protectorate 1921. Gaborone: Botswana National Archives.

BNA S.86/15. Census in Bechuanaland Protectorate 1936. Correspondence with resident magistrate, Francistown. Gaborone: Botswana National Archives.

BNA S.87/3. Census in the Bechuanaland Protectorate, Francistown. Gaborone: Botswana National Archives.

BNA S.130/6. Tati Native Reserve: Taxpayers under each chief. Cattle and small stock owned by followers of each chief. Gaborone: Botswana National Archives.

BNA S.238/4. Tati Native Reserve: Position of natives in, Congested state of, Question of settling natives on Nata Crown Lands. Gaborone: Botswana National Archives.

BNA S.238/6. Tati Native Reserve: Congested state of natives in, Question of settlement of certain sections on Nata Crown Lands. Including: Resident Commissioner's Report on his tour of the Bechuanaland Protectorate, 31 May to 19 June 1936. Gaborone: Botswana National Archives.

BNA S.238/7/1. Tati Native Resreve: Congested state of natives in, Question of settlement of certain sections on Nata Crown Lands. Includes: 'A report on the grazing condition of stock and stock water in the Tati Native Reserve, The G.V.O's Office, Francistown, 13/IX/46'. Gaborone: Botswana National Archives.

BNA S.238/7/2 Tati Native Reserve: Congested state of natives in, Question of settlement of certain sections on Nata Crown Lands. Gaborone: Botswana National Archives.

BNA S.238/8. Tati Native Reserve: Congested state of Tati Native Reserve. Prof. I. Schapera's report on. Gaborone: Botswana National Archives.

BNA S.238/9. Tati Native Reserve: Congested state of, Question of lease of grazing land from Tati Company, and others till such time as permanent arrangements can be made (purely temporary measure). Gaborone: Botswana National Archives.

Botswana 1975. Quarter degree sheet 2027C with parts of 2027D and 2027B, Sebina–Tshesebe (1:125 000, surveyed 1971–72), Geological Survey of Botswana. Gaborone: Geological Survey Department, Ministry of Mineral Resources and Water Affairs. Republic of Botswana.

Botswana 1984. Ramokgwebana sheet (1:250 000). Soil map. Soil Mapping and Advisory Service Project Bot/80/003. Gaborone: FAO/UNDP/Government of Botswana.

Botswana 1989. Ramokgwebana sheet (1:250 000). Land suitability for improved traditional dryland farming. Soil Mapping and Advisory Service Project Bot/85/011. Gaborone: FAO/UNDP/Government of Botswana.

Botswana 1980–91. Botswana Agricultural Statistics 1980–91. Planning and Statistics Division (published annually), Ministry of Agriculture. Gaborone: Government of Botswana,.

Casenave, A. and Valentin, C. 1992. A runoff capability classification system based on surface features criteria in semi-arid areas of West Africa. *Journal of Hydrology* **130**: 231–249.

CSO 1973. Guide to the villages of Botswana 1971. Central Statistics Office, Ministry of Finance and Development Planning, Government of Botswana.

CSO 1982: Summary statistics on small areas. 1981 Population and Housing Census. Central Statistical Office, Ministry of Finance and Development Planning. Gaborone: Government of Botswana.

CSO 1983. Guide to the villages and towns of Botswana. 1981 Population and Housing Census. Central Statistical Office, Ministry of Finance and Development Planning. Gaborone: Government of Botswana.

CSO 1991. 1991 population census – Preliminary results. Stats Brief No 91/4, Central Statistics Office. Gaborone: Republic of Botswana.

CSO 1992. Population of towns, villages and associated localities in August 1991. Central Statistics Office, Ministry of Finance and Development Planning. Gaborone: Government of Botswana.

CSO 1994. Data from 1991 population census, computer printouts, transcripts. Central Statistics Office, Ministry of Finance and Development Planning. Gaborone: Government of Botswana.

Dahlberg,A.[C.] 1994. Contesting views and changing paradigms. The land degradation debate in Southern Africa. Discussion Paper 6. Uppsala: Scandinavian Institute of African Studies.

Dahlberg,A.C. 1995. On interpreting environmental change: Time, space and perception in the case of Kalakamate, North East District, Botswana. In: Ganry, F. and Campbell, B., (eds), *Sustainable land management in African semi-arid and subhumid regions*, pp.257–271. Proceedings of the SCOPE Workshop, 15–19 Nov 1993, Dakar, Senegal. Montpellier: SCOPE/UNEP/CIRAD.

Dahlberg, A.C. 1996. Interpretations of environmental change and diversity. A study from North East District, Botswana. PhD thesis. Dissertation Series 7. Stockholm: Department of Physical Geography, Stockholm University.

DTRP 1987. National Conservation Strategy: Household opinion survey. Final Report. Department of Town and Regional Planning, Ministry of Local Governments and Lands. Gaborone: Republic of Botswana.

Egner, E.B. 1971. Interim report of the Tati Settlement Project. Francistown. App. C. File BNB # 8897. Gaborone: Botswana National Archives.

Ellis, J.E. and Swift, D.M. 1988. Stability of African pastoral ecosystems: Alternate paradigms and implications for development. *Journal of Range Management* **41**: 450–459.

Ellis, J.E., Coughenour, M.B. and Swift, D.M. 1993. Climate variability, ecosystem stability, and the implications for range and livestock development. In R.H. Behnke Jr., I. Scoones and C. Kerven (eds) *Range ecology at disequilibrium: New models of natural variability and pastoral adaptation in African savannas*, pp. 31-41. London: Overseas Development Institute.

Fortmann, L. 1989. Peasant and official views of rangeland use in Botswana. Fifty years of devastation? *Land Use Policy* **6**: 197–202.

Fortmann, L., Gobotswang, K.E., Edzani, U., Woto, T., Magama, A. and Motswogole, L. 1983. Local institutions, village development, and resource management: Case studies from North East District, Botswana. Applied Research Unit, Ministry of Local Government and Lands, Government of Botswana, with the Land Tenure Center, University of Wisconsin. (Earlier versions of the reports of this project were published in Botswana by the Ministry, in limited numbers.)

Garanganga, B., Muchinda, M. and Timberlake, J. 1994. The climate factor. In: Chenje, M. and Johnson, P., (eds), *State of the environment in Southern Africa*, pp. 87–104. Harare: SARDC (IUCN, SADC).

Greenhow, T.A. 1976. The Tati Resettlement Scheme of Botswana: A study in natural resource planning and implementation within a social, economic and political framework. Unpublished MA thesis, University of Windsor.

Gulbrandsen, Ø. 1984. Access to agricultural land and communal land management in Eastern Botswana. Applied Research Unit, Ministry of Local Government and Lands. Gaborone: Government of Botswana.

Hesselberg, J. 1985. *The Third World in transition. The case of the peasantry in Botswana*. Uppsala: The Scandinavian Institute of African Studies.

Hesselberg, J. 1993. Food security in Botswana. *Norsk Geografisk Tidsskrift* (Norwegian Journal of Geography) **47**(4): 183–195.

Hesselberg, J. and Wikan, G. 1982. The impact of absenteeism on crop production and standard of living in two villages in Botswana. *Botswana Notes and Records* **14**: 69–73.

Holling, C.S. 1973. Resilience and stability of ecological systems. *Annual Review of Ecology and Systematics* **4**: 1–23.

Hoof, P.J.M. van and Maas, H. van der 1991. Land use, settlements and the rural poor – whither rural development in the North East District? In C. van Waarden (ed) *Kalanga retrospect and prospect*, pp. 9–18. Gaborone: The Botswana Society.

Huntley, B.J. 1982. Southern African savannas. In B.J. Huntley and B.H. Walker (eds) *Ecology of tropical savannas*, pp. 101–119. Berlin: Springer Verlag.

Jackson, I.J. 1989. *Climate, water and agriculture in the tropics*. Second edition. England: Longman.

Kinlund, P. 1995. Personal communication. Stockholm University.

Larsson, R.Å. and Strömquist, L. 1991. Air photo analysis for environmental monitoring in the SADCC Region. SADCC Environmental Monitoring Systems, Monitoring Techniques Series 2. Lesotho: SADCC, Ministry of Agriculture and Marketing, and Sweden: Landfocus AB.

Leach, M., Mearns, R. and Scoones, I. 1997. Environmental entitlements: A framework for understanding institutional dynamics of environmental change. IDS Discussion Paper 359. UK: Institute of Development Studies, Sussex.

Lindesay, J.A. and Vogel, C.H. 1990. Historical evidence for southern oscillation-southern African rainfall relationships. *International Journal of Climatology* **10**: 679–689.

Litherland, M. 1975. The geology of the area around Maitengwe, Sebina and Tshesebe, North East and Central Districts, Botswana. District Memoir 2, Geological Survey of Botswana. Geological Survey Department, Ministry of Mineral Resources and Water Affairs. Gaborone: Republic of Botswana.

Maas, H. van der and Jansen, R. 1990. North East District CFDA. Demography, farming systems and employment. Inventories for land use planning, Annex 2: 71–114. In Annexes – Vol 1: Soils, demography, farming systems and employment inventories, Government Employment Programmes Inventory. University of Utrecht, with Applied Research Unit, Ministry of Local Government and Lands. Gaborone: Government of Botswana.

NEDC 1986. North East District Development Plan 1986–1989. North East District Council, North East District Development Committee, Ministry of Local Government and Lands. Gaborone: Government of Botswana.

Phillips, M. 1991. *A guide to the weeds of Botswana*. Department of Agricultural Research, Ministry of Agriculture. Gaborone: Government of Botswana.

PRO DO 35/4494. Bechuanaland Protectorate: affairs of the Tati Co., possible termination of concessions, dissatisfaction with company under new control of Glazer Bros. of Johannesburg. Public Records Office. Kew, England.

PRO DO 102/37. Bechuanaland Protectorate: Dependencies Report 1960/61. Public Records Office. Kew, England.

PRO DO 112/31. Bechuanaland Protectorate: Annual Report of the Department of Agriculture, 1957. Public Records Office. Kew, England.

Radcliffe, D.J. (ed) 1990. The soils of North Eastern Botswana. Soil Mapping and Advisory Services, Field Document 17, FAO, UNDP. Gaborone: Government of Botswana.

RAO R12D. The stocking densities in North East District communal areas 1981–82. Regional Agricultural Office. Francistown, Botswana.

Roe, E. 1980. Development of livestock, agriculture and water supplies in Eastern Botswana before Independence: A short history and policy analysis. Occasional Paper 10, Rural Development Committee. Ithaca: Cornell University.

Rosenau, P.M. 1992. *Post-modernism and the social sciences: Insights, inroads, and intrusions*. Princeton: Princeton University Press.

Sandford, S. 1980. Keeping an eye on TGLP [Tribal Grazing Land Policy]. Working Paper 31, Gaborone: National Institute of Development and Cultural Research (NIR).

Sandford, S. 1983. *Management of pastoral development in the Third World*. Chichester: Wiley.

Schapera, I. 1947. *Migrant labour and tribal life: A study of conditions in the Bechuanaland Protectorate*. London: Oxford Univ. Press.

Schapera, I. 1971. The native land problem in the Tati District. Gaborone: *Botswana Notes and Records* **3**: 219–268.

Scoones, I. (ed) 1995. *Living with uncertainty. New directions in pastoral development in Africa*. International Institute for Environment and Development. London: Intermediate Technology Publications.

Scoones, I. and Thompson, J. (eds) 1994. *Beyond farmer first. Rural people's knowledge, agricultural research and extension practice*. Institute for Environment and Development. London: Intermediate Technology Publications.

Shepherd, N. and Caughley, G. 1987. Options for management of kangaroos. In G. Caughley, N. Shepherd and Short, J. (eds) *Kangaroos: Their ecology and management in in the sheep rangelands of Australia*. London: Cambridge University Press.

Silitshena, R.M.K. 1979. Chiefly authority and the organization of space in Botswana: Towards an exploration of nucleated settlements among the Tswana. *Botswana Notes and Records* **11**: 55–67.

Silitshena, R.M.K. 1982. Migration and permanent settlement at the lands areas. In R.R. Hitchcock and M.R. Smith (eds) *Proceedings of the symposium on Settlement in Botswana–The historical development of a human landscape*, pp. 220–231. London: Heinemann Educational Books.

Skarpe, C. 1992. Dynamics of savanna systems. *Journal of Vegetation Science* **3**: 293–300.

Skerman, P.J. and Riveros, F. 1989. *Tropical grasses*. FAO Plant Production and Protection Series 23: 310–315. Rome: FAO.

Stafford Smith, M. and Pickup, G. 1993. Out of Africa, looking in: Understanding vegetation change. In R.H. Behnke Jr., I. Scoones and C. Kerven (eds) *Range ecology at disequilibrium. New models of natural variability and pastoral adaptation in African savannas*, pp. 196–226. London: Overseas Development Institute.

Steenhuijsen Piters, B. de 1995. Diversity of fields and farmers. Explaining yield variations in northern Cameroon. PhD thesis. The Netherlands: Wageningen Agricultural University.

Tapela, H.M. 1982. Movement and settlement in the Tati Region: A historical survey. In R.R. Hitchcock and M.R. Smith (eds) *Proceedings of the symposium on Settlement in Botswana – The historical development of a human landscape*, pp. 174–188. London: Heinemann Educational Books.

Thomas, D.S.G. and Middleton, N.J. 1994. *Desertification: Exploding the myth*. Chichester: Wiley.

Thomas and Sporton, 1997. Understanding the dynamics of social and environmental variability. *Applied Geography* **17**:11-17.

Thornton, R.W. 1943. Report on the land leased from the Tati Company by the Bechuanaland Protectorate Government as a Native Reserve. In: Tati Native Reserve: Congested state of natives in, Question of settlement of certain sections on Nata Crown Lands. File # S.238/7/1. Gaborone: Botswana National Archives.

Tiffen, M., Mortimore, M. and Gichuki, F. 1994. *More people, less erosion. Environmental recovery in Kenya*. Chichester: Wiley.

Tinley, K.L. 1982. The influence of soil moisture balance on ecosystem patterns in Southern Africa. In B.J. Huntley and B.H. Walker (eds) *Ecology of tropical savannas*, pp. 175–192. Berlin: Springer Verlag.

Tumkaya, N. 1987. Botswana's population trends: Past and future. *Botswana Notes and Records* **19**: 113–128.

Turner II, B.L., Hydén, G. and Kates, R. (eds) 1993. *Population growth and agricultural change in Africa*. Gainesville: University Press of Florida.

Tyson, P.D. 1979. Southern African rainfall: Past, present and future. In M.T. Hinchey (ed) *Proceedings of the symposium on drought in Botswana*, pp. 45–52. Botswana Society. Clark Univ. Press.

Tyson, P.D. 1986. *Climatic change and variability in Southern Africa*. London: Oxford University Press.

Vanderpost, C. 1992. The 1991 census and Botswana's population problem. *Botswana Notes and Records* **24**: 39–48.

Vegten, J.A. van 1979. Some aspects of African ecology with special reference to Botswana. Working Paper 26. Gaborone: National Institute of Development and Cultural Research (NIR).

Veterinary Office 1992. Cattle numbers for Kalakamate and area formerly called Tati Native Reserve. Based on cattle crush counts, for period 1984–1991. Botswana: Veterinary Office Francistown.

Vossen, P. 1987. An analysis of agricultural livestock and traditional crop production statistics as a function of total annual and early, mid and late rainy season rainfall in Botswana. Department of Meteorological Services. Gaborone: Government of Botswana.

Walker, B.H. 1993. Rangeland ecology: Understanding and managing change. *Ambio* **22**(2–3): 80–87.

Walter, H. 1985. *The vegetation of the earth and ecological systems of the geo-biosphere*, pp. 75–98. Berlin: Springer Verlag.

Werbner, R.P. 1970. Land and chiefship in the Tati Concession. *Botswana Notes and Records* **2**: 6–13.

Werbner, R.P. 1975. Land, movement, and status among Kalanga of Botswana. In M. Fortes and S. Patterson (eds) *Studies in African Social Anthropology*, pp. 95–120. London: Academic Press.

Westoby, M., Walker B.H. and Noy–Meir, I. 1989. Opportunistic management for rangelands not at equilibrium. *Journal of Range Management* **42**: 266–274.

White, R. 1993. *Livestock development and pastoral production on communal rangeland in Botswana*. Gaborone: The Botswana Society.

Bamboo, Rats and Famines:
Famine Relief and Perceptions of British Paternalism in the Mizo Hills (India)

Sajal Nag

As the British entered the Mizo hills (part of the Indo-Burmese range of hills, then known as the Lushai hills) to chase the headhunting tribal raiders and try to gain control over them by securing a foothold in the heart of the hills at Aizawl, they witnessed an amazing ecological phenomenon: a severe famine apparently caused by rats. The Mizo hills are covered extensively by various species of bamboo, which periodically rot, flower and seed. The bamboo seeds appeared to be a delicious food item for jungle rats, which emerged in massive numbers to devour them, and the consumption of bamboo seeds seemed to produce a vast increase in the rodent population. Once the millions of rats had exhausted the bamboo seed, they began to attack the standing crops in the fields. As they devoured the grains the resulting scarcity of food led to massive hardship, starvation and deaths.

THE FAMINES

In the famine of 1881, which was the first to happen under British rule, about 15,000 people perished.[1] In 1912 another famine resulting from the same circumstances took place, affecting a region covering the Mizo hills, Chin hills, Chittagong hills, and the Chin hills falling under Burmese jurisdiction. The Government of Burma organised a great battle against the rodents and destroyed scores of thousands of them.[2] In the Mizo hills, on the initiative of the administration, the tribals set and reset traps in their fields. Individual farmers could traps as many as 500 rats in a single night, and were often seen with basketfuls of dead or flattened rats on their backs, which they had taken out of their long log-traps early in the morning.[3]

The Mizos ate rats. Trapped rats would be dried over the fire and then use as food. But the abundant supply of rats at these times would have diminished their value,[4] moreover, a diet of rats would hardly make up for the loss of rice, which was their staple food. Some of the tribals, who had rice left from the last harvest, struggled to protect it from the invading rats. The unfortunate remainder, who constituted the majority, would search the forest for roots, jungle yams and other wild produce.[5] Wild sago palm was collected from the forests, dried, pounded and its pith sifted, the powder being made into a kind of dumpling that

Environment and History **5** (1999): 245–252.

was wrapped in a leaf and boiled. The resulting food for the family was a very sticky insipid mass, full of gritty particles.

Others ate a kind of wild yam found in the jungle.[6] The plant itself was a creeper. The upper part of its root was inedible, but lower down it changed into a long tuber rich in starch and somewhat resembling a potato in taste. The root was vertical and often very long, so to get out the tuber the tribals frequently had to dig to a great depth in very hard soil. Tragic instances were related of tribals searching for these wild produce to satisfy their hunger.[7] It was reported that the entire forest in many parts of the country was honeycombed with yam pits – most of them four to ten feet deep and large enough to admit the body of a man. One heartbreaking scene is recorded, of a grown up man sitting near one of the holes and crying like a child, because after toiling for hours tracing the roots of a yam he found his way blocked by a huge rock, and therefore he would have to return to his family empty-handed. At another site there was a widow with her baby on her back, working with all her feeble strength to extract the tuber. Often she would become so exhausted that she would lie down to rest, only to find insects crawling all over her – and if she did not get out of the jungle before the dusk, the wolves would devour her.

ANTICIPATION OF FAMINES

The colonial administrators found it interesting that the tribals could correctly predict the next famine from indications in their surroundings. This was on the basis of their past experience. The Mizos had for ages gone through the ravages of the bamboo flowering, and dreaded its impact on their lives. They had observed that there were two distinct varieties of bamboos in their region, which they named as *Mau* and *Thing*.[8] The colonial botanists found that the *Mau* variety was known to European botany as *Melocanna bambu soidef* and the *Thing* as *Bambusa tulda*. Both these varieties had periodic reproductive blooming; in others words, they rotted, flowered and set their seeds every thirty to fifty years. It was during these times that the devastation described by the Mizos as *Tom* occurred. With the aid of the Mizo elders the colonial administrators constructed a record of the past famines, and on that basis could predict the approximate years of the impending series of famines. For example,

Mautam	1862
Thingtam	1881
Mautam	1911
Thingtam	1929
Mautam	1956
Thingtam	1977
Mautam	2007

On the basis of this calculation, the administrators had made advance preparation for the impending 1929-30 famine. Indeed, by 1925 the signs of bamboo flowering were already visible. This time the administrators had an active ally in combating the natural calamity – the missionaries.

The Baptist Mission Society was one of the first groups of missionaries to arrive in the Mizo hills. They had witnessed the ravages of the 1881 famine, and had been active in reducing the suffering of the people in the famine of 1912. This time they began preparation to counter the natural disaster that the Mizos were about to experience. Rev. J. H. Lorrain of the Baptist Mission post at Lungleh wrote to the Superintendent of Lushai Hills,

> I am taking this liberty of writing to you regarding the expected thingtam famine and although I have no connection with the government I trust the expression of my own opinion as to the means which might be employed successfully to counteract the effects of such a visitation will not be unwelcome to one like yourself who has the welfare of the Lushai people so much at heart.[9]

Lorrain then went on to suggest measures to counter the catastrophe. These were 1) ordering the tribals to save a little grain in rat-proof baskets, and 2) application of Liverpool virus to spread a deadly epidemic amongst the rodents, which would destroy them totally.

The Government appreciated the initiatives of the missionaries but found the measures impracticable on the following grounds: 1) the tribals themselves produced a bare subsistence.[10] Most of them did not have a full year's rice, hence to enforce compulsory saving might create more hardship for them and even provoke reactions. 2) No virus had been found to be effective in controlling rat population in other parts of the country. Moreover, the most deadly and rapid spreading virus, the plague bacillus, had had little effect on the rat population of north and western India during the past 28 years.[11] This rendered the application of Liverpool virus out of question. The Superintendent suggested the use of rat-traps and a poison (barium carbonate) instead. The latter would be most effective as well as easy to apply. But the most important task was to reduce the food supply available to the rats: thus the bamboo forests were to be burnt, and standing crops in the fields were to be protected.

THE RAT-BAMBOO CONNECTION

By 1925, the bamboos had started to flower, and signs of fear among the people were visible. Soon there was a massive increase in the number of jungle rats. The people had already begun to destroy them. In December 1924, 45,000 to 50,000 rats were killed in Aizawl subdivision alone.[12] A war against the rats had begun, in which the people, the administrative machinery and non-governmental agencies like the Church combined to fight the menace.

There was no doubt that the famine was caused by the rats invading the standing crops. What remained a mystery was the rapid multiplication of the rats after they had consumed the bamboo seeds. Alexander Mackenzie wrote in 1884, 'the famine arose according to the concurrent testimony of all persons concerned, from the depredation of rats. In the previous season bamboos had seeded; the supply of food thus provided caused an immense increase in multiplication of rats'.[13] There was corroboration of this from missionary witnesses, and like others they were also perplexed by the possible connections between the bamboo seed and the multiplication of rats. The Baptist Mission report stated,

> The periodical flowering, seeding and dying down of certain species of bamboo all over these hills was followed last autumn by an enormous increase in the number of jungle rats ... the connection between the flowering of bamboos and invasion of rats is a disputed point, but the theory which seems to be most satisfactory is that the bamboo fruits has the property of making the rats which eat it, extraordinarily prolific. Whatever may have been the cause directly, the bamboos had seeded and the rats begun to increase and swarm everywhere.[14]

In a letter to the administration, Rev. Lorrain wrote,

> It appeared that the rats began to get more than extraordinarily troublesome years before the simultaneous seeding of the raw-thing bamboos but as soon as the seeding was over they increased to such an extent that no human power could save the crops from their depradation.[15]

Although it was established that bamboo seeds had something to do with the increase in the rat population, no one was sure of the explanation. They felt that perhaps the seeds had some properties that made the rats extraordinarily prolific in terms of reproduction.[16] Perhaps there was some hormonal change in the rats due to the excessive protein that the bamboo seed contained, enabling the female rat to produce a litter much earlier in age than in normal circumstances.[17] Others brushed aside this theory, saying that whenever there is an increase in the supply of food it is normal to find an increase in the rat population.[18] Perhaps rats were migrating from deficit areas to areas of abundance. The third theory was that it was only a visible increase, not a real one. Generally the rats lived in their holes, but with the bamboo flowering they came above ground to eat the delicacy and became more visible to the people. This would be misconstrued by the people as an increase in the number of rats, as they were not used to seeing so many at a time.

The theory that gained most credence was the first one. Although the people, the administration and the Missionaries were firm in their belief of this theory, the administration made no attempt to establish its scientific basis. They concentrated on relief and rehabilitation.

FAMINE RELIEF AND THE IMAGE OF THE RAJ

The Mizo tribals had experienced many such famines, and were used to the hardships, starvation and death that accompany the phenomenon. What they were not used to was being assisted in such times of crisis: this was the difference that the British made to the tribals. Their first taste of British aid was when they began to migrate towards the plains. The tribals would not normally descend to the plains except for raiding purposes or trading, but the hardship due to scarcity of food pushed them down to the plains. In the first batch, about eighty families from the village Kalgom, followed by the eastern Chiefs and then the western Chiefs, migrated to the Dhaleshwari river valley via Jhalnacherra.[19] This caused alarm among the tea labourers of the plains, as they feared the tribals, but the administration apprised them of the situation and calmed them. The distressed tribals were desperately looking for food and livelihood till the famine subsided. They were willing to sell their labour and trade forest products which they had brought along. The administration facilitated their entrepreneurship by temporarily abolishing the duty charged on such products at forest toll stations. They were offered employment in clearing the jungle and felling of trees – jobs to which they were well suited. Within the hills, the administration realised the importance of having a communication network for taking relief to the tribals. So they employed the tribal manpower to construct roads and railways. The missionaries also employed them to construct houses, clear jungle, prepare gardens, etc.

But the problem of food supply still remained. About 18,000 *maunds* of rice and 2,000 *maunds* of paddy was exported to the interiors of Mizo hills in 1881-82 alone.[20] During that crisis the total expenditure in famine relief stood at Rs. 2,240. Of this 1,100 was used for the purchase of paddy and rice, and Rs. 1040 for hire of boats to transport the supply. The missionaries and the administration also supplied cooked food to the hungry. Private traders were encouraged to send rice up to the main markets of Tipaimukh on the east and Changsil on the west.[21] They were asked to open their storehouses of rice and paddy, and they were also provided with Frontier Police protection against possible attack from the tribals. The Government itself opened two storehouses at Tipaimukh and Guturmukh; these were not to compete with the private traders, but to act as a reserve. In addition, government officials visited the affected areas. In 1881 Rai Han Charan Bahadur, the Special Extra Assistant Commissioner, accompanied by Mr. Place, Subdivisional Officer of Hailakandi, visited the frontier areas. Bengalee doctors from Silchar and Chittagong were impressed to treat the sick. In 1911-12, W. N. Kennedy of the Lushai hills borrowed a sum of Rs. 80,000 from the British Government to help the Mautam famine victims.[22]

The administration also took initiatives to reduce the spread of the famine. Since invasion of rats was the main cause of the famine two methods were applied to combat their attack. One was to destroy the rats in large numbers. The

Government provided the people with rat traps, designed specially on the French model and further developed by a Dr. Chitre. They also used log traps around the paddy field, and rewards were announced for killing rats. In one night alone farmers trapped about 500 rats in one trap by setting and resetting it. But the destruction of rats in such massive numbers did not seem to make any impact on the exploding rat population.

The other initiative was to encourage the people to save: this came mainly from the missionaries. People made large rice bins with clappers attached to the bed by a string. During the night some member of family who was awake would occasionally pull the string to flap the clapper and make a sound to scare the rats. It worked for a time, but soon the hungry rats got used to the sound and were not afraid any more. Nor did other saving devices work for long, as the tribal economy was basically a subsistence economy, and they hardly had any surplus to save, except for the Chief and his patrons. The Government did not want to antagonise the tribals by making it compulsory to save, but it arranged to see that those who had surplus food shared with those less fortunate. Tribals were employed to descend to Demagiri and bring back sacks of rice to the hills.

Thus the combined efforts of the Colonial administration and the Church were able to relieve the distress of the famine affected people to a considerable extent. Significantly, this effected a metamorphosis of the image of the Raj in the minds of the tribals. The British first came into contact with the tribals of Mizo hills in 1826, when the latter raided the Sylhet plains and perpetrated headhunting and kidnapping. This began a long story of confrontation, warfare and punitive expeditions. When the British annexed the Cachar plains they also confronted the Mizos, who regularly made raids there for purposes of headhunting and kidnapping. After the discovery of tea in Assam there was a rush to acquire foothill lands for starting tea gardens in the Cachar area as well. This threatened the tribals, who feared that soon the Europeans would invade the hills and deprive them of their homeland. Since then they had led a valiant fight against the white men, resisting every advance of the British towards their hills. In fact, they would often attack the plains, loot settlements, kidnap people, and practise headhunting on the British subjects. This was to register their protest and to scare the Europeans from invading the Hills. The white-skinned Europeans were objects of hatred for the Mizos. They were also amazed at the physical look of these newcomers. 'These enemies [British] are different from other people we have ever seen. They are white as goats. They clothe themselves from head to feet. They cover their feet with leather and we believe they will not able to climb the slopes of the hills'.[23] The Europeans were also ridiculed for their white skins, as 'half-cooked' people. But the same Europeans came across as kind and helpful people during the successive famine-related hardships, as Church records testify:

> In many ways we have been able to alleviate the want and distress around us and gratitude of the poor people has been most pleasing to witness. Scores of men

and women who had no food to eat have been enabled to go down to Demagiri to a fresh supply of food by the loan of a few pounds of rice apiece. Many others have been kept from want by being employed in building, road-making, jungle cutting, gardening and other works about the compound. While not a few who have been unable to work have been assisted with gifts of rice. It has been a peculiar privilege to be living in the Lusai hills this year and thus be able to help the people in their hour of need. They have always looked upon us as their friends and at such times as this, the poor especially find our presence a source of comfort and strength for they feel that they come to us in their extremity and be sure of a helping hand.[24]

The same was true of the administration too. The same report further stated,

Whatever feelings of resentment may have lingered in the hearts of some of these hill people against those who have occupied their country in order to prevent a repetition of their headhunting raids upon the peaceful inhabitants of the plains, this famine must have surely dispelled it. For there are hundreds who would have starved to death this year but for the kindly help rendered by the Government in bringing up thousands of sacks of rice to supply their need.[25]

This report stated the situation after the second famine under the British rule.

Since then, three more famines have stalked the hills of Mizoram. The relief measures provided by the Raj had a profound effect on the overall image of the Raj in the minds of people, who began to look up to the Raj as a kind and merciful system manned by white-skinned Europeans. The administration was paternalistic, and the white men were now addressed as Saab-Pa (white-father), Mirang Bawipa, Mikang Topa or Mirang Topa or Mirang Lalpa, meaning white master, nice white people, or even the white Lord. One British officer, Lewin, was so popular among the Mizos that he was known to the villagers as Thangli-ana – a Mizo name. While the administration attended to the requirements of the people and their needs, the administrators merged totally with the people, learning their languages and within a short time participating in their festivals, rituals and even their routine social life.

The impact of famine relief on the image of the Raj can be better understood when contrasted with the indifference of the Indian State when a similar famine occurred in 1958. People constantly referred to the benevolence and kindness of the British, and recalled how they were better off during British rule; when their distresses were cared for if natural disasters took place. It may be mentioned that the insurgency in Mizoram started only after the 1958 famine, and that the Mizo National Front which led the secessionist movement was originally a voluntary organisation called the Mizo National Famine Front, formed to co-ordinate famine relief efforts. The lamentation of a Mizo poet after the Indian government launched repression against the Mizos, gives poignant expression to the peoples' perception of the Raj:

I dare not contemplate this grief of our land
Departed are our civilised white skinned masters
Oh, God who succour the poor, I pray thee
Set the tottering land on its feet once again.[26]

NOTES

[1] Suhas Chatterje, *Mizo Chiefs and the Chiefdom* (New Delhi, 1995), p. 13
[2] Report for 1912 of the Baptist Mission Society Mission in the South Lushai Hills, Assam.
[3] Ibid.
[4] Ibid.
[5] Ibid.
[6] Ibid.
[7] Ibid.
[8] Lalbiakthanga, *The Mizos: A Study in Racial personality* (New Delhi, 1978).
[9] Rev. Lorraine to the Superintendent of Lushai hills, 17 January 1925.
[10] J. Needham, Sub divisional Officer, Lungleh to the Superintendent Lushai HIlls, 5 February 1925.
[11] Report of Col. Hodgson, director of Pasteur Institute, attached to the letter to N. E. Parry, Suptd. Lushai Hills from J. E. Webster, 19 March 1925.
[12] N. E. Perry, Suptd. of Lushai hills to the Commissioner Surma Valley and Hill Division, Silchar 19 January 1925.
[13] Alexander Mackenzie, *History of the Relations of Government with the Hill Tribes of North Eastern Frontier of Bengal* (Calcutta, 1884). Reprinted as *The North East Frontier of India* (New Delhi, 1994), pp. 325-6
[14] Report for 1912.
[15] Lorraine, 1925.
[16] Report for 1912.
[17] Dr. S. Trivedi, Dept. of Forests, Govt of Arunachal Pradesh and Prof. H. Y. Mohanram, Department of Botany, Delhi university have conducted research on the phenomenon.
[18] Hodgson, 1925.
[19] Mackenzie, 1884.
[20] Ibid.
[21] Ibid.
[22] Vumson *Zo History* (Aizawl, n.d.), p. 139.
[23] Ibid., p. 116.
[24] Report for 1912.
[25] Ibid.
[26] Cited in V Venkata Rao et al., *A Century of Government and Politics in North East India: Mizoram* (New Delhi, 1987), p. 270.

Swidden Farming as an Agent of Environmental Change:
Ecological Myth and Historical Reality in Indonesia

David Henley

This article uses historical evidence from northern Sulawesi (Indonesia) to question the currently popular view of swidden farming as a form of forest management or even conservation. The argument is structured as follows. The first section traces the emergence, in the late twentieth century, of the modern view of swidden farming as an environmentally benign practice. The second introduces the regional setting and the related literature. The third examines historical evidence regarding the duration of the swidden cycle and its impact on the natural vegetation in those cases where the rotation appears to have been sustainable. The fourth highlights evidence for wholly unsustainable swidden practices and for their role in the creation and progressive extension of fire-climax grasslands. The fifth briefly examines why recent writers have tended to forget facts that were once well known about the historical prevalence of short fallow cycles and unsustainable swidden variants. The sixth and concluding section outlines some implications for environmental historiography and conservation policy.

1. A REVOLUTION IN ATTITUDES TO SWIDDEN FARMING: FROM PROBLEM TO SOLUTION

The second half of the twentieth century saw a dramatic reassessment, in academic and policy-making circles, of the environmental impact of swidden farming. In 1966, J.E. Spencer, in his landmark survey *Shifting Cultivation in Southeastern Asia*, could still portray the dominant attitude to this type of agriculture as one of condemnation: swidden farming was held responsible for the 'destruction and waste' of timber, soils, flora, fauna, and 'wild landscapes'.[1] Yet less than thirty years later, an official World Bank country study on environment and development in Indonesia concluded that because swidden agriculture in its traditional form involves 'long term rotation cycles that allow for forest regeneration and soil rebuilding', its environmental impact 'should probably not be considered deforestation at all'.[2]

It was Harold Conklin's *Hanunóo Agriculture: A Report on an Integral System of Shifting Cultivation in the Philippines*, published in 1957, which led the reaction against the colonial and early postcolonial stereotype of swidden cultivation as something 'axiomatically destructive of plant growth, soil, and

Environment and History **17** (2011): 525–554.

other resources'.[3] In this meticulous field study Conklin revealed, essentially for the first time in detail, the ecological sustainability, as well as the complexity, productivity and labour-efficiency, of a traditional bush-fallow foodcrop farming system operating at an appropriate population density. Although Conklin's study was commissioned by the Food and Agriculture Organization of the UN in the context of an international investigation into shifting cultivation as a problematic type of land use, his conclusion, as the FAO noted in its preface, was that in the Hanunóo case there was in fact no problem to be addressed.

> It will perhaps come as a surprise to some readers that Dr. Conklin has not concluded his work by suggesting possible ways of improving the standards of living of the group he has studied. It is felt, however, that in this particular case there was no urgent need for such suggestions. It is a case of almost perfect equilibrium between man and his environment, and if there is any deterioration on either side it is an extremely slow process.[4]

Not all forms of shifting cultivation, Conklin acknowledged, were this benign. Some involved progressive expansion of the deforested area, rather than continuous *in situ* rotation. These unstable forms, however, were characterised in his study as 'incipient', 'pioneer', or 'partial' variants, practised by recent and/ or unskilled migrants from more densely populated areas. As such they fell outside the category of 'integral' (established and traditional) swidden systems exemplified by that of the Hanunóo.[5]

Conklin's work was a key source of inspiration and data for Clifford Geertz's equally influential 1963 book *Agricultural Involution: The Processes of Ecological Change in Indonesia*. Geertz took Conklin's ecological approach one step further by characterising swidden farming as 'the imitation of a tropical forest'.[6] The variety of cultivars planted in the swidden echoed the natural biodiversity of the rainforest, while the fallow stage resembled the process of regrowth following the death of a single tree under natural conditions. Geertz, however, did also emphasise the instability of the system under conditions of population growth, which beyond a critical threshold would lead either to an unproductive grassland climax or to to the adoption of an area-intensive farming system such as wet rice cultivation. This kind of demographically-driven intensification was explicitly modelled by Ester Boserup in *The Conditions of Agricultural Growth: The Economics of Agrarian Change Under Population Pressure* (1965). Boserup proposed a universal transition from forest-fallow cultivation, via bush-fallow and short-fallow, to annual and multi-cropping systems such as irrigated rice farming. In this scheme each transition led to higher areal productivity but at the same time to lower labour-efficiency; hence the preference of sparse populations for swidden farming, which yielded more food for less work than any of its alternatives. What for centuries had been thought of as the most primitive form of agriculture now came increasingly to be seen as the most efficient, and also as 'the ideal way to exploit the tropical environment while conserving it'.[7]

David Henley

During the 1960s and 70s the idea of swidden farming as a close, and under appropriate conditions benevolent, adaptation to forest nature percolated steadily into conventional wisdom among scholars of Southeast Asia, particularly Indonesia. At the same time there was a tendency toward liberal estimation of the customary fallow period: 'an interval that varies from 15 to 20 years'; 'a fallow period of anything from a few years to more than twenty years, during which the vegetation regenerates'.[8] Conklin, by comparison, had reported an average fallow interval of just 8–10 years among the Hanunóo.[9] Geertz, for his part, refrained – prudently, it is tempting to add, in view of his argument that a fallowed swidden was a kind of simulated rainforest – from citing this (or any other) specific figure.

In the 1980s and 90s, concern over rapid deforestation as a result of commercial logging, settler migration and the spread of oil palm and pulpwood plantations gave additional reason to appreciate the ecological virtues of swidden cultivators, whose own practices were indisputably less destructive and who in many cases were being dispossessed and marginalised as the forest frontier retreated. At the same time, and partly for the same reason, there was also rising interest in the smallholder arboriculture or 'agroforestry' systems, based on commercial tree products (including rubber, copra, resins and in some cases also fruits, timber and firewood for local markets), which had emerged in many places as developments of traditional swidden farming via the planting of economically useful trees on abandoned swiddens. Typically far more biodiverse than any estate arboriculture, these systems were hailed as 'recreating the forest' in the form of 'forest gardens'.[10] Some authors even began to describe rotational swidden cultivation itself as a form of agroforestry.[11]

Those who have studied the development of smallholder agroforestry on former swidden land have understood that unlike swidden farming, it is entirely predicated on commerce.[12] They have also understood that the longevity of the tree crops involved means that these cannot generally be integrated into a closed fallow cycle and so must either replace swidden farming proper or, if they continue to coexist with it, lead to additional clearance of natural forest.[13] Nevertheless the reinterpretation of swidden-derived commercial arboriculture as 'ethnoconservation', and of swidden farmers as 'managers of the forest', has helped to inspire an idealistic conviction among anthropologists, ecologists and forest scientists that deforestation in Southeast Asia has always been the result of external pressures or interference, in the absence of which local populations would have continued to live in harmony with the forest.[14]

> While forests were utilized and managed by local communities, no irreversible changes occurred. Tropical forests began to deteriorate as modern state formations introduced centralized management regimes, which triggered the loss of local autonomy and control over self-support systems ...[15]

The same kind of idealism is also evident in the work of historians of the region. Anthony Reid, writing in this journal, has portrayed the relationship between farmers and forests in pre-colonial Southeast Asia as one of idyllic symbiosis.

> Until about fifteen centuries ago the interaction of humans with the Southeast Asian rainforest was primarily one of interdependence. Trees were felled for food and aromatic woods, and in dryer zones to burn in a process of shifting cultivation, but population pressures were low enough for routine regeneration.[16]

Before the modern era of plantation agriculture and mechanised logging, in this view, the only important agents of permanent environmental change were wet rice farming and the cultivation of smallholder cash crops, particularly pepper.[17] Swidden food-crop farming was more a part of the natural landscape than an intrusion into it. Robert Cribb's generally authoritative *Historical Atlas of Indonesia* presents a similar picture: under traditional swidden cultivation, abandoned plots were allowed to 'revert to jungle over a period of perhaps twenty to thirty years'. While this 'certainly affected the structure of tropical forests', Cribb doubts whether it was 'any more significant than natural destructive forces such as landslides or lightning strikes'.[18] Peter Boomgaard, in his standard work *Southeast Asia: An Environmental History*, is more circumspect on this point but nevertheless reports that in traditional swidden farming the fallow phase 'may take as long as thirty years' and that if it falls below eight years, the system tends to become unsustainable.[19]

Given the magnitude of the assumptions that have become entrenched over the past half century, it is high time to reassess the environmental impact of swidden cultivation in a systematic way on the basis of concrete historical evidence. In the remainder of the present article, I attempt to do this in relation to an Indonesian region on which I have made a detailed study encompassing environmental and demographic history.[20]

2. SWIDDEN FARMING IN NORTHERN SULAWESI

'Northern Sulawesi' refers here to the northern half of that island (and its smaller outliers), equivalent to the late-colonial 'residency of Manado' or the combined present-day provinces of North Sulawesi, Gorontalo and Central Sulawesi (Map 1). It covers a land area of some 90,000 square kilometres (roughly the size of Scotland). At the time of the first reasonably comprehensive census in 1930, its population was a little over 1.1 million; a century earlier, in 1830, probably about 700,000. In the nineteenth century its inhabitants were mostly farmers growing subsistence crops on swidden fields and obtaining the limited quantity of trade goods which they consumed (principally textiles and iron) by selling coconut oil, cacao, coffee, surplus rice, or gold dust and forest products collected during the agricultural off-season.

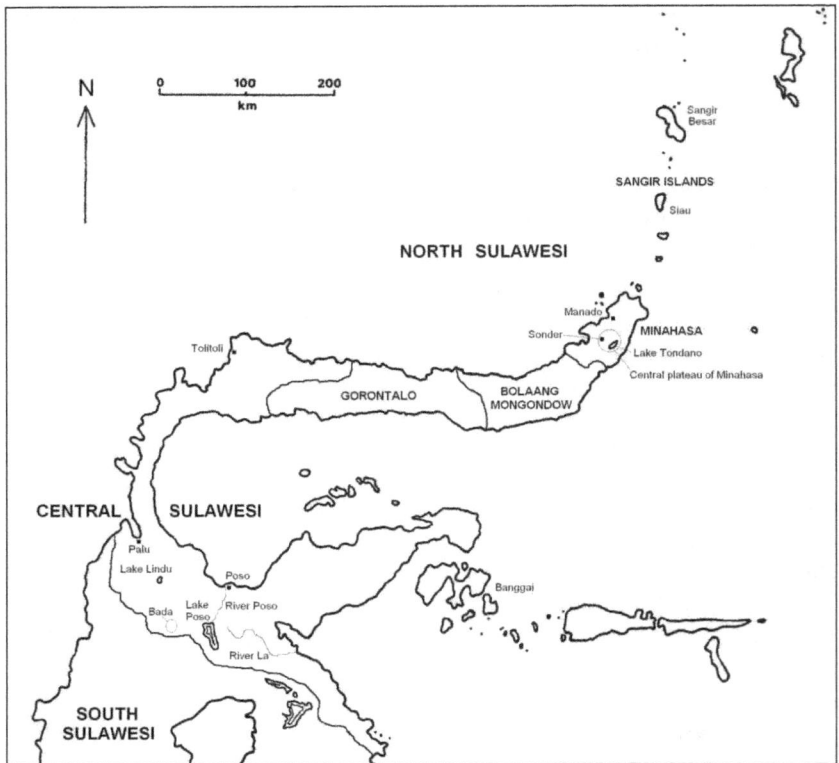

MAP 1. Northern (North and Central) Sulawesi, showing the areas and places mentioned in the text.

Northern Sulawesi is a region for which extensive data relating to swidden farming practices are available in Dutch sources covering the period from the beginning of the nineteenth century up to 1950. Recent literature, by contrast, is of little help on this topic, at least from a historical perspective. The standard reference work *The Ecology of Sulawesi*, by Whitten, Mustafa and Henderson, provides a good illustration of how recent writing on the ecology of swidden cultivation tends to reflect assumptions and extrapolations rather than empirical research. Both in its original (1987) and its second (2002) editions, this book describes swidden agriculture as a sustainable system 'characterised by long fallow periods between short periods of intensive production'. Following temporary cultivation, fields are said to be 'abandoned for 15–30 years during which time the soil recovers'.[21]

Only two references are cited by the authors in support this statement. Of these, one is a general source and the other refers to a forest reserve in West

Java. Neither includes evidence from Sulawesi itself, whether contemporary or historical. Ironically, Whitten and his co-authors do note that in parts of Central Sulawesi there are 'vast areas of grassland caused by shifting cultivation on poor soils'. What 'went wrong' here, they speculate – along the same lines as Conklin, but again without citing any local evidence to support their theory – was that the farmers responsible were immigrants who 'originally worked land in the richer lowlands but were forced by various pressures to move to the poorer soils of the hills where their techniques were inappropriate'. As a result they were forced to abandon their original sustainable swidden rotation in favour of an unsustainable, itinerant practice of continuously shifting cultivation, leading to progressive deforestation and grassland creation.[22]

How do these retrospective speculations compare with the contemporary data on swidden farming practices found in historical sources? The discussion below is divided into two parts. Section 3 looks at what contemporary reports say about length of the swidden cycle and its implications in terms of how much primary forest is permanently replaced by what kind of secondary vegetation, assuming that the rotation is indefinitely sustainable. Section 4 highlights some evidence for wholly unsustainable swidden practices, and for their role in the creation and progressive extension of fire-climax grasslands.

3. THE SWIDDEN CYCLE: DURATION AND IMPACT UNDER SUSTAINABLE CONDITIONS

Table 1 shows all the quantitative information on cultivation periods and fallow intervals that I have been able to find in Dutch sources from the colonial period. The earliest figures date from the 1820s, the latest from the 1940s. The majority refer to one of two areas: Minahasa, the mountainous, volcanic, agriculturally productive area at the tip of the northern peninsula of Sulawesi where Dutch involvement was longest and most intensive; and the almost equally fertile and populous Sangir islands to the north of Minahasa. However there are also some data from less densely populated areas, with more typical, infertile tropical soils, further south in Bolaang Mongondow, Gorontalo, and Central Sulawesi. What immediately stands out here is the general shortness of the recorded swidden cycles. Of the 41 entries in this table, only nine (22 per cent) mention a fallow period of more than eight years. Unexpectedly, the fallow intervals from the less fertile areas outside Minahasa and Sangir (1857, 1888, 1912b, 1917, 1927, 1931a, 1931b, 1932, 1939) are no longer than the rest. Only two (5 per cent) of the entries – both of them, ironically, from Minahasa – seem to refer to what might be described as long fallow cycles: respectively up to fifteen years (1846c), and up to 25 years (1895a). Unlike the majority of the data, both of these high figures are of doubtful reliability since they are actually estimates

David Henley

by visitors of the age of briefly observed secondary vegetation, rather than descriptions by resident officials of farming practices observed over long periods.

TABLE 1: Swidden cycle data in Dutch sources on northern Sulawesi, 1821–1949

Year	Locality	Fallow interval (years)	Main crop cultivation period (years)	Source
1821	Minahasa	5–6	1	Reinwardt 1858, 585
1824	Minahasa	5	1	Olivier 1834–37, II, 35
1825	Minahasa	at least 3, often 5 or more	not specified	Riedel 1872, 539, 540
1825	Sangir islands	4–5	not specified	Van Delden 1844, 18
1833	Minahasa	2–4	not specified	AV Manado 1833 (ANRI Manado 48)
1840	Minahasa	3–5	not specified	Van Doren 1857–60, II, 362
1846a	Minahasa	at least 3	1–2	Francis 1860, 349, 350
1846b	Minahasa (central uplands)	3–5	2	Grudelbach 1849, 406
1846c	Minahasa (northwest coast)	6–15	not specified	E. Francis, Aantekeningen, 12 juni 1846 (ANRI Manado 50)
1855	Minahasa (west coast)	5	not specified	Fragment 1856, 150
1857	Bolaang Mongondow (north coast)	3–5	not specified	Riedel 1864, 272
1860	Minahasa (west coast)	7–10	1–2	Teysmann 1861, 344
1861	Minahasa (central uplands)	4–6	not specified	CV Manado 1861 (ANRI Manado 95)
1863	Minahasa	3–10	not specified	CV Manado 1863 (ANRI Manado 52)
1864a	Minahasa	3–10	not specified	CV Manado 1864 (ANRI Manado 39)
1864b	Minahasa	2–5	1	Graafland 1864, 8
1866	Minahasa	4–7	not specified	CV Manado 1866 (ANRI Manado 52)
1869	Minahasa (southwest coast)	5–8	2	De Clercq 1870, 526–7
1870	Minahasa	2–4 or more	1	N.P. Wilken 1870, 374
1872a	Minahasa	2–4 or more	1	G.A. Wilken 1873, 134
1872b	Minahasa (southeast coast)	9–10 or more	not specified	G.A. Wilken 1873, 134
1875	Minahasa	2–10 or more	2–4	Edeling 1919, 48

Year	Locality	Fallow interval (years)	Main crop cultivation period (years)	Source
1879	Minahasa	3–10	not specified	CV Manado 1879 (ANRI Manado 86)
1881	Minahasa	at least 3, 'usually much longer'	not specified	Matthes 1881
1888	Central Sulawesi (Palu Bay)	2–3	not specified	Landschap Donggala 1905, 522
1895a	Minahasa	up to 25	not specified	Koorders 1898 (vegetation map)
1895b	Minahasa (central uplands)	6–8	2–3	Koorders 1898, 26
1895c	Central Sulawesi (Poso area)	6–8	not specified	A.C. Kruyt 1895–97, II, 117
1895d	Central Sulawesi (Poso area)	3–4	not specified	A.C. Kruyt 1895–97, III, 131
1900	Central Sulawesi (western highlands)	5–6	not specified	A.C. Kruyt 1938, IV, 35
1912a	Minahasa	2–6	1	Dirkzwager 1912, 1165
1912b	Gorontalo (south coast)	5 or more	1–2	Regeeringsrapport Boalemo 1914, 169
1917	Gorontalo (north coast)	4–10	1–3	Van Andel and Monsjou 1919, 118–19
1927	Central Sulawesi (Tolitoli, west coast)	2–3	2–3	Kortleven 1927, 88
1931a	Central Sulawesi (Palu Bay)	3–6	1	Dutrieux 1931, 3
1931b	Central Sulawesi (western highlands)	5–8	1	Dutrieux 1931, 3
1932	Central Sulawesi (Banggai, east coast)	about 5	not specified	A.C. Kruyt 1932, 477
1938	Minahasa	3–5	1–2	Weg 1938, 147
1939	Central Sulawesi (Lindu, western highlands)	6	1	Bloembergen 1940, 390
1941	Sangir (Siau)	5–8	3–5	M. van Rhijn 1941, 42
1949	Sangir (Sangir Besar)	5–7	1	Blankhart 1951, 86

A second striking feature of the swidden cycle data is that the stated fallow periods show no discernible downward trend over time: in the early nineteenth century they are already just as short as in the early twentieth. Figure 1 illustrates graphically the average reported fallow intervals listed in Table 1. Here the trend line lies virtually flat, over the whole period from 1820 to 1950, at an average fallow period of between five and six years. This line takes no account of locality or of changes in the cultivation (as opposed to fallow) period,

David Henley

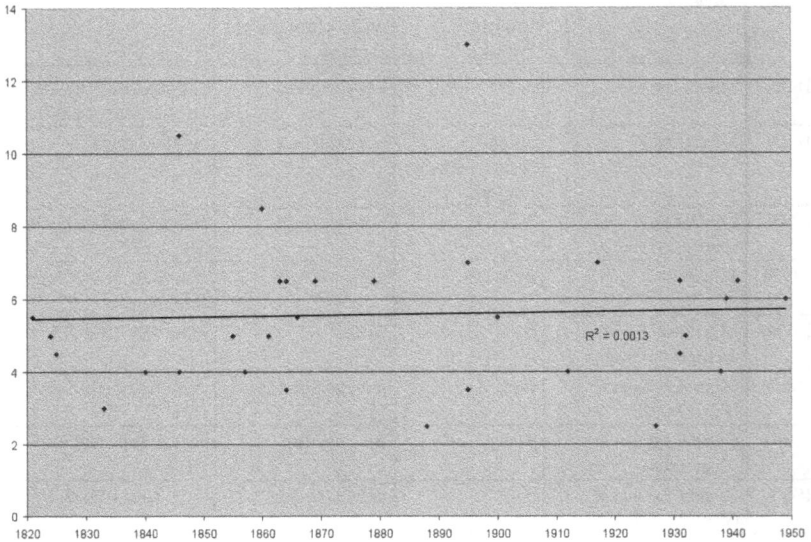

FIGURE 1. Average reported fallow interval (years) for swidden cultivation systems
in northern Sulawesi, 1821–1949. Sources: see Table 1.

and does not include those fallow interval data which specify no upper limit. However, the data from the two localities (Minahasa and Sangir) for which a considerable series of fallow period indications is available do not suggest local shortening either; the limited information on cultivation periods likewise implies no systematic change; and the open-ended fallow period indications (1825, 1846a, 1870, 1872a, 1872b, 1875, 1879, 1881, 1912b) mostly date from the middle rather than the beginning of the documented period, which again does not imply a consistent chronological trend.

The absence of any observable tendency for swidden cycles to become shorter between 1820 and 1950 is remarkable, given that this was a long period of strong population growth. In Minahasa, the area to which most of the swidden data refer, the population more than tripled from about 100,000 in 1850 to over 300,000 in 1930, raising the average population density from about twenty persons per square kilometre to more than sixty. The additional inhabitants were supported not by means of an intensification of swidden agriculture, but by a direct transition in many places from rotational to permanent-field farming (and to some extent by food imports). The proportion of the Minahasan rice crop grown on irrigated rather than dry (mostly swidden) fields grew from less than a quarter in 1850 to over sixty per cent in 1930.[23]

FIGURE 2. Wet ricefields in Central Sulawesi (Kulawi, near Lake Lindu), with swidden fallow vegetation on adjacent slopes, circa 1913. Source: Abendanon 1915–18, II, Plate CLVIII.

The paradoxical combination, at the beginning of our period, of a relatively low population density with a short average swidden cycle is clarified by the fact that the distribution of the population was highly uneven. In the early nineteenth century the population of Minahasa was concentrated on the fertile central plateau around Lake Tondano, where its density was at least seventy persons per square kilometre.[24] The rest of the country was largely uninhabited except for a few small coastal harbour settlements, so that the average population density for Minahasa as a whole was only around twenty persons per square kilometre. Elsewhere in the region too, small, populous enclaves of intensive short-fallow swidden farming, often located in high mountain valleys where they were supplemented by even smaller areas of wet rice cultivation on bunded or terraced fields (Figure 2), alternated with wide expanses of uninhabited wilderness (Map 2). Almost nowhere was the population sparsely and evenly enough distributed to permit long-fallow swidden rotations.

The prevalence of very short swidden cycles, even at the beginning of our documented period, was partly a direct result of the geographical concentration of the population, which in turn reflected the limited availability of high-quality farmland and the need for defensive security in the form of relatively large nuclear settlements. However, there was also a positive preference among farmers for opening their swiddens in secondary vegetation that was still young, and therefore relatively easy to cut and clear.[25] Fallow vegetation more than eight

David Henley

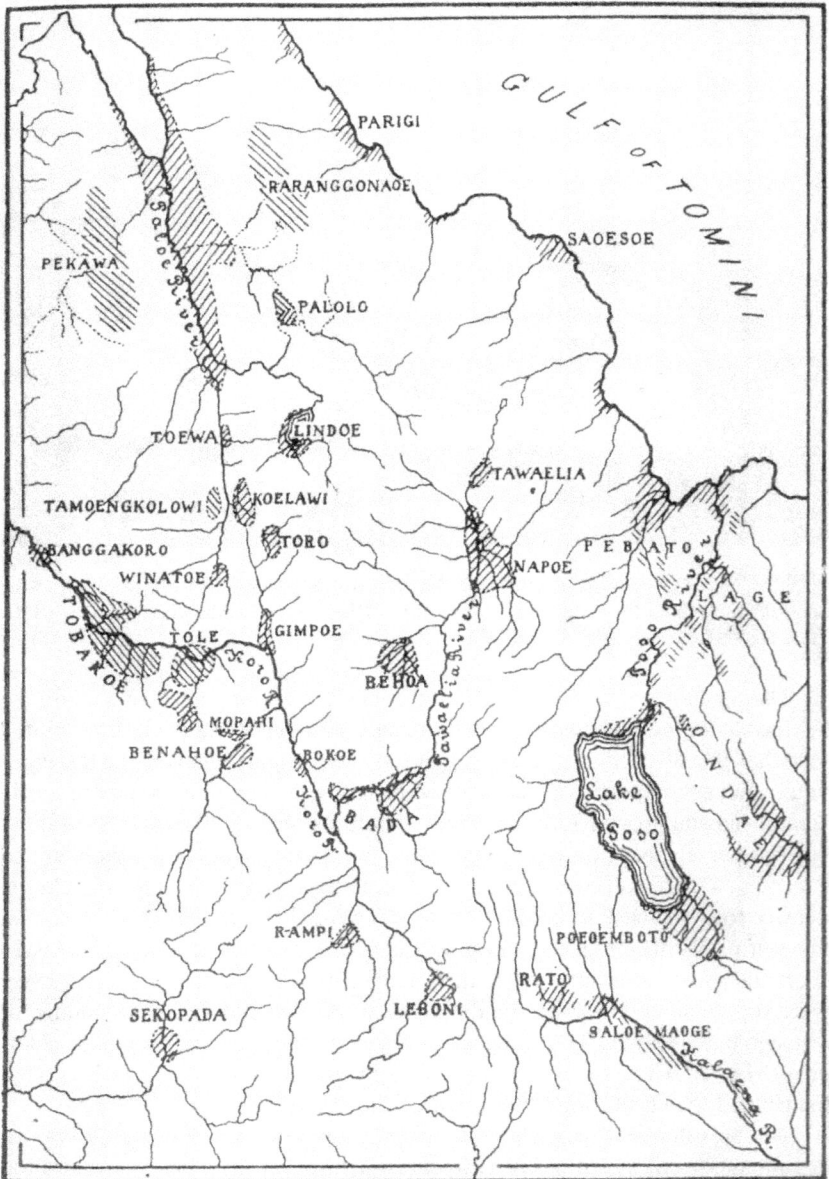

MAP 2. Western Central Sulawesi: inhabited (shaded) and uninhabited (unshaded) areas, circa 1920. Source: Kaudern 1925, 33.

Swidden Farming

years old, as Seavoy has noted, can no longer be felled with a bush knife, only with an axe.[26] Besides being much more laborious to fell, older forest is also much more difficult to burn, entailing a risk that the cleared plot will have to be abandoned until the large trunks have rotted.[27] The fact that the productivity of short-fallowed swiddens was evidently not so poor as to outweigh these disadvantages of long fallowing is consistent with recent empirical evidence that fallow length is a weak predictor of crop yields in swidden cultivation.[28]

With swidden fallow intervals from the outset averaging only five to six years, it is obvious that within the populated areas the swidden landscape (Figures 3 and 4) never bore much resemblance to the natural rainforest it replaced, in which the canopy trees were typically centuries old and over thirty metres tall. The man-made character of the fallow vegetation was particularly evident where it consisted partly of deliberately planted fruit trees and palms.

> This fallow land ... has a monotonous appearance. Invariably it is characterized by a great number of ferns and copses of bamboo together with papaya and *tagalolo* [*Ficus septica*] trees, and by *lingkuwas* [*Alpinia galanga*] and *galoba* [*Costus* species] bushes, among which a coconut palm or a banana tree rises up here and there.[29]

FIGURE 3. Swidden landscape, Minahasa (village of Sonder, central plateau), 1824.
Source: RMV, Collectie A. Payen, Calpin A, Sketch 36.

David Henley

FIGURE 4. Swidden landscape (with swidden houses), Poso area, circa 1905. Source: Adriani and Kruyt 1912–14, Plate 47.

Even in those places where the secondary woodland was denser and contained fewer cultivars, the stem diameter of its constituent trees was typically still only five to six centimetres, its canopy height five to six metres. And instead of the thirty to fifty species of large tree typical of virgin hill forest, here there were just four or five.[30]

With a ratio of fallowed to cultivated land area often as low as 3:1, the swidden landscape was characterised not by occasional clearings scattered through high forest, but rather by a continuous patchwork of contiguous blocks of anthropogenic vegetation, some under current cultivation and others in various stages of more or less woody regrowth.

> The land in the neighbourhood of the village is usually divided into sections, the number of which depends on the number of years for which the land is left fallow, such that each farmer always returns to the same location after the cycle has elapsed. Usually there are five sections, sometimes more, sometimes fewer … The larger landowners have land of their own in each of these sections.[31]

These blocks of adjoining, simultaneously cultivated swiddens were sometimes very large, bearing during the burning and planting phase a striking resemblance (except for the absence of large-diameter tree stumps) to the landscapes created by modern logging operations (Figure 5).

FIGURE 5. Planting rice on a large swidden complex, Central Sulawesi, 1912.
Source: ARZ photograph collection, Land- en Volkenkunde Celebes, 403.

During the pre-colonial and colonial periods, the fact that the population was still small by modern standards meant that the proportion of the total natural forest cover lost to swidden cultivation (and other human land uses) was still limited: just over a quarter in 1941.[32] Map 3 shows the approximate distribution of the deforested areas around that date. It is also true that even within the

David Henley

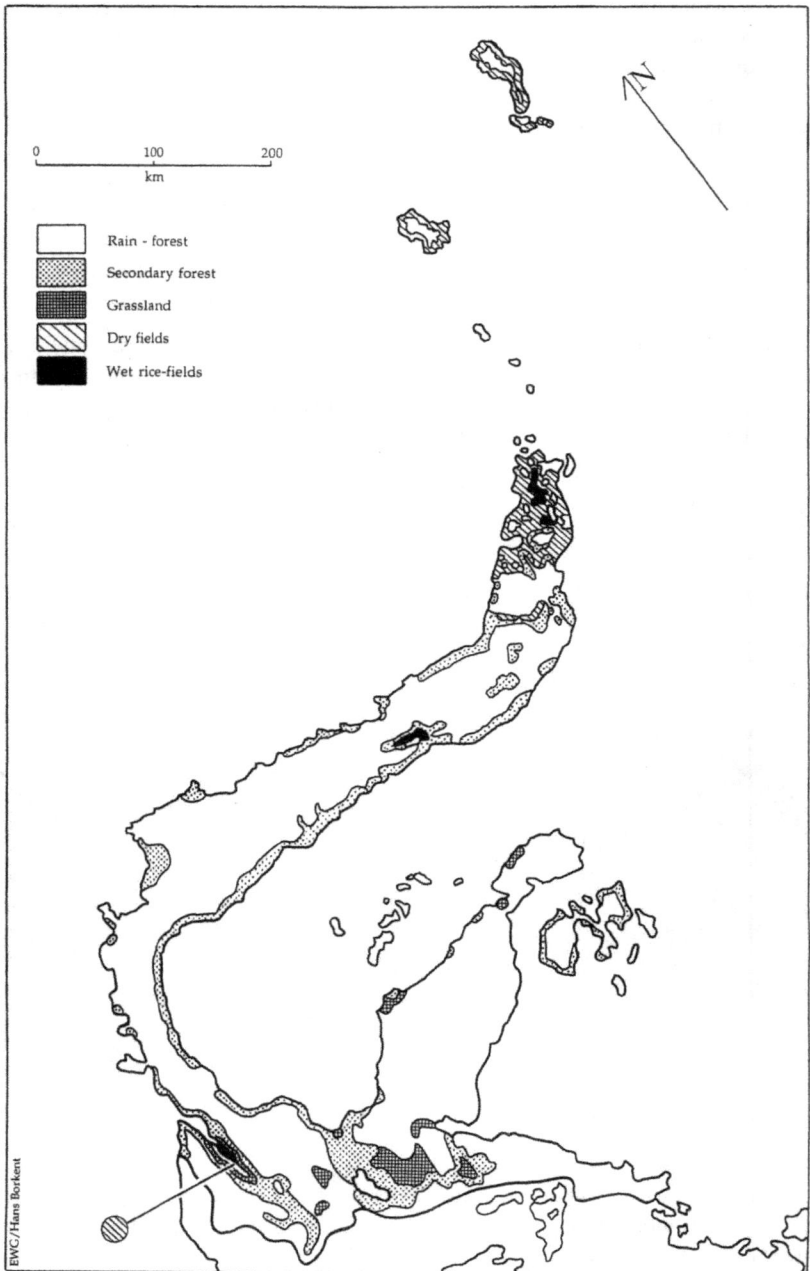

MAP 3. Northern Sulawesi: vegetation cover circa 1950. Source: redrawn from Van Steenis 1958.

populated enclaves, small areas of permanent semi-natural forest were sometimes preserved for ritual purposes or as timber reserves.[33] Nevertheless, traditional swidden farming practices in northern Sulawesi clearly did not involve any kind of symbiosis with the natural forest. On the contrary, they involved the destruction of that forest and its replacement by completely different, man-made ecosystems that were much less rich and diverse.

4. UNSUSTAINABLE SWIDDEN PRACTICES AND PROGRESSIVE DEFORESTATION

So far it has been assumed that the swidden rotations documented in Table 1 were stable and sustainable, in the sense of not entailing progressive destruction of natural forest beyond the boundaries of an established complex of periodically fallowed and recultivated swidden plots. In most areas, as even sceptical Dutch colonial observers conceded, this was indeed the case.[34] There were, however, exceptions. Early descriptions of farming and settlement in the sparsely populated hinterland of Poso in Central Sulawesi, for instance, reveal a deviant pattern. Food-crop agriculture, here as in many other areas, was based on rice and maize, and took place exclusively on dry swidden fields. The fallow period, at three to eight years (Table 1, 1895c, 1895d), was also unexceptional. In Poso, however, this rotation was reportedly unsustainable, so that whole villages, and not just their outlying swidden houses, were periodically forced to shift to new locations.

> When the area of land that such a village needed to feed itself had become completely deforested and exhausted, the villagers chose a site for a new settlement within the territory of their tribe and founded a new village there. The old village was abandoned to its fate; the houses quickly became derelict, the protective pallisade decayed and only the coconut palms remained to mark the hill as the site of a former village. The traveller in this land saw at least as many such abandoned villages as inhabited ones.[35]

The existence of itinerant farming systems in Poso and adjacent areas had to do with the fact that farmers here combined swidden agriculture with extensive animal husbandry and hunting.[36] For this purpose they routinely set their abandoned swiddens on fire to promote the growth of young grass shoots for grazing by semi-domesticated water buffalo and wild deer. Because of the limited amount of fertile ash which the grass yielded after burning, and because of the heavy labour involved in extracting grass root mats from the soil or weeding their fast-growing shoots, grasslands created in this way were difficult to reincorporate into the swidden farming cycle.[37] As the grass spread, moreover, fires raging across it became increasingly difficult to control, accelerating the emergence of a permanent or semi-permanent fire-climax grassland (Figure 6).

It was this blocking and displacement of the swidden cycle by progressive

David Henley

FIGURE 6. Grasslands in Central Sulawesi (Bada), circa 1912. Source: ARZ photograph collection, Land- en Volkenkunde Celebes, 39.

grassland formation which caused farmers in eastern Central Sulawesi to become
shifting rather than rotational cultivators, periodically moving their villages
on to new forest lands and leaving open expanses of sword grass or savanna
parkland in their wake. The approximate distribution of these grasslands in the
mid-twentieth century is shown in Map 3. The biggest of them, in the La river
basin between Lake Poso and the east coast, stretched unbroken over an area
of several hundred square kilometres, its treeless expanses almost uninhabited
except for herds of deer and water buffalo, but still periodically set on fire in
order to prevent vegetation succession and keep the livestock density high.[38]
Contemporary observers noted that such landscapes could not be created by
swidden agriculture alone, without the additional factor of repeated burning.

> When trees are felled to clear a swidden and the land is left to its fate once more
> after the harvest, young forest grows up again spontaneously, and there is no
> question of *alang-alang* fields being formed. This species of tough grass, which
> has a subterranean rhizome, does come up, but is subsequently overwhelmed by
> shrubs and trees. If people burn off the *alang-alang*, however, they destroy the
> young trees while actually promoting the growth of the grass, which burns only
> to ground level and then immediately puts up young shoots from its rhizomes,
> whereas the bushes and trees need a much longer recovery period.[39]

If human settlement became so sparse or distant that the burning stopped altogether, then the tree cover would slowly regenerate, eventually making the land
suitable for settlement and cultivation once again.[40]

Large parts of eastern Central Sulawesi are naturally infertile, consisting either of limestone overlain by shallow calcareous soils, or ultrabasic rocks overlain by acid soils.[41] This helps to explain why farmers there were unusually willing to convert swiddens into permanent pasture, even when doing so would eventually make it necessary for them to relocate their villages. There is also some evidence in the historical record, and in oral tradition, for relocation as a direct result of declining soil fertility rather than exhaustion of forest and fallow woodland.[42] Nevertheless there is no evidence that itinerant cultivation and progressive grassland formation in Central Sulawesi were the results of 'incipient', 'pioneer' or 'partial' swiddening systems practised by inexperienced migrants from more densely populated areas, as suggested by Whitten, Mustafa and Henderson.[43] The areas concerned had a long history of settlement; the water buffalo herds central to the expansionary dynamic were not a recent or commercial innovation but a traditional element of the subsistence farming system, their importance underpinned by ritual and by competitive feasting practices involving animal slaughter (Figure 7).

FIGURE 7. Funeral feast in the Poso area with animals awaiting slaughter, circa 1918. Source: A.C. Kruyt 1919, 55.

Progressive expansion of grassland and savannah at the expense of cultivated or cultivable land could not, of course, have continued indefinitely at a constant population density. The fact that the population of eastern Central Sulawesi underwent a secular decline during the late nineteenth century, together with reports of sustained emigration from one area due to 'overpopulation and the resulting lack of arable land', confirms that the demographic-ecological system was not in equilibrium.[44] But it would be wrong to conclude from this that the farming practices involved were novel, eccentric or unusual. Protracted episodes of ecological disequilibrium have been common throughout history, sometimes with fatal consequences for whole civilisations.[45]

Half a century ago, the antiquity and extent of grassland formation as a consequence of the combination of swidden farming with animal husbandry was well known to scholars of Indonesian agriculture.

> [T]his farm system and the corresponding extensive grassy plains are to be met with in the wet districts of Indonesia, e.g. the Gajo, Alas and Batak lands, in Upper Palembang (Pasumah) and the Toradja district and ... the north coast of Atjeh, the Padang Lawas south of the Batak district, in the lower districts of Palembang and the Lampongs, in southern Priangan, and in the whole of southern Celebes ... It is obvious that this farm system is very soon bound to create extensive grasslands in the drier districts, and this used to be seen in the north of Indramaju and is still found sporadically in eastern Java, although to a greater extent on Sumbawa, Sumba, Flores and Timor and the smaller islands where cattle are tended ...[46]

Likewise, the prevalence of short swidden cycles in Southeast Asia is also frequently noted in older literature.[47] R.D. Hill, in his historical geography of rice cultivation in Malaya, went so far as to regard all swidden systems involving fallow periods of more than eight years as recent deviations from a traditional short-fallow norm.[48] Why, then, have many recent writers lost sight of the fact that swidden farming was always an important agent of permanent deforestation?

5. LOSING SIGHT OF THE PAST: REASONS FOR FORGETTING

One reason for this scholarly amnesia is the disproportionate emphasis in recent literature on Borneo, the front line of current biodiversity destruction and the site of some of the best-known studies of indigenous agriculture. Although Borneo always had its share of short-fallow systems, its exceptionally infertile soils and low population densities also gave rise to some singularly unintensive agriculture – notably the itinerant shifting cultivation, influentially described by Freeman (1955), of the Iban, whose traditional preference for felling virgin rather than secondary forest was the more remarkable given that they were neither livestock farmers nor grassland burners. Genuinely rotational long-fallow systems also seem to have been more widespread in Borneo than elsewhere, and some are still in operation there today.[49]

A second explanation for the neglect of historical evidence for short-fallow and unsustainable swidden farming is that many of the relevant sources, especially for regions outside Kalimantan, have over the years become obscure and inaccessible. Certainly few of the Dutch publications cited in the present article, and none of the archive materials, were readily available to Whitten, Mustafa and Henderson when they were compiling their handbook on *The Ecology of Sulawesi*. Given that swidden farming was not a central concern of their study, under these circumstances it no doubt seemed expedient to deal with the subject by extrapolating from recent and general sources.

In addition, however, it is hard to avoid the impression that even authors who do possess good information on this topic sometimes prefer not to address the questions which that information poses. A table of 'features of swidden agriculture of various groups in Indonesia', presented by De Jong *et al.* in a special issue of the *Journal of Tropical Forest Science* devoted to 'Secondary Forests in Asia', includes three groups in Sulawesi which employ fallow periods of between two and four years – too short for the growth of anything that could be called 'secondary forest'.[50] In fact none of the fallow periods given in this table, the two from Kalimantan not excepted, exceed ten years, and only one (from Malay Sumatra) exceeds seven years – although several are accompanied by the suspiciously parenthetical note: '(previous 15–20)'. No comment on the shortness of the actually observed cycles appears in the accompanying text, which instead proceeds to focus on 'emerging tree crops', particularly rubber, that are increasingly replacing swidden systems 'in stands that may be considered secondary forest gardens'.[51] In other words it is to permanent agroforests, not to fallow vegetation, that the term 'secondary forest' refers here.

By not highlighting this distinction, these and many other authors give the impression that it is the traditional practice of the farmers in question, rather than their conversion to commercial arboriculture, which has led them to engage in 'recreating the forest'.[52] The characteristic conflation here of smallholder arboriculture, swidden farming and forest conservation is not, of course, intended to mislead. But it does reflect an ethically, politically and romantically inspired insistence on the ecological virtue of Indonesia's tribal and post-tribal 'indigenous peoples', most of whom live close to forested areas and were swidden farmers in the past.[53]

6. IMPLICATIONS FOR SCHOLARSHIP AND POLICY

The portrayal of swidden farmers as guardians of the forest is, in effect, a variant on the myth of the 'ecologically noble savage'.[54] Swidden farming is a vanishing way of life in Southeast Asia, and there is no doubt that the transformation of its practitioners into smallholder tree crop planters is a better ending to its story, both for the farmers and for their environment, than their displacement or proletarianisation by big business in the form of logging and plantation concerns.[55] De Jong is also right to point out that in the past, many swidden farming communities did preserve small areas of primary (or at least old secondary) forest within their territories for ritual reasons and as reserves of timber, rattan, medicinal plants and game.[56] But when considering the historical relationship between swidden farming and deforestation, it remains important not to miss – if the pun can be excused – the wood for the trees. Even if we accept that modern smallholder tree plantations are forests, they are still dependent on access to markets, without which tree crops could not be sold, imported foodstuffs could not be purchased, and more land would have to be cleared of trees for local food production. In the past, when commerce was less highly developed, 'integral' swidden farmers like those described by Conklin may have cultivated some fruit and other trees for their own use; but their primary focus was on subsistence field crops like rice, which could never be grown under a forest canopy. It follows that on balance, they had a much greater interest in felling trees than in planting them.

Swidden farmers did, of course, allow young uncultivated trees to grow on their fallowed fields. But this does not mean that they lived in a 'forest'. In *The Conditions of Agricultural Growth*, Ester Boserup distinguished between two types of shifting cultivation. The first was 'forest-fallow', with fallow intervals of at least twenty years, allowing the development of 'secondary forest'. The second was 'bush-fallow', with intervals 'usually between six and ten years'. 'No true forest', Boserup noted, 'can grow up in so short a period, but the land left fallow is gradually covered with brush and sometimes also with small trees'.[57] Most traditional Indonesian swidden farming practices fell into this second, bush-fallow category. The associated fallow vegetation – even at its maximum development, in which form it covered only a small part of the landscape – consisted precisely of 'small trees'. It was also far less complex in composition than the Southeast Asian rainforest that it replaced, the biodiversity of which is greater than that of any other ecosystem on earth.[58]

The impact of swidden farming on the natural vegetation, then, was neither slight nor temporary. The system's relatively profligate use of land meant that even when it took sustainable forms, compared to wet rice cultivation and other non-rotational systems it was still responsible for much more deforestation per head of the human population which it supported. Of all the sustainable options, the preferred bush-fallow swidden variant was in a sense the most destructive possible choice: neither intensive enough to spare the forest, nor extensive enough to incorporate and recreate it.

A final point to reiterate is that not all swidden farming was sustainable anyway, in the sense of involving a stable *in situ* fallow rotation. Some swidden farmers were always shifting rather than rotational cultivators, periodically moving their villages on to new forest lands and leaving uninhabited expanses of grassland or savanna in their wake. This practice was associated with the deliberate but poorly controlled use of fire to stimulate the growth of young pasture for livestock. The evidence from Sulawesi does not suggest that it resulted in any direct way from commerce, migration or other external influences. The unstable combination of shifting cultivation with animal husbandry and grassland burning which produced the most spectacular deforestation in pre-colonial Indonesia must be regarded as an indigenous system, no less 'traditional' in nature than its (relatively) less environmentally destructive counterparts.

These facts require scholars to dispense once and for all with the comforting modern myth that at one time, when swiddening was the dominant form of agriculture, 'the interaction of humans with the Southeast Asian rainforest was primarily one of interdependence', and that deforestation began with commerce, 'settled cultivation' and 'the loss of local autonomy' to 'modern state formations' and 'centralized management regimes'.[59] On a more purely practical note, taking a realistic view of swidden farmers as agents of forest destruction may also mean reconsidering calculations of recent deforestation rates which underestimate the early extent of bush-fallow and fire-climax vegetation.[60]

Does the revisionist view of swidden cultivation presented above also have policy implications for environmental protection and conservation? In a time when swidden farming as such has almost disappeared, any such implications are bound to be limited. In northern Sulawesi today, cocoa planting – largely by migrants from other regions who settle for that purpose on the forest frontier – has long eclipsed swidden farming as an agent of deforestation.[61] Nevertheless there have been instances in which indigenous farmers, caught between the migrant expansion and the boundaries of the few remaining protected nature reserves, have been permitted to occupy land within a national park on the grounds that their traditional land use practices, including swidden cultivation, enable them to 'manage park resources in a sustainable fashion'.[62]

While the human (as opposed to environmental) ethics of this decision are beyond the scope of the present discussion, its premise – that ecological sustainability is equivalent to, or at least compatible with, nature conservation – is flawed. Ultimately that premise proved irrelevant to the outcome of the experiment, which was that the beneficiaries of the dispensation planted cocoa trees just as migrant settlers would have done. But even if they had kept the promise given by their representatives that they would practice only traditional agriculture, their presence would have been inconsistent with any idea of preserving natural ecosystems. The conservation of tropical rainforests is incompatible with agriculture – traditional or otherwise, and notwithstanding insistent claims to the contrary.[63]

This is as true of swidden cultivation as it is of other farming methods: more true, in fact, since swiddening, even when it takes sustainable forms, is uniquely profligate in its use of land. With the age of mature trees in virgin Southeast Asian rainforests ranging between 200 and 700 years, it would still be true even if the twenty- and thirty-year cultivation cycles mentioned in the literature were common in reality.[64] But with the real norm, even in the distant past, lying between five and eight years and with unsustainable, expansionary systems being common at all documented periods, any idea of swidden farming

David Henley

as a form of rainforest conservation is clearly far-fetched. The support or consent of nearby swidden farming populations, if it can be obtained, may facilitate the protection of nature reserves. But what is certain is that in the area they actually farm, there will be no more rainforest.

LIST OF ABBREVIATIONS

ANRI Manado [...] Arsip Nasional Republik Indonesia (Indonesian National Archive, Jakarta), Manado residency collection (1677–1914), [archive bundle number]

ARZ Archief van de Raad voor de Zending der Nederlandse Hervormde Kerk (Archive of the Council for Missions of the Dutch Reformed Church), Het Utrechtse Archief, Utrecht.

AV Algemeen Verslag (General Report)

CV Cultuur Verslag (Cultivation Report)

NA MvO [...] Nationaal Archief (Dutch National Archive, The Hague), Indonesian Memories van Overgave collection, [document number]

RMV Rijksmuseum voor Volkenkunde (National Museum of Ethnology), Leiden

NOTES

[1] Spencer 1966, p. 3.
[2] World Bank 1994, p. 52.
[3] Spencer 1966, p. 3.
[4] Conklin 1957, p. v.
[5] Conklin 1957, pp. 2–4, 154–5.
[6] Geertz 1963, p. 31.
[7] Russell 1988: 92.
[8] Hardjono 1971, p. 133; Missen 1972, p. 34.
[9] Conklin 1957, p. 145.
[10] De Jong 1995; Salafsky 1994.
[11] Colfer, Peluso and Chin 1997, p.156; Sinclair 1999: 175.
[12] Gourou 1940, p. 348; Michon and de Foresta 1995: 94.
[13] Dove 1993: 145; Tammes 1949, p. 7.
[14] De Jong 1995; Colfer and Dudley 1993.
[15] De Jong, Lye and Abe 2003, p. 19.
[16] Reid 1995: 93.
[17] Reid 1995: 93, 101–103.
[18] Cribb 2000, p. 23.
[19] Boomgaard 2007, pp. 220, 222.
[20] Henley 2005b.

[21] Whitten, Mustafa and Henderson 1987, pp. 575, 577; 2002, 570.

[22] Whitten, Mustafa and Henderson 2002, p. 571.

[23] Henley 2005b, pp. 405, 408, 526.

[24] Henley 2005a: 160.

[25] Adriani and Kruyt 1912–14, II, p. 239; Graafland 1864, p. 20.

[26] Seavoy 1973: 219.

[27] Li 1991, pp. 37, 40.

[28] Mertz *et al.* 2008.

[29] De Clercq 1870: 527.

[30] Koorders 1898, p. 26.

[31] Edeling 1919: 51.

[32] Boomgaard 1996, p. 166.

[33] Henley 2005b, pp. 578–9.

[34] Henley 2005b, pp. 567–70.

[35] Adriani 1919, pp. 9–10.

[36] Henley 2005b, pp. 534–7.

[37] Adriani and Kruyt 1912–14, II, p. 239.

[38] Henley 2005b, pp. 477–9.

[39] J. Kruyt 1933: 38.

[40] Adriani and Kruyt 1950–51, I, p. 167.

[41] RePPProT 1990, Map 8.

[42] Henley 2005b, pp. 572–3.

[43] Whitten, Mustafa and Henderson 2002, p. 571.

[44] A.C. Kruyt 1903: 203–4; 1899: 608.

[45] Diamond 2005.

[46] Terra 1958: 171.

[47] Seavoy 1973: 219; Van Steenis 1937: 638.

[48] Hill 1977, p. 183.

[49] Knapen 2001, p. 248; Mertz *et al.* 2008: 79.

[50] De Jong *et al.* 2001: 710.

[51] De Jong *et al.* 2001: 710, 715.

[52] De Jong 1995.

[53] Henley and Davidson 2007, pp. 34–5.

[54] Redford 1991.

[55] Padoch *et al.* 2007.

[56] De Jong 1997: 193.

[57] Boserup 1965, p. 15.

[58] Whitmore 1995, p. 7.

[59] Reid 1995: 93; Cribb 2000, p. 23; De Jong, Lye and Abe 2003, p. 19.

[60] Henley 2005b, pp. 475, 484–9.

[61] Ruf and Yoddang 2004.

[62] Li 2007, p. 147.

[63] Colfer and Dudley 1993; Colfer, Peluso and Chin 1997.

[64] Age range for virgin forest: Nicholson 1965, p. 82.

REFERENCES

Abendanon, E.C. 1915–18. *Geologische en Geographische Doorkruisingen van Midden-Celebes (1909–1910)*. Leiden: Brill. 4 vols.

Adriani, N. [1919] *Posso (Midden-Celebes)*. Den Haag: Zendingsstudie-Raad. [Onze Zendingsvelden 2.]

Adriani, N. and Alb. C. Kruyt. 1912–14. *De Bare'e-sprekende Toradja's van Midden-Celebes*. Batavia: Landsdrukkerij. 3 vols.

Adriani, N. and Alb. C. Kruyt. 1950–51. *De Bare'e Sprekende Toradjas van Midden-Celebes (de Oost-Toradjas)*. Amsterdam: Noord-Hollandsche. 3 vols.

Allied Geographical Section, Southwest Pacific Area. 1944. *Menado (Celebes)*. [Terrain Study 83.]

Andel, W.J.D. van and M.A. Monsjou. 1919. 'Bestuursinrichting en Grondenrecht in de Afdeeling Gorontalo (1917)'. *Adatrechtbundel* **17**: 114–22.

Blankhart, David Meskes. 1951. *Voeding en Leverziekten op het Eiland Sangir in Indonesië*. [Doctoral thesis, Rijksuniversiteit Utrecht.]

Bloembergen, S. 1940. 'Verslag van een Exploratie-Tocht naar Midden-Celebes (Lindoe-Meer en Goenoeng Ngilalaki ten Zuiden van Paloe) in Juli 1939'. *Tectona* **33**: 377–418.

Boomgaard, Peter. 1996. *Forests and Forestry 1823–1941*. Amsterdam: Royal Tropical Institute. [Changing Economy in Indonesia, 16.]

Boomgaard, Peter. 2007. *Southeast Asia: An Environmental History*. Santa Barbara, California: ABC Clio.

Boserup, Ester. 1965. *The Conditions of Agricultural Growth: The Economics of Agrarian Change Under Population Pressure*. London: George Allen and Unwin.

Clercq, F.S.A. de. 1870. 'De Overzijde der Ranojapo'. *Tijdschrift voor Indische Taal-, Land- en Volkenkunde* **19**: 521–39.

Colfer, Carol J. Pierce, and Richard Dudley. [1993] *Shifting Cultivators of Indonesia: Marauders or Managers of the forest? Rice Production and Forest Use Among the Uma' Jalan of East Kalimantan*. Rome: Food and Agriculture Organization of the United Nations. [Community Forestry Case Study 6.]

Colfer, Carol J. Pierce, Nancy Peluso, and Chin See Chung. 1997. *Beyond Slash and Burn: Building on Indigenous Management of Borneo's Tropical Rain Forests*. New York: The New York Botanical Garden. [Advances in Economic Botany 2.]

Conklin, Harold C. 1957. *Hanunóo Agriculture: A Report on an Integral System of Shifting Cultivation in the Philippines*. Rome: Food and Agriculture Organization of the United Nations.

Cribb, Robert. 2000. *Historical Atlas of Indonesia*. Richmond, Surrey: Curzon.

[Delden, A.J. van] 1844. 'De Sangir-Eilanden in 1825'. *Indisch Magazijn* **1** (4–6): 356–383; **1** (7–9): 1–32.

Diamond, Jared. 2005. *Collapse: How Societies Choose to Fail or Survive*. New York: Allen Lane.

Dirkzwager, N. 1912. 'Het Nederlandsch-Indisch Gouvernement en Zijne Tekortkomingen Tegenover de Minahassa'. *De Indische Gids* **34**: 1160–73.

Doren, J.B.J. van. 1857–60. *Herinneringen en Schetsen van Nederlands Oost-Indië*. Amsterdam: J.D. Sybrandi. 2 vols.

Dove, Michael R. 1993. 'Smallholder Rubber and Swidden Agriculture in Borneo: A Sustainable Adaptation to the Ecology and Economy of the Tropical Forest'. *Economic Botany* **47**: 136–47.

Dutrieux, F.B. 1931. *Toelichtingen op het Ladangvraagstuk Betreffende de Onderafdeeling Donggala*. [NA MvO KIT 1197, Bijlage B.]

Edeling, A.C.J. 1919. 'Memorie omtrent de Minahasa', *Adatrechtbundel* **16**: 5–95.

Fragment. 1856. 'Fragment uit een Reisverhaal'. *Tijdschrift voor Nederlandsch Indië* **18** (1): 391–432; **18** (2): 1–38, 69–100, 141–60.

Francis, E. 1856–60. *Herinneringen uit den Levensloop van een' Indisch' Ambtenaar van 1815 tot 1851*. Batavia: H.M. van Dorp. 3 vols.

Freeman, J.D. 1955. *Iban Agriculture: A Report on the Shifting Cultivation of Hill Rice by the Iban of Sarawak*. London: Her Majesty's Stationery Office.

Geertz, Clifford. 1963. *Agricultural Involution: The Process of Ecological Change in Indonesia*. Berkeley: University of California Press.

Gourou, Pierre. 1940. *L'utilisation du Sol en Indochine française*. Paris: Centre d'études de Politique étrangère. [Travaux des Groupes d'études, Publication XIV.]

Graafland, N. 1864. 'Fragment eener Onuitgegevene Beschrijving van de Minahassa'. *Mededeelingen vanwege het Nederlandsch Zendelinggenootschap* **8**: 1–23.

Grudelbach, J. 1849. 'Het Meer van Tondano en Omstreken'. *Indisch Archief* **1** (1): 399–407.

Hardjono, J. 1971. *Indonesia, Land and People*. Jakarta: Gunung Agung.

Henley, David. 2005a. 'Agrarian Change and Diversity in the Light of Brookfield, Boserup and Malthus: Historical Illustrations from Sulawesi, Indonesia'. *Asia Pacific Viewpoint* **46**: 153–72.

Henley, David. 2005b. *Fertility, Food and Fever: Population, Economy and Environment in North and Central Sulawesi, 1600–1930*. Leiden: KITLV Press. [Verhandelingen KITLV, 201.]

Henley, David and Jamie S. Davidson. 2007. 'Introduction: Radical Conservatism – The Protean Politics of Adat', in Jamie S. Davidson and David Henley (eds.) *The Revival of Tradition in Indonesian Politics: The Deployment of Adat from Colonialism to Indigenism*. London: Routledge.

Hill, R.D. 1977. *Rice in Malaya: A Study in Historical Geography*. Kuala Lumpur: Oxford University Press.

Jong, W. de. 1995. 'Recreating the Forest: Successful Examples of Ethnoconservation Among Dayak Groups in Central Kalimantan', in O. Sandbukt (ed.) *Management of Tropical Forests: Towards an Integrated Perspective*. Oslo: Centre for Development and the Environment, University of Oslo. [SUM Occasional Papers, 1/95.]

Jong, W. de. 1997. 'Developing Swidden Agriculture and the Threat of Biodiversity Loss'. *Agriculture, Ecosystems and Environment* **62**: 187–97.

Jong, Wil de, Lye Tuck-Po, and Abe Ken-ichi. 2003. 'The Political Ecology of Tropical Forests in Southeast Asia: Historical Roots of Modern Problems', in Lye Tuck-Po, Wil de Jong, and Abe Ken-ichi (eds.) *The Political Ecology of Tropical Forests in Southeast Asia: Historical Perspectives*. Kyoto: Kyoto University Press. [Kyoto Area Studies on Asia, 6.]

Jong, W. de, M. van Noordwijk, M. Sirait, N. Liswanti and Suyanto. 2001. 'Farming Secondary Forests in Indonesia'. *Journal of Tropical Forest Science* **13**: 705–26.

David Henley

Kaudern, Walter. 1925. *Structures and Settlements in Central Celebes*. Göteborg: Elanders. [Ethnographical Studies in Celebes; Results of the Author's Expedition to Celebes 1917–1920, I.]

Knapen, Han. 2001. *Forests of Fortune: The Environmental History of Southeast Borneo, 1600–1800*. Leiden: KITLV Press. [Verhandelingen KITLV 189.]

Koorders, S.H. 1898. *Verslag Eener Botanische Dienstreis Door de Minahasa: Tevens Eerste Overzicht der Flora van N.O. Celebes Uit een Wetenschappelijk en Praktisch Oogpunt*. Batavia and 's-Gravenhage: G. Kolff. [Mededeelingen van 's Lands Plantentuin 19.]

Kortleven, K.S. 1927. *Nota van Toelichting Betreffende het Zelfbesturend Landschap Toli-Toli*. [NA MvO KIT 1199.]

Kruyt, Alb. C. 1895–97. 'Een en Ander Aangaande het Geestelijk en Maatschappelijk Leven van den Poso-Alfoer'. *Mededeelingen van wege het Nederlandsch Zendelinggenootschap* **39**: 1–36 (I), 106–128 (II), 129–53 (III); **40**: 121–60 (IV); **41**: 1–22 (V), 23–42 (VI), 42–52 (VII).

Kruyt, Alb. C. 1899. 'Het Stroomgebied van de Tomasa-Rivier'. *Tijdschrift van het Koninklijk Nederlandsch Aardrijkskundig Genootschap* (second series) **16**: 593–618.

Kruyt, Alb. C. 1903. 'Gegevens voor het Bevolkingsvraagstuk van een Gedeelte van Midden-Celebes'. *Tijdschrift van het Koninklijk Nederlandsch Aardrijkskundig Genootschap* (2nd series) **20**: 190–205.

Kruyt, Alb. C. 1919. 'Het Gedenken der Dooden op de Tweede Paaschdag'. *Maandblad der Samenwerkende Zendingscorporaties* **2** (4): 55.

Kruyt, Alb. C. 1932. 'De Landbouw in den Banggai-Archipel'. *Koloniaal Tijdschrift* **21**: 473–92.

Kruyt, Alb. C. 1938. *De West-Toradjas op Midden-Celebes*. Amsterdam: Noord-Hollandsche. 4 vols.

Kruyt, J. 1933. 'Een Reis door To Wana'. *Mededeelingen: Tijdschrift voor Zendingswetenschap* **77**: 19–48.

Landschap Donggala. 1905. 'Het Landschap Donggala of Banawa'. *Bijdragen tot de Taal-, Land- en Volkenkunde van Nederlandsch-Indië* **58**: 514–531.

Li, Tania Murray. 1991. *Culture, Ecology and Livelihood in the Tinombo Region of Central Sulawesi*. Jakarta: Environmental Management Development in Indonesia Project. [EMDI Environmental Report 6.]

Li, Tania Murray. 2007. *The Will to Improve: Governmentality, Development, and the Practice of Politics*. Durham, North Carolina: Duke University Press.

Matthes, P.A. 1881. *Memorie van Overgave van de Residentie Menado*. [NA MvO MMK 300.]

Mertz, Ole, Reed L. Wadley, Uffe Nielsen, Thilde B. Bruun, Carol J.P. Colfer, Andreas de Neergaard, Martin R. Jepsen, Torben Martinussen, Qiang Zhao, Gabriel T. Noweg and Jakob Magid. 2008. 'A Fresh Look at Shifting Cultivation: Fallow Length an Uncertain Indicator of Productivity'. *Agricultural Systems* **96**: 75–84.

Michon, Geneviève, and Hubert de Foresta. 1995. 'The Indonesian Agro-Forest Model: Forest Resource Management and Biodiversity Conservation', in Patricia Halladay and D.A. Gilmour (eds.) *Conserving Biodiversity Outside Protected Areas: The Role of Traditional Agro-Ecosystems*. Gland, Switzerland: International Union for the Conservation of Nature and Natural Resources.

Missen, G.J. 1972. *Viewpoint on Indonesia: A Geographical Study*. Melbourne: Thomas Nelson.

Nicholson, D. I. 1965. 'A Study of Virgin Forest Near Sandakan, North Borneo', in *Symposium on Ecological Research in Humid Tropics Vegetation, Kuching, Sarawak: July 1963*. Kuching: Government of Sarawak and UNESCO Science Cooperation Office for Southeast Asia.

Olivier, J. 1834–37. *Reizen in den Molukschen Archipel naar Makassar, enz. in het Gevolg van den Gouverneur-Generaal van Nederland's Indië, in 1824 Gedaan*. Amsterdam: G.J.A. Beijerinck. 2 vols.

Padoch, Christine, Kevin Coffey, Ole Mertz, Stephen J. Leisz, Jefferson Fox and Reed L. Wadley. 2007. 'The Demise of Swidden in Southeast Asia? Local Realities and Regional Ambiguities'. *Geografisk Tidsskrift, Danish Journal of Geography* **107**: 29–41.

Redford, Kent H. 1991. 'The Ecologically Noble Savage'. *Cultural Survival Quarterly* **15**: 46–8.

Regeeringsrapport Boalemo. 1914. 'Regeeringsrapport Nopens de Toestanden in de Onderafdeeling Boalemo der Afdeeling Gorontalo (1912)'. *Adatrechtbundel* **9**: 169.

RePPProT [Regional Physical Planning Programme for Transmigration] 1990. *The Land Resources of Indonesia: A National Overview* [atlas]. Jakarta: Government of the Republic of Indonesia/Overseas Development Administration, Foreign and Commonwealth Office, U.K.

Reid, Anthony. 1995. 'Humans and Forests in Pre-colonial Southeast Asia'. *Environment and History* **1**: 93–110.

Reinwardt, C.G.C. 1858. *Reinwardt's Reis Naar het Oostelijk Gedeelte van den Indischen Archipel, in het jaar 1821*, ed. W.H. de Vriese. Amsterdam: Frederik Muller.

Rhijn, M. Van. 1941. *Memorie van Overgave van de Residentie Menado*. [KITLV H1179.]

Riedel, J.G.F. 1864. 'Het landschap Bolaäng-Mongondouw'. *Tijdschrift voor Indische Taal-, Land- en Volkenkunde* **13**: 266–84.

Riedel, J.G.F. 1872. 'De Minahasa in 1825. Bijdrage tot de Kennis van Noord-Selebes'. *Tijdschrift voor Indische Taal-, Land- en Volkenkunde* **18**: 458–568.

Ruf, François and Yoddang 2004. 'The Sulawesi Case: Deforestation, Pre-cocoa and Cocoa Migrations', in Didier Babin (ed.) *Beyond Tropical Deforestation*. Paris: United Nations Educational, Scientific and Cultural Organization (UNESCO), Centre de Coopération Internationale en Recherche Agronomique pour le Développement (CIRAD).

Russell, W.M.S. 1988. 'Population, Swidden Farming and the Tropical Environment'. *Population and Environment: A Journal of Interdisciplinary Studies* **10**: 77–94.

Salafsky, N. 1994. 'Forest Gardens in the Gunung Palung Region of West Kalimantan, Indonesia: Defining a Locally-developed, Market-oriented Agroforestry System'. *Agroforestry Systems* **28**: 237–68.

Seavoy, Ronald E. 1973. 'The Transition to Continuous Rice Cultivation in Kalimantan'. *Annals of the Association of American Geographers* **63**: 218–25.

Sinclair, Fergus L. 1999. 'A General Classification of Agroforestry Practice'. *Agroforestry Systems* **46**: 161–80.

Spencer, J. E. 1966. *Shifting Cultivation in Southeastern Asia*. Berkeley: University of California Press. [University of California Publications in Geography, 19.]

358

David Henley

Steenis, C.G.G.J. van. 1937. 'De Invloed van den Mensch op het Bosch'. *Tectona* **30**: 634–53.

Steenis, C.G.G.J. van. 1958. *Vegetation Map of Malaysia*. N.p.: UNESCO. [Flora Malesiana Series I, Vol. 2.]

Tammes, P.M.L. [1949] *De Bevolkingscultuur van Klapper in het Bijzonder in Oost Indonesië*. [Mededelingen van het Departement van Economische Zaken in Nederlandsch-Indië,11.]

Terra, G.J.A. 1958. 'Farm Systems in South-East Asia'. *Netherlands Journal of Agricultural Science* **6**: 157–82.

Teysmann, J.E. 1861. 'Verslag van den Honorair-Inspecteur van Kultures J.E. Teysmann, Over de Door Z.Ed. in 1860 Gedane Reize in de Molukken'. *Natuurkundig Tijdschrift voor Nederlandsch-Indië* **23**: 290–369.

Weg, U.J. 1938. *Memorie van Overgave van de Afdeling Menado*. [NA MvO KIT 1178.]

Whitten, Tony, Mustafa Muslimin, and Gregory S. Henderson. 1987. *The Ecology of Sulawesi*. Yogyakarta: Gadjah Mada University Press.

Whitten, Tony, Mustafa Muslimin, and Gregory S. Henderson. 2002. *The Ecology of Sulawesi* [second edition]. Hong Kong: Periplus. [The Ecology of Indonesia Series, IV.]

Whitmore, T.C. 1995. 'Comparing Southeast Asian and Other Tropical Rainforests', in Richard B. Primack and Thomas E. Lovejoy (eds.) *Ecology, Conservation, and Management of Southeast Asian Rainforests*. New Haven: Yale University Press.

Wilken, G.A. 1873. 'Het Landbezit in de Minahasa'. *Mededeelingen van wege het Nederlandsch Zendelinggenootschap* 17: 107–37.

Wilken, N.P. 1870. 'Iets over den Landbouw in de Minahassa en de Daarbij Gebruikelijke Benamingen'. *Tijdschrift voor Nederlandsch Indië* 4: 373–85.

World Bank. 1994. *Indonesia: Environment and Development*. Washington, DC: The World Bank. [A World Bank Country Study.]

Index

360

E

ecosystems vi, xi, xvii, 8, 35, 71, 77–8,
80, 85, 89, 94, 102, 121, 129,
168, 194, 276, 294–5, 313, 345,
350–1

erosion 54, 157–83, 276–7, 306, 309–10

essentialism vi, 1–3, 14–16, 45, 230

F

famine xiv, xvii, 166, 190–2, 195, 197–9,
202–3, 207–8, 223–4, 322–9

farming; famers; agriculture *see also*
cultivation, shifting; herding,
herders vii, x, xiii–xviii, 6, 26,
41–2, 50–4, 71, 72, 75, 82, 91,
103–4, 110, 113, 120–2, 137–9,
142–6, 148, 161–7, 171–5, 188,
192, 194, 204, 211–14, 222–3,
226–30, 236, 275–7, 279–88,
293–5, 299–313, 322, 327,
330–58

fire xi, xii, 12, 72, 92–3, 96, 108, 111,
157–83, 276, 293, 306, 330, 335,
345–6, 350–1

forest *see also* rainforest vi, x–xiii, xv,
xvii–xviii, 1–20, 25–9, 36, 71–
87, 88–101, 114, 121, 157–83,
186, 194, 200, 204, 219–44,
322–9, 330–58

G

globalisation; global knowledge systems;
global forces v–vii, xiv, xv–xvi,
xviii, 2, 3, 7, 8, 10–16, 21–5,
33–4, 42, 44, 136–7, 186, 188,
194, 201, 206, 220, 237, 257

god(s) v, x, xv, 71–87, 103, 105, 142,
145, 150, 226, 237

government vii, xvii, 6, 15, 31, 40, 41,
51, 57, 79, 82, 91–2, 96, 105,
108, 113, 118, 124, 126, 148,
157–7, 173, 185, 189–91,
195, 199, 202–4, 222, 224, 233,
236–8, 292, 293, 311, 322–8

groves, sacred x–xi, xv–xvi, 71–87,
89–90, 112–13, 226–8, 235–9,

H

Hawaii v, 225

herding, herders 26–9, 49–50, 75, 80,
123, 167–8, 172, 174 191, 279,
282, 284–5, 287, 293, 295, 307,
310, 347

Hobsbawm, Eric 219, 235

hybridisation (of knowledge / discourse)
vii, xiv–xv, 16, 45, 201

I

imagination xiv, 173, 228, 236

imperialism, post-imperialism see also
colonialism 1, 4, 47, 190, 224,
233, 253

idealism vi, xvi, xvii, 93, 107, 332–3

India vi, x–xiii, 42, 50, 60, 71–87, 144,
219–44, 322–9

Indonesia vi, 330–58

Iran xiii, 136–53

irrigation xiii, 54, 61, 136–53, 158, 164,
169, 222, 331, 338

L

landscape x, xv, xvi, 1, 4, 9, 11, 12, 15,
25–6, 56, 60, 74, 77, 85, 96, 118,
121, 163–5, 169, 175, 185–6,
192, 196, 198, 200, 204, 207,
209, 219–30, 235, 237, 239,
275–321, 333, 341–3, 346, 350

Index

www.ingramcontent.com/pod-product-compliance
Lightning Source LLC
Chambersburg PA
CBHW020237290326
41929CB00044B/91